THIS WILD ABYSS

THIS WILD ABYSS
THE STORY OF THE MEN WHO MADE MODERN ASTRONOMY

Gale E. Christianson

THE FREE PRESS

A Division of Macmillan Publishing Co., Inc.

New York

Collier Macmillan Publishers

London

The Free Press
A Division of Macmillan Publishing Co., Inc.
866 Third Avenue, New York, N.Y. 10022

Collier Macmillan Canada, Ltd.

Library of Congress Catalog Card Number: 77-81428

Printed in the United States of America

printing number

1 2 3 4 5 6 7 8 9 10

Library of Congress Cataloging in Publication Data

Christianson, Gale E
  This wild abyss.

  Bibliography: p.
  Includes index. .
  1. Astronomy--History. I. Title.
QB15.C44   520'.9   77-81428
ISBN 0-02-905380-3

We are grateful to the following institutions and libraries
who have given us permission to reprint the photographs
in this book:
                Kunsthistorisches Museum, Vienna
                Lilly Library, Indiana University, Bloomington, Indiana
                Oxford University Press, Oxford
                Yerkes Observatory, Williams Bay, Wisconsin

*To Howard who is—as e.e. cummings
once said of his most respected mentor—
one of those blessing and blessed spirits
who deserve the name of teacher—
predicates who are utterly in love with their
subject and who
because they would gladly die for it
are living for it
gladly.*

Beware when God lets loose a thinker.
Emerson

Into this wild abyss,
The womb of Nature and perhaps her grave,
Of neither sea, nor shore, nor air, nor fire,
But all these in their pregnant causes mixed
Confusedly, and which thus must ever fight,
Unless the Almighty Maker them ordain
His dark materials to create more worlds,
Into this wild abyss the wary Fiend
Stood on the brink of Hell and looked a while,
Pondering his voyage; for no narrow frith
He had to cross.
Milton, Paradise Lost, Book II

Contents

Preface

ALTHOUGH I HAVE PERSONALLY spent a good deal of time studying the lives and contributions of the men who made the first scientific revolution—that in astronomy—I must admit that I have never been able to grasp, to my complete satisfaction, the totality of that magnificent intellectual achievement. I am particularly conscious of this when, on a crisp and cloudless autumn night, I wander about my back yard picking out the planets and the constellations as they move across the seemingly limitless heavens. How, I repeatedly ask myself, did Pythagoras, Aristotle, Ptolemy, Copernicus, Brahe, Kepler, Galileo, and Newton know what they were doing? What were the exact thoughts that crossed their inventive minds, and exactly how were those thoughts ordered? How, indeed, can the mind of man truly grasp such concepts as "timelessness" and "infinity?" Such a handful of men and such an immense universe; there is no comparable experience in all of human history. Alas, I have finally acknowledged the fact that I will never know with certainty the answers to these and many other questions my study of the subject has raised, but neither will I ever become completely reconciled to this realization. All of the above-mentioned individuals were men of genius, and perhaps only a genius can truly understand a like spirit. Even Newton, a man possessed of unmatched intellectual gifts, felt constrained to observe: "If I have seen farther than other men it is because I stood on the shoulders of giants."

I am also painfully aware of the fact that the life of every single individual discussed in this work is better known to many others than to me. Scarcely a day passed during the research and writing of this volume that I did not seriously question the wisdom of its undertaking. At one point I actually terminated the study for a period of several months. However, I eventually found in the example of Bertrand Russell the inspiration to see the project through. In the preface to his *History of Western Philosophy* Russell remarks that "It is obviously impossible to know as much about every philosopher as can be known about him by a man whose field is less wide." Nevertheless, "If this were considered sufficient reason for respectful silence, it would follow that no man should undertake to treat of more than some narrow strip of history." As an enthusiastic admirer of Russell's, I have taken heart from this observation, for it applies equally as well, I believe, to the study of the great astronomers as to the great philosophers. I must also ask the indulgence of those readers who find fault with my knowledge in the treatment of the areas in which their expertise lies. Error, of course, is error no matter how laudable the intentions of the writer, and it should not be permitted to become a part of the historical record. On the other hand, I think it fair to point out that this work was not conceived with the expert in the history of science directly in mind. Rather it is a different type of audience that I hope to reach.

Nearly twenty years ago the distinguished British historian, Herbert Butterfield, in his splendid study, *The Origins of Modern Science*, observed that the scientific revolution of the sixteenth and seventeenth centuries "reduces the Renaissance and Reformation to the rank of mere episode, mere internal displacements, within the system of medieval Christendom." In fact, wrote Butterfield, the scientific revolution "looms so large as the real origin both of the modern world and of the modern mentality that our customary periodization of European history has become an anachronism and an encumbrance."

Many scholars, myself included, are in agreement with Butterfield's bold analysis, and even those who are not would surely grant that few if any historical processes have had greater impact upon modern society —the world over—than the ongoing revolution in science. Certainly we are far beyond the point in historical studies where the influence of science upon the past and present can be comfortably ignored; yet it often is.

This unfortunate situation is at least partially attributable to the sobering fact that we live in an age of professional polarization in which the historian, like his colleagues in the sciences, has come to regard his particular area of study as somehow more important than those pursued by his counterparts in other disciplines. I do not refer here to the justifiable pride which most professionals quite naturally take in their chosen field after long and frequently difficult years of study and reflection. Rather I allude to the overweening pride which the ancient Greeks called hubris. This inordinately large sense of self-importance and the resulting cleavage it has produced between intellectuals of the scientific community, on the one hand, and the intellectuals who pursue nonscientific studies, on the other, is brilliantly analyzed by the Cambridge scientist and novelist C. P. Snow in his provocative little book *The Two Cultures and a Second Look.* Snow contends that within the intellectual community the nonscientists share the impression that scientists "are shallowly optimistic and unaware of man's condition," while the scientists see literary intellectuals as "totally lacking in foresight, peculiarly unconcerned with their brother man . . . and anxious to restrict art and thought to the existential moment." Both groups have a distorted image of the other—so distorted, in fact, that even on the emotional level they can find little common ground. And due to the increasing demands of narrow specialization the gap is unfortunately wider now than it was a generation ago, when their college-trained parents had at least some basic understanding of several divergent fields. If Snow's is a basically accurate analysis of current conditions within the scholarly community, which I believe it to be, then one can only guess at the situation which abides among educated laymen, the majority of whom have been trained by specialists.

Snow further maintains that university-trained nonscientists (historians included) can be compared to the Luddites of early nineteenth-century England. They do not, of course, dash wildly about smashing the machinery of industry, as did the dispossessed workers of over a century and a half ago; but the nonscientist all too often closes his mind to the contributions which science and technology have made and continue to make to our culture, an exercise tantamount to Luddism.

Although much has been written about the brilliant personages and events surrounding the scientific revolution in astronomy, I have as yet been unable to find a single, comprehensive work that satisfactorily encompasses the subject for the general reader. On the one hand, much of the scholarship is far too technical for the nonspecialist. On the other, when a good general work is available it frequently deals only with the life of a single individual or limited aspects of the revolution in astronomy. As a result, I undertook to write a volume that would cover the middle ground, so to speak—one that would not be so narrow and complex that the beginning student or educated layman would flounder from the outset; yet a work that will hopefully challenge and stimulate the individual who demands something more than superficiality from the books he chooses to advance his knowledge of the past. Moreover, I have attempted to interweave science and history in a manner which I hope will provoke the nonspecialist's curiosity while deepening his appreciation for a series of intellectual accomplishments far too little discussed both within and outside of the classroom.

I have also tried to produce a complete work. By complete I do not mean a book that covers everything, but one that touches the major themes and historical figures in the history of astronomy while doing so against the far broader background of social and cultural developments. For this reason, the study begins with man before the dawn of civilization in the society of the hunter–gatherer. To some, this will undoubtedly appear odd, as might the rather extensive treatment accorded to both barbarian hegemony during the European Dark Ages and to medieval Christianity: Copernicus does not make his appearance until chapter four. I hope, however, that as the reader follows the unfolding of historical and intellectual events, he will discover in his own way the reasons for my particular choice of organization and emphasis.

Furthermore, by beginning with early man and moving beyond the mathematician–astronomers of the seventeenth century, I have attempted to present a broader picture of the first scientific revolution than has generally been done in a single volume. Both my academic training and my sense of esthetics have created in me a desire for completeness. Obviously, the final chapters on the scientific revolution have not yet been written, and, if the optimists are correct, they never will be. (Of course, if the pessimists are correct, they will never be written either; our destructive capacity is both too swift and too great.) Still, as regards the historical background and implications of the first scientific revolution, I have attempted to make my analysis as complete and as lively for the general reader as I know how. Nor, when the occasion seemed appropriate, have I refrained from introducing certain controversial issues—the hydraulic theory of Karl A. Wittfogel and Lewis Mumford, for example, and the possibility that Copernicus seriously contemplated an infinite universe. In such instances, both sides of the issue have been presented; thus, my views notwithstanding, the reader is free to draw his own conclusions. Hopefully he will be sufficiently stimulated to undertake further study on his own.

Lastly, it might well be argued that it would have been better had a widely known and highly respected scholar in the history of astronomy undertaken this study. Perhaps the reason none has chosen to do so is that the topic seems too broad, the potential rewards too small, and the possibility of seriously damaging a hard-earned reputation too great. I run no such risk, for at this early stage in my career I have no major reputation to gamble away. Yet even if I had, it would have been a chance worth the taking. I will consider my efforts to have been well worthwhile if, from the following pages, the reader finds that in his study of the scientific revolution in astronomy he has gained a little better understanding of the human condition, and perhaps a little wisdom as well.

I have become indebted to so many persons during the research and writing of this book that it is impossible to acknowledge each of them individually. However individual mention must be made of a special few. Richard S. Westfall, Distinguished Professor of History and Philosophy of Science at Indiana University, gave unselfishly of his time and expertise by reading all of the work in manuscript, even though we had never met before I requested him to undertake what was in many ways a most thankless task. Owen Gingerich, Professor of Astronomy and of the History of Science, Harvard University, and Astrophysicist, Smithsonian Astrophysical Observatory, also read much of the manuscript. His comments on the chapters dealing with Copernicus, Brahe, and Kepler have proven invaluable. My mentor and good friend Howard Thompson, Professor of European Intellectual History at the University of Northern Iowa, performed the same indispensable service on the nonscientific side. It goes without saying, of course, that the responsibility for any remaining errors is mine alone. The books, documents, special collections, and services of the libraries at Indiana State University and Indiana University, in addition to the Lilly Library, have been generously placed at my disposal. Librarians Karen Chittick and June Dunbar diligently assisted in tracking down and making available difficult to obtain research materials. To secretaries Martha Layton and Rita Phillips goes a special note of thanks for their editorial assistance and handling of the typescript. I am also deeply indebted to Dr. Barbara A. Chernow and Elly Dickason, my excellent editors at The Free Press. Their skillful handling of the manuscript prior to publication dispelled my unfounded fears that editors have little sympathy toward those whose work they must proof and criticize. Finally, but most importantly, my wife, Brenda, whose patience, encouragement, and technical assistance with the manuscript can never be repaid, contributed far more than any other individual to the completion of this study.

I. The Dawn of Human Consciousness

Beyond Memory

BETWEEN 1,750,000 AND 10,000 B.C. MAN'S KNOWLEDGE of the world was the limited knowledge of the wandering hunter, who migrated with the seasons and who left no record of his existence save discarded spears and bleached bones. During this period, hunters lived in small groups of between twenty and sixty individuals. That their life was precarious at best is attested to by the animal figures so skillfully painted on cave walls at Lascaux, France and Altimira, Spain, reminding us that human existence about 16,000 B.C. was entirely dependent upon game, which in turn depended upon the balance of nature. With tools of polished stone, antlers, bone, and ivory, the hunter moved ceaselessly throughout the year in search of prey, probably returning to his cave or other natural shelter only during the most severe winter months.

As with most of the animal species that they hunted, leadership in these small Paleolithic groups very likely fell upon a single individual whose physical strength, predatory skills, and native intelligence won him a position of dominance and authority. There probably existed a loose network of stylized relationships among the various hunting groups which established acknowledged limits on their hunting grounds. This pattern, known as territoriality, contributed to the maintenance of order and harmony by assuring that violent confrontation between rival bands would be the exception rather than the rule. Only when one group of hunters encroached upon the territory of another, either because of dwindling supplies of game or an expanding population, did aggression openly manifest itself. Because man's numbers were small, sufficient quantities of game were usually available, although successfully hunting it down was another matter. Certainly the concern for ecology—the relationship of man and animals to their environment—was every bit as important to the hunter as it is to citizens of contemporary society. Moreover, hunting societies experienced first hand, with sometimes fatal consequences, the adverse effects of overkilling or overgathering—some-

1

thing modern people, isolated as they are from nature by the streets of sprawling cities and a complex network of social and economic institutions, can too easily disregard.

Because of the relentless preoccupation with the primary biological drive of feeding himself and the other members of the group, the hunter–gatherer was in no position to develop a complex and continually changing culture. Constantly on the move in pursuit of the migratory herds, they carried on their backs their few meager possessions. Inventions, even of the most simple kind, if unrelated to the hunt, were rarely if ever contemplated because they would have no immediate value: large pottery, weaving looms, and furniture were all superfluous to a nomadic tribe or clan perpetually on the move. Nor was there the time or stimulation for the hunter to develop a written language with which to pass on the experience of one generation to the next. Using a limited vocabulary and a large number of symbolic gestures, the adults taught the young virtually all of the accumulated wisdom of their group within a few years after childhood. Very little, if anything, new was added by successive generations, so that the life of each individual was patterned after those who had gone before. The past, the present, and the future were quite literally one, as if a man or woman who had died a thousand years earlier could exchange places with one of a much later generation and recognize no fundamental difference in life styles. (Compare this with the disorientation that would result if two individuals separated by only a single decade were to change places in contemporary society, and the implications of the accelerative thrust of modern civilization become much clearer.)

Of primary concern to our understanding of the hunter–gatherer was his radically different conception of time compared to that of modern man. Nothing more clearly separates us from our Paleolithic hunter ancestors than the tiny machines, strapped around our wrists, whose ticking regulates nearly every facet of contemporary life. The so-called "primitive" or "pagan view of time," as understood by Paleolithic man, is evoked by the scientist–writer Loren Eiseley in his penetrating study of the history of evolution, *Darwin's Century*. Eiseley writes of the hunter:

> Among these men gray-headed elders may speak of many grasses or, perchance, of innumerable leaf falls, and then, speechless, they can only make a gesture and refer to the 'dream time' or the 'old ones'. It is on this level of society man feels the touch of time emotionally but he cannot implement his feelings nor grasp the full significance of that vast waste across which today the astronomer and the geologist peer.[1]

The individual possessed of the primitive consciousness of time was thus circumscribed by the experiences of his own generation and perhaps a few scattered reminders of his father's time. He had no written records to draw upon, none to leave, and precious few unique experiences to record, even if he had possessed knowledge of the written word. The hunter knew only that the seasons change, the herds move on, and winter is nigh; each year the time inevitably arrived to retreat into the mouth of the cave until three new moons had come and gone. It was all one needed or could ever hope to know of time.

The Beginnings

Because of his frail knowledge, a meager cultural legacy, and the absence of a written account of his experiences, it has become quite popular to think of Paleolithic man as being intellectually inferior to his modern progeny. How natural it seemed to earlier scholars to picture him as a skulking, stoop-shouldered, hairy biped, communicating in guttural grunts, and possessed of a childlike intelligence that was ever subservient to instinctual passions and unbridled brutality. This popular stereotype can perhaps be best termed "the myth of primitive man," and it was cultivated, among many others, by Darwin himself. Only recently has a new reality, fostered by the painstaking work of anthropologists and scientists in related fields, emerged to challenge this view.

Cro-Magnon man, the earliest representative of Homo sapiens, or modern man, appeared on the scene between 50,000 and 100,000 years ago. After three billion years of biological evolution, he had become the sole inheritor of nature's most remarkably complex and least understood organ—the human brain. Virtually all scientific evidence indicates that by 50,000 B.C. it was a brain which in size and shape was very much like, if not identical to, our own. What is so puzzling to the scientist is that the expansion of the human neocranium to almost double its former size took place at a rate far greater than that of any other evolutionary process of which we have knowledge. This literal explosion of the brain began with and was completed during the million years or so of Ice Age time, and it was the major factor that assured human survival, while such overspecialized species as the hairy mammoth perished because of their inability to adapt to the changing climatological conditions. His enlarged brain provided Cro-Magnon man and successive lines of Homo sapiens with a generalized capacity to survive by adapting to the vicissitudes of a changing environment in a variety of ways. Man could not run as swiftly as the deer, see with the clarity of the hawk, smell or hear as well as the fox. But he could make weapons, fashion rude clothing from animal skins, and more than make up in cunning and intelligence for what those he preyed upon possessed in instinct, swiftness, and physical strength. Homo sapiens had become a generalist with an adaptive capacity denied to every other species deprived of the enlarged brain. Thus, from the time they first gazed across the endless plains, or watched the moon die, only to be reborn more brilliant than before, for the past 100,000 years or so, Homo sapiens possessed a brain little, if any, different from our own.

The question has been frequently asked: Just how intelligent were our early modern ancestors? There is, unfortunately, no simple answer, for our knowledge on the subject will always be limited. In this instance, scientists can never recreate in the laboratory what has been lost for many thousands of years. The greater weight of the evidence tends to indicate, however, that the similar size and structure of their brains endowed our Paleolithic ancestors with an intelligence quite comparable to that of our contemporaries. The tools they fashioned, the weapons they left, and most important, their capacity to survive the rigors of an endlessly hostile environment, provide strong support for this conclusion. Of course, other members of the scientific community challenge this view by pointing out that in the intervening thousands of years between the appearance of Cro-Magnon man and the rise of major world civilizations, the intellectual capacity of the species has continued to improve as a result of the ongoing process of natural selection. Furthermore, the size and weight of the brain are not the only determinants of intellectual capacity. The quality of the cerebrum is dependent upon both the number of neurons and the quantity of dendritic spines, which link the brain cells into a complete information-processing matrix. Obviously, the availability of such data will remain forever beyond our reach. We shall never know for certain whether Paleolithic man fathered an Einstein of the hunt or a Newton of the hearth, although it seems highly probable that he did. In any case, recent research has provided us with enough sound evidence to demonstrate that the gifted humans who painted pictures of deer and wild bulls on the walls of caves in western Europe some 20,000 years ago were every bit as capable as contemporary people of carrying out the higher intellectual functions normally associated with modern society. Thus, when we apply the word "primitive" to our Paleolithic ancestors, it must no longer be a euphemism for such pejorative terms as brutal, unintelligent, and unfeeling.

Despite the explosive growth of his neocortex, Homo sapiens remained a cave-dweller and a hunter–gatherer for several tens of thousands of years. During this entire period, the organ which would take us to the moon and perhaps one day even beyond that lay waiting to be tapped inside his skull. The tentative beginnings of the process of developing the brain's capacities cannot be pinpointed with great exactitude, but they date from the period 12,000–8,000 B.C. By this time, man had become master of the animal kingdom insofar as he was the major and most efficient of the predators. Yet even with his tools, his cultural development, and his social organization, he remained strictly dependent upon the balance of nature. His life lacked a sense of permanence and attachment, without which any true feeling of security remained nothing more than an elusive dream.

While Paleolithic man played with fire and became a specialist in one area of adaptation—predation—he generally took his surroundings as he found them. Living conditions during the Ice Age were so adverse that it was impossible to undertake major constructive changes in the environment. The first giant step in human cultural development paralleled the appearance of an increasingly warmer climate, the subsequent melting and flooding of massive glacial ranges, and the drying up of vast swamplands in the continental interiors. During the Mesolithic period (c. 15,000–10,000 B.C.) and the Neolithic period (c. 9500 B.C. onward), Homo sapiens first discovered the methods and means with which to alter the natural environment to suit basic human needs. With axe in hand man felled great forests, constructed reservoirs and vast networks of irrigation ditches, staked out fields and pastures, erected wood and stone edifices, and, at the same time, shaped the matrix of a highly complex cultural apparatus. The farmer ultimately achieved what the hunter could not: the capacity to support large numbers of people at relatively small permanent locations in an environment that was increasingly humanized. The domestication of plants and animals, permitting man as it did to gradually escape the limits imposed by his predatory past, constituted the most fundamental of all revolutions, one without which later civilization would have been inconceivable.

Unlike the swift and often violent political upheavals commonly referred to as "revolutions" in the parlance of political scientists and historians, the Neolithic, or Agricultural Revolution, as it is commonly called, was a long-term process as opposed to a brief series of rapidly occurring events. It is probable that agriculture was invented more than once and in widely scattered locations including Asia, Africa, the Americas, and the Middle East. Like the ripples from a stone cast into a quiet pond, husbandry slowly spread outward from its small village centers until it united significant numbers of individuals into a common social and cultural framework sufficiently complex to be called "civilization."

By 6500 B.C. man had already tamed the wolf for purposes of self-protection and to assist in the hunt, and the first steps in the domestication of wild sheep, pigs, cattle, and goats had been taken. For countless centuries women had been searching out patches of wild grasses, including wheat and barley, using them to supplement the diet of their families when supplies of game became scarce, or when the annual inundations of the great rivers made hunting impossible. In these first human communities, the hunter still retained his position of dominance, but perhaps he had already experienced a vague yet troubling awareness that his days as a predator were numbered. At first by accident, and later by design, women found ways of assisting the growth of plants. During the summer, some of the plants were uprooted to permit the full maturity of others, and each autumn a portion of the crop was left unharvested to insure a supply of seed for the following spring. A major break-through occurred when the discovery was made that by scattering seed on previously prepared ground fields of grain could be grown where none had previously existed. As agricultural techniques im-proved through the process of trial and error, coupled with the intro-duction of simple plows and cutting tools, the population of the Neolithic communities gradually but steadily increased. As the same time the hunter's prey, once within easy reach, moved farther away as the village fields expanded to meet the increased demands of a growing population.

At this time (c. 4000 B.C.) agriculture was still primarily women's work; for while farming greatly supplemented the consumable resources of the community, it did not as yet completely sustain it. Still, their in-creasingly important contribution to the welfare of the village saw the woman's position elevated to that of a major provider, in addition to her traditional maternal role as the bearer of new generations. In some areas, the heightened status of women earned them a position of such domi-nance that the traditional patrilineal family, subject to total male control, gave way to the female-dominated or matrilineal family. It was also dur-ing this period that the relationship between child bearing and agri-cultural fertility was established, an act of recognition which ultimately led to the rise of the great Axial religions of the Middle East, whose central deities became the great Mother Goddesses: Tiamat, Cybele, Ishtar, and Astarte. The later appearances of the maternal influence in Western religions, such as Demeter, the Greek Mother of Harvests, and Mary, Mother of Christ, is directly traceable to the elevation in the status of women during the early Neolithic era.

Of Villages

It is worthwhile to have a basic understanding both of the life within and of the broader cultural contributions made by the Neolithic com-munity as the foundation of higher civilization, for although the seven-teenth-century scientific revolution in astronomy was removed from this period by several millennia, its origins are rooted in the developing con-sciousness of at least some Middle Eastern villagers that people could—with the aid of generous and benevolent gods—exert a degree of control over their own destiny.

Regardless of whether or not we are speaking of one of the hundreds, perhaps thousands, of Neolithic villages located in the lower Mesopotamian valleys of the Tigris and Euphrates or along the banks of the serpentine gorge of the Nile, upon close examination we will encounter very similar conditions. The physical characteristics of the villages are described by historian Lewis Mumford in his seminal work, *The City in History*:

> A heap of mud huts, baked, or of mud-and-reed construction, cramped in size, at first little better than a beaver's lodge. Around these villages lie garden plots and patches, all the dimensions modest; not yet the broad but bounded fields, rectangles in shape, that come in with the plow. Nearby in swamp and river, are birds to snare, fish to net, extra food to tide over a bad crop or enrich the daily diet.[2]

Once agriculture had established itself, the range of material goods and equipment, so limited before by the unrelenting demand of mobility in the nomadic society of the hunter, increased at a rapid rate. This is not to say that invention succeeded invention with anything approaching the pace of technological innovation in highly industrialized societies today; it obviously did not. But for the first time in the record of man's existence, his natural resistance to change was sufficiently reduced to open up major opportunities for innovation and adaptive behavior which in turn contributed to the stability of the community at large.

Now that they were securely anchored to their environment, villagers were able to acquire things both bulky and fragile for use in the home and in their work. Of primary importance were the clay pots, so much better suited as containers than the skins and baskets employed by hunters. Grain for eating and later cultivation could be kept fresh, out of the reach of such predators as rats, and the techniques of cooking must have undergone considerable improvement when clay vessels were made fireproof. Historian William McNeill points out that, "Meat could be roasted on a wooden spit; but to cook cereals required a container both fireproof and waterproof: required, in short, a pot."[3] In fact, with the increased dominance of women, Neolithic society was geared to the creation of containers both large and small. They ranged in size from the smallest everyday pottery vessels to cisterns, houses, granaries, and the later massive collective containers including the communal temple, the pyramid, and man-made lakes. Indeed, the process undertaken by women of collecting grain, paralleled by the development of the container, could be considered an outward manifestation of the internal encasement of the fetus within the womb, and perhaps as a subconscious stimulus toward the social and religious elevation of the woman in Neolithic society.

With a ready supply of domesticated animals such as the goat and the sheep, hair and wool were woven into cloth for the first time, and the wooden apparatus employed in weaving was rapidly improved, as later, with the cultivation of flax, was that used in spinning. The art of brewing was soon mastered, along with significantly improved techniques of tool manufacture. Speaking the same tongue, working fields of a similar size, meeting together in a common place, each family followed an almost identical way of life, from worshipping the gods of the hearth to sharing a common burial ground.

Any fundamental historical changes like those wrought during the Agricultural Revolution involve both gain and loss. We have already noted that throughout the entire history of civilization people have been dependent upon an abundant supply of food produced through agriculture and bolstered by the domestication of animals. Without such a surplus no civilization could support the doctor, poet, scribe, artisan, priest or philosopher. Yet as the ongoing routine of Neolithic agriculture produced increasingly successful results in the form of a more stable and enduring society, life probably became less spontaneous and increasingly conservative. There were limits to the degree of social change and technological innovation possible within the village context, and apparently these limits had been approached by about 3000 B.C. As Mumford points out, "conformity, repetition, patience, were the keys to their culture, once it solidified. Doubtless it took thousands of years for the Neolithic community to establish its limits, but once it reached them, it had little impulse to further development." [4] The normal resistance to innovation, reduced for a time during the unsettling transformation from predation to agriculture and animal domestication, had reasserted itself. In the wake of a number of major technical innovations, acceptable limits of routine and stability were again established. The impulse to further developmental change was muted: The formula for survival became, "Look no further, keep what we have." Slowly, and with consequences that were largely imperceptible to the countless forgotten generations that had made the Neolithic Revolution possible, the instinctual satisfactions of the hunt and the exhilarating experience of moving across unmapped plains and through the dark untrodden forests, in symbiotic union with nature, passed beyond the memory of the villager, who had forsaken the sport and uncertainty of the spear and arrow for the greater security and daily routine of the plow. Doubtless the reason why so many nomadic bands stubbornly refused to abandon their hunting, fishing, and herding activities—even to the point of enduring prolonged periods of famine—when they could have made the transition to agriculture, was their desire to conserve the nonmaterial satisfactions and values associated with the hunter's life. They shunned the increased material security provided through cultivation because of its demands of greater political, social, and economic submission to authority, whether natural or man-made. This decision, for better or worse, predestined the hunter and herdsman to a gradual extinction, the final chapter of which is being written in our own time.

The very routinization required of a society dependent on agricultural production once again initiated a basic change in the respective roles played by men and women in Neolithic culture. Once the community had developed a primary dependence on cereal grains it became tied to a fixed location. With the invention of the traction plow, sometime around 3000 B.C., men began to take over many of the agricultural duties previously handled by women. This was partly due to the fact that plowing required considerable physical strength and, in a society where the male was thought physically superior, it was natural for him to undertake these and similar tasks. More important, however, was the realization, directly resulting from the pioneering efforts of women, that agriculture, rather than hunting, would provide a more secure subsistence. Like the once wild animals he had tamed, man, too, became the unwitting domestic of social and cultural forces that would exert an ever increasing control over his life.

Yet if the hunter's predatory habits had been tamed by the process of domestication, his desire to subdue and conquer had not. Whereas he had once sought to master the stag, the wild boar, and the lion, man now sought dominion over the dog, pig, horse, elephant and, above all, over the land itself. In this respect his predatory instincts were sublimated for the welfare of family and community. However, there unfortunately remained a latent excess of the predatory drive which was to later manifest itself in the often violent quest for power over other human groups. The strength of woman had lain in the special qualities of her sex: in the menstrual cycle, in childbirth, and in lactation. Ultimately these mysteries were overshadowed by the sheer force of male aggression, in his desire to overcome natural obstacles, and the will of other men, in his readiness to kill in order to dominate. The Great Mother Goddesses were gradually consigned to the background as champions of the male ego, such as the fierce and powerful deities Horus, Atum, and Ptah, made their violent appearance.

The Paleolithic hunter had been able to give vent to his aggressive compulsions in the act of pursuit and in the killing of his prey. But with the cultivation and harvesting of grain, the herding of animals, the construction of containers large and small, dyeing, tanning, brewing, weaving, and countless other tasks, the ritualization of work became the basic human activity. The concept of work most certainly came into being in the Neolithic village community. By "work" I refer to the direction of physical and mental activity toward the accomplishment of a major task. In the Neolithic context this meant the application of repetitive acts on a routine basis for the purpose of enhancing the welfare of the group. It increased the supply of daily bread, which in turn insured the multiplication and survival of more individuals, a process that must have limited the imagination to a certain extent while fostering the degree of conformity essential to a stable society.

Social controls in the form of government existed only in the most rudimentary form in the early villages of the ancient Middle East. The decision-making process had not yet been institutionalized in bureaucratic terms due to the simple social and economic structure of the community. Whereas a system of patrilineal families, molded into kinship groups under the authority of a chieftain had made the daily decisions among the hunters and pastoralists, such rigidity was alien to the first tillers of the soil. As traditions of violence and predation faded, the decision-making process was undertaken by a Council of Elders, for age commended respect and was venerated as a virtue unto itself. Life expectancy was almost certainly much shorter than today, and those few who survived beyond the age of fifty were the only tangible link to the unrecorded past. The elders no doubt met at certain designated times during the year, but probably on an infrequent basis, at least in the beginning. There were few, if any, day-to-day decisions to be made. Only when new land was to be distributed, sacrifices made, or defenses against outside threats planned, did the village leaders assemble. Even when they did meet, such was the size of the community that every man knew the attitudes and opinions of his neighbors. Neolithic man did not practice democracy in the modern sense of the word, but the smallness of the village permitted its members to know each other well. Archeological remains from these villages suggest surprisingly peaceful societies. As long as there was sufficient land to till and agricultural production was fairly equalized among households, there was little reason or opportunity for widespread conflict. Each family was perennially engaged in securing an adequate food supply for its members. Only with the rise of a leisured class, supported by the enforced toil of others, was this balance upset. For many centuries, perhaps for several thousands of years, Neolithic man lived relatively free from the violence, both internal and external, that has become a plague on modern civilization. For all of its technical limitations, its boring routine, and cultural simplicity, the life style of the citizens of ancient Middle Eastern agricultural communities held significant advantages over that of its hunter–gatherer forebearers: Food was much more abundant and varied, violence considerably less common, and people could exist—for the first time—on a limited amount of land. No one as yet had dreamed that one day greedy strangers, dissatisfied and searching for more, would appear, and that they would have no qualms about taking by force that which they desired.

Kingdoms Come

During the late fourth and early third millennia B.C., major cultural changes took place in the valleys of the great river systems of the Middle East. At the center of this transformation was the extension of human control over the environment through the technique of irrigation. Each year, following the melting of the mountain snows, the Nile and Euphrates overflowed their banks and covered the surrounding plains with several feet of water for as far as the eye could see. In the wake of these great floods, extremely rich deposits of silt remained, so that the land was never exhausted, as it would be in medieval Europe, by relentless cropping absent of such rejuvenating agricultural techniques as fertilization. However, this annual cycle of inundation was beneficial to the human community only to the extent that it could be controlled through the development of a systematic network of dikes, irrigation ditches, dams, and reservoirs. The annual rainfall in Egypt was so sparse that farming proved untenable in most regions unless water could be stored in the wake of the yearly floods. At first attempted along the banks of lesser streams, where flood and drought were more easily coped with, the feasibility of growing crops sustained by an artificial supply of water was later transferred to the great desert valleys of the Nile, Tigris, and Euphrates. As they gained confidence in their newly mastered engineering skills, the farmers of the Middle East moved farther and farther away from the rivers by constructing an ever more elaborate system of artificial waterways. These conditions called for, and were capable of supporting, much larger social units than the traditional Neolithic villages. The increasingly abundant agricultural surplus provided by irrigation opened the way for the rapid growth of relatively large and territorially defined communities under the control of specialists. Kingship emerged for the first time around 3500 B.C. and gradually supplanted the oldest mode of social control—government through kinship. The development of kingship was paralleled by the rapid elaboration of ideas, institutions, and ceremonies, whose existence was recorded for the first time in the form of a transmissible written record of events: history. By 3000 B.C. the wealth, complexity, and radically transformed cultural patterns of Mesopotamia and Egypt justified, for the first time in history, the use of the rubric "higher civilization."

Once again, in return for its continued advancement, society had made a trade-off: in place of the simplicity, routine, and rough equality of Neolithic village life, a new dynamism asserted itself. Monumental art, massive architecture, and unsurpassed engineering feats were all undertaken within the context of the new city civilizations. Human horizons were extended far beyond the imagination of the inhabitants of the down-to-earth villages of the Agricultural Revolution. But in undertaking this new style of life, the individual became part of a collective work force, often several thousand strong, whose slavish existence was highly regimented and carefully monitored by a class of professional managers, planners, and overseers, all in the service of the king. For the first time in human history, a huge gulf, created by the immense wealth of a few and the impoverishment of the many, separated one individual from another. What became and still remains an historical constant in most areas of the world had gained general acceptance by 3000 B.C.—the division of the human community into upper and lower classes. Wealthy nobles inhabited splendid villas surrounded by shady groves and fertile gardens. Their food, dress, and material goods encompassed all the richness and variety society could provide. In contrast, the poor endured wretched conditions as history's first congested slums, composed of mud-brick hovels, appeared in the cities. Their meager furnishings, if indeed they existed at all, were of the rudest kind, while the diet and dress of the lower class was simple and monotonous.

The recently developed institution of kingship became the dominant force behind this new outburst of cultural energy and its accompanying condition of human degradation. Every facet of daily life was subject to a set of rigid institutional controls and physical sanctions entirely new to human experience. These sanctions were based upon religious myth and ideology whose roots antedated the Neolithic Revolution, reaching as far back as the magical rites conducted in the protective shadows of the Paleolithic cave. Only a strong central authority could guarantee internal peace and economic stability. Water had to be continually directed, rivers deepened, dikes checked, canals dug, and conflicting interests reconciled. When these functions were neglected, the kingdom literally disintegrated as the man-made waterways fell into disrepair, the earth dried up or flooded beyond control, and the people starved to death. The wrath of the gods then lay upon the land.

The centralization of power fulfilled another vital function by maintaining the defense of the low lying plains against the more aggressive occupants of the nearby deserts or mountains. These rivals scratched out a tenuous existence through a combination of herding animals and marginal cultivation of the soil. From economic necessity, rapacious design, or both, these seminomadic peoples were periodically drawn to the fertile valleys, where they plundered and exacted tribute from their more prosperous neighbors. In order to meet these threats, a warrior caste developed under the command of a chieftain or king. It was out of this relatively simple system of military organization that there first emerged a ruling class, an elite able to control the agricultural surplus of the community through the use of force. At other times, aggressive neighbors turned from looters into conquerors by establishing themselves as an aristocracy in the midst of their victims, taking on the reverse role of protectors and overlords. In either case, the end result was the emergence of a strong centralized state under authoritarian rule. From that time until the present, the overwhelming majority of all the people of the world have known no other.

The Stars

Perhaps from the time the human mind had first risen above the animal level, and most certainly by the time the tool-making stage had been reached, the general permanence as well as the nonrandom motions and basic rhythm of the heavens had been observed. For both the Paleolithic hunter and the Neolithic farmer, the sun was the most common object of veneration, for it took relatively little sophistication to realize that this brilliant object was responsible for seasonal changes and was, in fact, the giver of life on earth. Nothing could have been more natural than to deify this powerful and benevolent force. And if the sun was a god, the moon had also to share in divinity, as, too, did the planets and the brighter stars.

For those of us who inhabit latitudes where clouds regularly obscure our vision of the heavens, or who reside where city lights hide the stellar glow even on clear nights, it is difficult to imagine the clarity and unswerving regularity with which the stars appeared to the Neolithic villager of the Middle East. The region bordering on the Persian Gulf, the very cradle of Mesopotamian civilization, is blessed with extremely clear skies the year round. For untold generations the priests of the kingdom of Babylonia meticulously observed and recorded the movements of the heavenly bodies. It was here that the discoveries were made which set the pattern for the development of man's first science: astronomy.

As early as 6000 years ago, the priestly class constructed terraced pyramid towers called ziggurats on the level areas of the countryside from which they scanned the skies and gathered information to be recorded on maps or used in the construction of ephemerides, time-tables of planetary motion. The priests guarded their secrets well by living highly secluded lives, usually in monasteries adjacent to their observatories. Every day they made careful notations of the previous evening's sightings and correlated them with terrestrial phenomena such as floods, famine, war, and rebellion. From earliest times, man had concluded that the forces that govern planetary and stellar movements must also control events on earth. The priests knew, for example, that the height of the ocean's tides corresponds to the variation of the moon's phases, that the menstrual cycle of women recurs at intervals roughly coincident with the length of the lunar month—about every 29 days—and that the sun is hottest when in the constellation Cancer and least hot when in Capricorn. Were not all earthly phenomena, then, the direct result of influences exerted by celestial bodies, whose appearance and movements are totally dependable and predictable? The priests believed this was so, and their judgment was accepted by the king. The reports compiled by the priestly observers were signed and presented to the ruler, who used the knowledge contained therein as a basis for making crucial decisions. Fortunately, the cuneiform tablets were preserved in the royal libraries, and it is from this fragmentary record—later gathered together by their Assyrian conquerors in a collection of writings known as the *Enuma Anu Ellil*—that modern archeologists have gained valuable insight into the origins of Babylonian astronomy.

The Babylonian priesthood was not only responsible for developing the roots of astronomy, it also invented the zodiacal system of astrology which has played a major part in astronomical tradition. Through careful calculation, the priests discovered that, in addition to the sun and moon, five other planets also occupy the heavens—the planets we now call Mercury, Venus, Mars, Jupiter, and Saturn, after Roman tradition.* They were also aware of the fact that there are approximately twelve lunar cycles in a solar cycle and subsequently divided the year into twelve fairly equal parts. The question next arose: How could the twelve parts of the year be best characterized? They finally established the method of identifying the twelve months with twelve stellar constellations. It was this system which developed into the now famous Signs of the Zodiac. They gave each of the constellations a name derived from existing mythological characters or symbols: for example, Cancer, or the Crab, Gemini, or the Twins, Leo, or the Lion, and Aries, or the Ram. These constellations bore no true resemblance to the individuals or animals after which they were named, a fact of which I myself became aware only after considerable pain and frustration as a child. (I remember how I struggled with my grade school science books to memorize the constellations so that, on the next cloudless night, I could identify the Roman archer Sagittarius or the Greek warrior Orion. Try as I might, however, success eluded me, although many of my friends, possessed no doubt of imaginations more vivid than my own, eagerly reported "authentic" sightings. It was only later—from a better informed source than my elementary teacher—that I learned the symbolic as opposed to the true nature of such representations.)

It was further believed by these early astronomers that as the sun passed through each of the constellations for a period of one month during the year, the particular constellation housing it at any given time

---

* To the Babylonians the five planets were Nebo, Ishtar, Nergal, Marduk and Ninib.

controlled events on earth. For hundreds, perhaps thousands of years, the Babylonian priestly caste employed astrology only for purposes of predicting natural events including storms, floods, and drought. The use of natural astrology, as it was subsequently called, became inseparable from the exercise of political and military power.* On the predictions of priests rested the fate of dynasties and empires, while the general populace stood in respectful awe of those who communicated in secret with the gods.

The astrologer–priests stood next in power only to the king himself —members of the highly privileged upper class whose status was maintained through the exercise of absolute power and the agricultural surplus accumulated by the labors of over 90 percent of the population. Yet theirs was also a profession that exposed the practitioner to certain inherent dangers: woe to the man whose predictions proved false. To invoke the wrath of the king was to suffer disgrace, banishment, or even death. With this thought ever in the back of his mind, it is no wonder the court astrologer of antiquity became as skillful at hedging his predictions or clouding them in a protective blanket of ambiguous phraseology and double-entendre, as he was in tracing the course of the planets and the stars. The embarrassing inaccuracies and ritualistic mumbo jumbo traditionally associated with the creed of the ancient astrologers notwithstanding, when laid bare of its emphasis on the unfounded and overly optimistic prognostications used to assuage anxious patrons, the detail, accuracy of observation, and methods of recording their findings place the Babylonian priests on the first frontier of scientific discovery.

The god Marduk, senior deity of the Babylonian pantheon, personified the power of nature at its ultimate. In the *Epic of Creation*, Marduk's preeminence is supported by his powers of giving life and decreeing death, in addition to countless lesser abilities. Under Marduk's control, the cycle of life and death was tangibly demonstrated by the annual flooding of the Tigris and the rhythmic agricultural process it initiated of planting, cultivating, and harvesting. This cyclical community activity further stimulated the growth of the new science of astronomy. As Dutch astronomer and historian of science Antonie Pannekoek observes: "It [astronomy] proceeded directly from the demands of time-reckoning, and more especially from the problem of adapting the moon calendar to the solar year." [5] This was a highly complicated undertaking, one that could be accomplished only after completing the laborious process of taking down and correlating an immense number of separate observations. Undoubtedly the technical problems encoun-

---

* It was not until the first millennium B.C. that judicial astrology—used to predict events in the lives of individuals through the casting of horoscopes—came into general practice.

tered in this exercise served to promote as a major goal the much more careful tracking and plotting of celestial phenomena.

Painstaking accuracy was not absolutely essential to the needs of agriculture, however. After all, cultivation had been undertaken long before an accurate method of calendric reckoning became available. Moreover, those civilizations that arose on the banks of the great river systems of the Middle East possessed a natural calendar, the annual beginning of which was marked by massive flooding. No astronomer was needed to tell the populace what operations to perform once the waters had receded. On the other hand, we must bear in mind the fact that agricultural activities were inseparable from religious and related social events. The priests, as interpreters of the dominant natural forces, determined what was to be done on the social level by issuing commandments from the gods. These commandments then became scrupulously fixed in the rites of the community. Everyone knew the gods were impatient, capricious, and vengeful; if not properly respected and appeased, their favor might be withdrawn, their wrath invoked. Under these circumstances it would not do to conduct religious rites at *approximately* the same time each year. The gods demanded that their human servants maintain an exact observance of ritual, thus making the calendar—founded on the extremely accurate recording of celestial phenomena—an indispensable part of ancient Middle Eastern life. Although it would not become a recognized part of the scientific tradition until the time of Galileo, the empirical method, guided by observation and practical experience, had already become an indispensable tool of the astronomer as he directed his gaze across the horizon above the sweeping plains of Babylon.

The Old Babylon, as it is usually called, of the great lawgiver Hammurabi, who ruled from 1792–1750 B.C., remained more or less the same until a new Semitic power arose at Ashur on the upper reaches of the Tigris to the north. This was the kingdom of the Assyrians, who steadily expanded their domain to the point where Babylon itself was taken and razed by King Aennacherib in 689 B.C. The Assyrians, following the pattern of past conquerors, abandoned many of the harsher aspects of their own culture for the more refined customs of their new subjects. Of primary importance is the fact that they left the Babylonian pantheon untouched in all important aspects save one: Ashur replaced Marduk as the god of gods, although the old patriarch of the Babylonian deistic hierarchy retained a place as a major god, a privilege rarely accorded the chief deity of a conquered people. This enlightened policy meant that the tradition of the ancient priesthood was preserved, so that those who had labored for the kings of Babylon now continued to do so unmolested in the service of the new rulers. Their records on astrology and religion were preserved as before, assuring that the scientific and cultural legacy of Babylon remained largely intact.

By the time of the Assyrian conquest, sufficient data had already been assembled over the centuries to assure accurate record keeping for the purposes of offering sacrifices and holding feasts in honor of the gods. Consequently, the calendar ceased to be the primary motive for celestial observation. Astrology came to the fore as heavenly phenomena were now studied for the purpose of detecting signs that could influence the fate of individual men, as well as the destiny of kings and empires. Luckily, the Assyrian king Ashurbanipal (668–626 B.C.) was possessed of a strong sense of history: When a library was established in the royal palace, he decreed that it contain copies of the texts and writings from the temples of Old Babylon. From these records, many of which were subsequently destroyed or lost, scholars have determined that the priestly class was less interested at this time in the sun and moon than in the planets and major constellations. It was now believed by the priest–astronomers that as the planets wandered among the stars, they held the key to human destiny. But to unlock the secret of man's fate was no simple matter, for the planets did not appear with the regularity of the sun and moon, nor did they seem to follow a constant course across the sky. Rather, they behaved in a most unorthodox manner, "sometimes by standing still or reversing their courses, sometimes combining among themselves or with bright stars into ever-changing configurations. They seemed like living beings spontaneously roaming through the starry landscape." [6] It was this seeming violation of the previously established pattern of predictability and order—a series of phenomena that, despite numerous and detailed attempts at explanation, evaded wholly satisfactory analysis until the time of Johannes Kepler—that increasingly commanded the astronomer's attention.

Under the Assyrians, celestial observation reached the point where a true science of astronomy could arise for the first time. Yet from the modern point of view, something essential was missing. Although the Babylonian priests possessed detailed knowledge of the heavenly phenomena, they lacked what Pannekoek refers to as "scientific purpose." In other words, the motivation behind their intensive labors differed significantly from that of the modern astronomer. "Throughout antiquity man's attitude to natural phenomena was marked not by causality but by finality." [7] The ancient priests sought ultimate answers to questions posed by anxious kings, in contrast to the modern scientific observer who is motivated by a desire to establish cause and effect relationships. This does not make the contributions of the ancients in the field of astronomy any less valid or important than those of the modern scientist; it does mean, however, that the priest of 1000 B.C. and the astronomer of today operate under a different set of premises, while employing factual data whose validity, by virtue of its demonstrated accuracy, has proven acceptable to both.

The observational abilities of the ancients were to have practical application beyond those of time-reckoning and attempts to predict future events. The regular pursuit of astronomy over several centuries had enabled the Egyptians, as it did the Babylonians of Mesopotamia and the Mayans of South America, to determine with remarkable accuracy the length of the year. They employed a calendar of twelve months of thirty days each, adding five days at the end of the twelfth month called *epagomenes,* the Greek word for "added days." This five-day period was devoted to a festival, perhaps because "somebody in authority had come to the conclusion that it would probably be unlucky to work during these days." [8] But after the development of the 365-day solar calendar, the Egyptians turned their attention to more pressing matters.

Running like a sinuous thread through the Libyian desert, the valley of the Nile, in contrast to the broad plains of lower Mesopotamia, is extremely narrow, reaching a maximum width of only ten miles. And although the distance between the upper delta and the first cataract, a natural barrier to further settlement in ancient times, is some 650 miles, the narrowness of the valley floor meant that less than ten thousand square miles of land were suitable for cultivation. Nor could the Egyptian farmers spread outward, for on either side of the valley are cliffs beyond which lay the no-man's-land of the Libyian and Sahara deserts. Living as they they did in a kind of geographical funnel, the Egyptians were more acutely aware than any other ancient people of the direct relationship between the annual floods and economic survival. The inundation of the Nile river is caused by the melting of snows deep in the African interior. The rushing waters of the spring freshets wash away the fertile loam of the Abbyssian highlands, carrying it down to the valley below where it is deposited to depths of from three to thirty feet. In the delta region of the north, an area approximately 100 miles in length, the soil deposits are considerably deeper and have risen over the silent remains of what were once flourishing centers of a great and vibrant culture.

After the spring floods, the need arose to restore the obscured boundaries of the annually submerged fields. Practical necessity turned the Egyptians into land surveyors, and they developed a profound knowledge of arithmetic and geometry. It was the Egyptians who devised the arithmetical operations of division, addition, and subtraction, and it was their mathematical scholars who initiated the decimal system. However, for some unknown reason they never learned to multiply, an inconvenience that forced them to substitute a series of additions in its place.

Perhaps this was because they had no symbol for "zero," an innovation made by Indian scholars and carried westward by the Arabs at a much later date.

Besides the application of geometry to the task of surveying, mathematics was employed in the construction of the most lasting of all the Egyptian monuments—the Great Pyramids. Repeated examples of the so-called "golden section" have been found in pyramid construction; that is, the division of a straight line into two unequal parts so that the ratio of the shorter to the longer part is the same as the ratio of the longer to the whole. This operation, traditionally credited to the third-century B.C. Greek mathematician Euclid, was mastered over 2000 years earlier, as evidenced in the sloping planes of mankind's greatest burial chambers, and in figures drawn on the walls surrounding the mummified remains of once all-powerful Pharaohs. In Egypt, then, both astronomy and mathematics were developed for practical ends: for calculating the time of the Nile's inundations and for laying out and executing the plans of massive pyramids, irrigation ditches, dams, and temples. The Egyptians were not scientists, but they, like their Babylonian neighbors to the east, solved difficult practical problems through the combined use of celestial observation and mathematical calculation. They did not attain the high degree of knowledge in astronomy acquired by the priests of Babylon, but the Egyptians were the better mathematicians and engineers.

Still, Egyptian cosmology was influenced no more than that of the Babylonians by their practical application of astronomy to construction and design. Unlike the European astronomers of the sixteenth and seventeenth centuries, they developed no new geometric vision of the universe. The astronomer–priest was not the ancient counterpart of the modern philosopher–mathematician René Descartes; rather, he remained wedded to traditional religious practices and rejected the prospect of developing new cosmic theories not in conformity with the holy doctrines. The universe of Egyptian civilization remained, from beginning to end, a box with the earth as its floor. The sky was a woman supporting herself on knees and elbows. Beneath her, around the inner walls of the box, flowed a river upon whose waters sailed the barques of the sun and moon gods. The planets, too, sailed their own boats while the stars were either lamps affixed to the female vault or carried in the hands of lesser deities. During the middle of each month the full moon was attacked by a colossal sow and ravenously devoured. Yet there was no reason to fear, for the moon god was reborn in all his splendor at the beginning of each month, and the endless cycle once again repeated itself. The other great civilizations of the Middle East had equally graphic visions of celestial occurrences, and, throughout their long and often splendid history, all things were done in the name of the gods.

Ancient Time

In that area of the Middle East where the first great human cultures rose and fell and rose again, a new conception of time permeated the consciousness of civilized man. In Mesopotamia and Egypt the ways of the Paleolithic and Mesolithic hunter, including his concept of primitive time, were abandoned. Men now kept records of their accomplishments through the invention of the written word, and the growth of this ac-cumulated wisdom profoundly influenced the human outlook on the world and the universe. History, the extension of human memory through time, is a contemplative luxury of advanced civilizations. The hunter had only his own memory and the oral record of a few elder tribesmen upon which to draw. Only the knowledge essential to the sur-vival of the species passed on. To the hunter, temporal or spatial mea-surements mattered very little, if at all.

All this changed in the agricultural civilizations, whose rhythm of existence was directly related to the functions of measurement and calcu-lation. Supported by the surplus resources of the community, the priest measured time, traced the course of celestial phenomena, and predicted future events. Time and space became part of the human dimension and, once mastered, were used to control the lives of great masses of men and women. To know when the next flood would take place, to predict the "death" and "rebirth" of the sun during an eclipse, to share the wisdom of the gods by reading their signs in the sky—all these magni-fied to an awe-inspiring degree the power exercised by the masters of the citadel. Loyalty to the gods demanded loyalty to their most favored servants in the form of kings and an elite and highly secretive priesthood. The new intellectual class that came into existence included among others physicians, scribes, engineers, magicians, and architects; and all took an oath, administered by the priest, to uphold the State gods which, if vio-lated, meant instant death.

Gradually, as a result of the fluctuating pattern of their own exis-tence, the intellectuals of Middle Eastern civilization formulated what historians now refer to as the "ancient concept of time": "notions of vast cycles and undulations in a time stream where things became and passed, perhaps only to come again." This cyclical view of the historical process differs from that of modern Western man in a most fundamental man-ner. Our view of the past, and particularly of the future, is linear. Ever since the Enlightenment of the eighteenth century, we have thought in terms of a precisely defined line of human progress, as opposed to the recurrent pattern of rise and fall discerned by the ancients, who held to their cyclical interpretation of the past and future throughout the Greek and Roman periods, after which it was replaced by the Christian philosophy of history and time. Perhaps Marcus Aurelius, Roman emperor and Stoic philosopher of the late second century A.D., best summarized the ancient view of time:

> The periodic movements are the same, up and down from age to age. Soon will the earth cover us all: then the earth, too, will change, and the things also which result from change will continue to change for ever, and these again for ever. For if a man reflects on the changes and transformations which follow one another like wave after wave and their rapidity, he will despise everything which is perishable.

One wonders if this vision of time and history made somewhat more palatable to the seers of ancient Babylon, Greece, and Rome the knowledge that just as the civilizations of which they were a part had risen upon the broken monuments and forgotten dreams of their predecessors, theirs, too, would one day decline and fall, only to be followed by some successor? Or did the belief in their inevitable demise subject them to a constant state of tension and anxiety? Whatever the case, we shall give the last word to the poet, who, reflecting no doubt upon the fate of the ancients, wrote, "Old gods wither on the stem but new gods will arise from them." It was in the nature of things, for time and tide would wait for no man.

Hydraulics and Megamechanics

In 1957, Karl A. Wittfogel published his scholarly, albeit highly controversial, study of the politics and social institutions of the great ancient Near Eastern and Eastern civilizations, *Oriental Despotism: A Comparative Study of Total Power*. Wittfogel contends there was a trait—common to a majority of these major civilizations—that bound them together in terms of their political and social organization: "The common substance in various Oriental societies appeared most conspicuously in the despotic strength of their political authority." From this observation Wittfogel goes on to develop his now famous "Hydraulic Theory" of ancient civilization.*

As the once independent peasant communities of the Middle East, with their own councils of elders or chieftains and their local gods, were merged into larger political units, the need first arose to develop an

---

* Subsequent criticism of Wittfogel's work has mainly centered on its application to ancient China, and his ideas on this score have been challenged by W. Eberhard, Jean Escarra, and Henri Maspero among others. My discussion of the "Hydraulic Theory" will be confined to the civilizations of the Middle East; i.e., Sumeria, Babylonia, and Egypt, where the roots of Western science began and where hydraulic engineering appears to have had a more direct, if not a determining influence on the subsequent development of authoritarian political institutions.

effective means of centralizing the regulation of the precious water resources upon whose conservation the welfare of the community depended. As has been noted above, the annual inundation of the Nile and Euphrates had to be continuously controlled by dikes, canals, and reservoirs. Group survival through irrigation led to the development of higher mathematical functions and constructional techniques associated with the science of hydraulic engineering. Wittfogel assigns the term "hydroagriculture" to the large-scale irrigation undertaken by the central governments of Mesopotamia and Egypt, and, for the first time, refers to them as "hydraulic societies." "By underlining the prominent role of the government, the term 'hydraulic,' as I define it, draws attention to the agromanagerial and agrobureaucratic character of these civilizations." [9]

It naturally followed that if governments were to manage such an ambitious and ongoing undertaking, a large number of specialists—highly skilled in mathematics (particularly geometry), construction, administration, and celestial observation—would be essential. In other words, by virtue of the decision to undertake hydroagricultural production, a ruling elite, possessed of skills far beyond those of the Neolithic villager, had to come into being. "It explains why, in hydraulic society, there exists a bureaucratic landlordism, a bureaucratic capitalism, and a bureaucratic gentry." Central authority and regulation, by necessity, had to prevail over local interests. "The effective management of these works involves an organizational web which covers either the whole, or at least the dynamic core, of the country's population. In consequence, those who control this network are uniquely prepared to wield supreme political power." [10] Wield it they did; the masters of the hydraulic "apparatus" that had been responsible for their elevation to positions of unchallenged authority in the first place now became the dominant force behind all important functions, not just those with a direct bearing on hydroagriculture.

In the shadow of this powerful bureaucratic elite cowered the anonymous mass of agrarian producers: "The bulk of the hydraulic workers are expected to remain peasants, and in most cases they are mobilized for a relatively short period only—at best for a few days, at worst for any time that will not destroy their agricultural usefulness." [11] They were divided into work gangs whose foremen usually undertook "no menial work at all." The threat of physical punishment or death was ever present, and it was invoked without hestitation against the recalcitrant and disrespectful. Any restraint in the exercise of authority in the face of disobedience was considered a sign of weakness. Dissidents whose conduct led to the open challenge of State authority could not be tolerated. This is the major reason why kings took on the trappings of gods or, as in the case of the Egyptian Pharaohs, claimed outright divinity for themselves. The meaning quickly became clear: the individual who challenged the leadership of a hydraulic society also challenged its gods—the ultimate "sin" whose commission deserved and received no quarter. Under the circumstances, it seems doubtful that open defiance was very common or that it was even contemplated, except in isolated instances of utter desperation. The vast majority of the populace accepted its fate as "willed by the gods" and interpreted by the priestly hierarchy. There was coercion to be sure—both implied and overt—but as Wittfogel points out: "It is the circumspection, resourcefulness, and integrative skill of the supreme leader and his aides which play the decisive role in initiating, accomplishing, and perpetuating the major works of hydraulic economy." [12] Order and group conformity through bureaucratic systematization became the rule of the day; disobedience and open rebellion the rare exception. In this manner the societies of the Middle East were able to intensify cultivation through the division and mass application of labor on a hitherto unimagined scale.

Exactly a decade after the publication of Wittfogel's major study, the first volume of Lewis Mumford's two-volume work, *Techniques and Human Development*, was brought to press under the title *The Myth of the Machine*. In this highly provocative, yet no less controversial work, Mumford carries the "Hydraulic Theory" far beyond the point of its originator. Like Wittfogel, Mumford adheres to the basic thesis that the rise of authoritarian government in the Middle East is the direct result of man's attempts to exert hydraulic controls over the periodic flooding of the natural waterways. Nor is the manner in which Mumford traces the historical roots of this development much different from that of Wittfogel. Mumford's major innovation is his conceptualization of the ancient hydraulic societies as "archetypal machines." He believes that hydroagricultural societies comprise "the earliest working model for all later complex machines," although in our own time there has been an emphasis away from human components to mechanical parts:

> Now to call these collective entities machines is no idle play on words. If a machine be defined, more or less in accord with the classic definition of Franz Reuleaux, as a combination of resistant parts, each specialized in function, operating under human control, then the great labor machine was in every aspect a genuine machine: all the more because its components, though made of human bone, nerve, and muscle, were reduced to their bare mechanical elements and rigidly standardized for the performance of their limited tasks. The taskmaster's lash ensured conformity. Such machines had already been assembled if not invented by kings in the early part of the Pyramid Age, from the end of the Fourth Millennium on.[13]

Mumford's "invisible machine" is the most highly organized of collective undertakings. Its components are none other than the economic, political, military, royal, and bureaucratic branches of the social order. This multidimensional composition of working parts can be best characterized as the "megamachine" or, more simply, the Big Machine; and the technical equipment derived therefrom becomes "megatechnics."

Mumford cites as the ancient example of the megamechanical process at its best and most efficient, the Great Pyramid Age of Egypt during the third millennium B.C. "In its elemental geometric form, in the exquisite accuracy of its measurements, in the organization of the entire working force, in the sheer mass of construction involved, the final pyramids demonstrate to perfection the unique properties of this new technical complex." [14] Thousands of unskilled laborers and specialized craftsmen alike worked in slavish devotion with a single common aim: the construction of an indestructable tomb to house the mummified remains of their Pharaoh for the purpose of securing his safe passage into the bliss of after-life. The entire project was undertaken and completed with the use of only the lever and the inclined plane, for as yet the wheel and pulley had not been invented: "These workers were stripped down to their reflexes in order to ensure a mechanically perfect performance." Human muscle took the place of wheels and cranes, battalions of men, under national conscription or slavery, did what no mechanical device could accomplish until the twentieth century. Mumford points out, for example, that the stone slab that covered the inner chamber of the Great Pyramid at Giza weighs upwards of fifty tons and that "an architect of today would think twice before calling for such a mechanical exploit." Without the use of a "human machine," whose individual components had been "mechanically conditioned" to execute their tasks in strict obedience to authority, this ultimate product of mankind's unbridled ego and hubris could have never been attempted. It is within the context of this high degree of communal regimentation that Wittfogel and Mumford both discover the seeds of modern totalitarianism based upon the cult of personality and total devotion to the State.

The "governors" of these human engines were god-like kings whose every act was sanctioned and amplified by a host of royal officials. Directly below the king was a coterie of intellectuals possessing minds of the highest order. They employed "a unique combination of theoretic analysis, practical grasp, and imaginative foresight." The surviving records indicate that their technical abilities and intellectual powers were equal to virtually any task of which their king might dream. Further down the line were the lesser officials who, like modern bureaucrats, simply take orders and pass them on. Except at the very highest levels, individual initiative had no place in the megamachine. Like its later mechanical counterparts, it performed well only when each component was integrated into a larger whole. Once again the human community had gained greater security and order than ever before; it had also developed the techniques of engineering and a grandiose artistic and architectural style that would have been impossible under less controlled conditions. But the wastage in human life and the monumental social perversions associated with the "cult of leadership" exacted a price that made the megamachine the most costly, in human terms, of all our race's many inventions.

If Mumford's hypothesis is valid, then when men sought to tame the raging rivers in the absence of sophisticated mechanical devices, they incurred the inevitable loss of their identity by becoming an expendable part of a great machine. Moreover, this human "engine" became the prototype for later machines, even though they would be constructed of largely inorganic materials. Of course, we will never know for certain if the power elite of ancient Mesopotamia and Egypt thought in mechanistic terms. Perhaps it is only our own inurement to the mechanical process, institutionalized by the Industrial Revolution, that has colored our vision, permitting us to read a major cultural development of our own back into the ancient past. It was one thing for Johannes Kepler or Isaac Newton to compare the universe to a great timepiece during an age when the clock was well known and the accurate measurement of time was of increasing importance to the new men of science. It is quite another to attribute a sense of mechanical consciousness to ancient pre-industrial civilization. Still, the slow process of mechanical development, paralleled by its penetration into the human consciousness, began somewhere. Perhaps the seeds of mechanical innovation are as deeply, albeit less visibly, rooted in the flood plains of ancient Babylon and Egypt as those of the cereal grains associated with the Agricultural Revolution. Mathematical calculation and meticulous astronomical observations were essential components of the process of planning and constructing the giant collective works of the ancient world. Mumford's observation is quite convincing: "The order that was transmitted from Heaven through the king was passed on to every part of · the machine, and in time created an underlying mechanical unity in other institutions and activities: they began to show the same regularity that characterized the movements of the heavenly bodies." [15] In any event, human consciousness had been permeated and deeply influenced by the order and regularity of the movements associated with celestial bodies, whose days as gods were numbered, but whose influence on man's cosmology had just begun.

2. The Ancient View of the Cosmos

The Background of Greek Astronomy

DURING THE LAST FEW CENTURIES before the birth of Christ, a new civilization arose in the West that took over where Babylonia and Egypt left off. In contrast to the extensive fertile plains that nurtured the ancient cultures of the Middle East, Greek civilization flourished on a mountainous peninsula interspersed by narrow valleys and long gulfs. Rivers are few and swift-flowing, while the arable plains, miniscule by Babylonian standards, are separated from one another by thickly wooded mountains or isolated by rocky outcroppings along the southern Gulf coast. Still, there was sufficient agricultural potential to attract the Hellenic tribes of the north, whose resourcefulness, first as farmers and later as merchant seamen, led to a rapid rise in population, part of which migrated to establish new Greek colonies on the Aegean islands, the Ionian coast of Asia Minor, southern Italy, and Sicily. Greece, as a whole, faces to the southeast so that her valleys, harbors, and mountain ranges pointed the ancient voyager toward Egypt and western Asia, journeys that could be completed in relative safety within a few weeks' time, and without the aid of the compass and advanced navigational skills. Next to agriculture, overseas trade became the Greeks' major livelihood. And the Athenians, who were the best at it of all the Hellenic traders, by establishing a maritime empire, were able to dominate commerce in the Aegean and become, at the same time, formidable competitors of the Phoenicians in the eastern Mediterranean. Moreover, these commercial contacts opened new and even more exciting vistas: The Hellenic mind became receptive to new and stimulating ideas, while the keen competition in the arts, crafts, and trade made the Greeks an inventive and resourceful people.

Of equal importance to the development of classical Greek culture was the climate. The winters are severe in the mountains, but elsewhere it is normally moderate and very sunny. Summer comes early and is hot, but the humidity does not stifle vigorous activity, except in the landlocked plains. Along the coastal areas where, with the exception of militaristic Sparta, the capitals of the major city-states were located, the heat is tempered by the daily influx of cool sea-breezes. Rain, while more common than in Egypt and Mesopotamia, is quite scarce in the summer and falls in quantity only during the autumn and late winter.

These conditions enabled the typical Greek to lead an active life, both vigorous and physical, in the out-of-doors. The main center of community activity was the open marketplace or agora, where the Greeks bartered their wares, met in political assembly, engaged, as did Socrates, in philosophical disquisitions, or simply gossipped with their neighbors. Nearby stood the gymnasium, where physical fitness was combined with intellectual discourse. It has, in fact, been pointed out by classical scholar H. D. F. Kitto, that neither Athenian democracy nor drama could have developed as they did, if the prevailing climate had made a roof and walls a daily necessity. The economic and social divisions between rich and poor were not as visible as in many of the contemporary Eastern societies, and appear to have been much less corrosive in fifth-century B.C. Athens, since all things could be open to everyone "because they could be open to the air and the sun." [1] Of course, it would be erroneous, as Kitto himself maintains, to explain Greek cultural development solely on the basis of the physical conditions in which it took place. Nevertheless, the climate and geography of the area exerted a strong, if not a deterministic influence, on the Greek temperament and character and on the subsequent development of Greek scientific thought as well.

In tracing the background of Greek science, we must also bear in mind the fact that in the region bordering on the Aegean Sea there was no concept of order and stability like that experienced by Mesopotamia and Egypt. Throughout their long history the civilizations of the Tigris, Euphrates, and Nile rivers remained integral parts of large, centralized land empires. Political control might shift from the Assyrians to the Babylonians or from the Medes to the Persians, but the area remained remarkedly unchanged. Even after the Roman and Arab conquests of the region, the bureaucratic apparatus and great network of irrigation canals and reservoirs continued intact. Only after some 3000 years, in the wake of the Mongol invasion of A.D. 1258, was the basic fabric of ancient Middle Eastern civilization rent beyond repair. Against this background, Mesopotamian society remained, as we have previously noted, basically conservative. There were few contacts with outside peoples to open new horizons; individualism itself was considered a major threat to the power of the State; while the preservation of tradition was deemed the highest good by the ruling elite. For these reasons the sciences of astronomy and mathematics were limited to a secondary role by a powerful priestly hierarchy that employed them to sustain a strong religious tradition, as opposed to undertaking their study as a legitimate independent intellectual function unto itself. Science had always been and continued to remain the handmaiden of the great Eastern religions.

The Greeks, on the other hand, were a fiercely independent people. They believed that the human family consisted of two major classes: the Barbarians and the Hellenes. The Greek word for barbarian, "bar-bar," simply meant one who did not speak Greek, and it was applied equally to the lowest of wilderness herdsmen and the most sophisticated residents of ancient Babylon. However, during the classical period the term "barbarian" took on a much deeper meaning, for it was used to designate those who did not live or think in the Greek manner.[2] It meant, quite simply, those people with a different attitude toward life. It is for this reason that citizenship in the polis could be attained by birth only: No matter how long he (for only males became citizens) or his ancestors might have lived in Corinth, Athens, or any of the other poleis, unless he was born of Greek parents living within the boundaries of the city-state in which he resided, the individual and his children were forever known as resident foreigners. The metic, as the resident-foreigner was called in Athens, was considered a freeman, but he was denied the constitutional rights of entering into the political life of the polis. Thus, for the first time in the history of Western thought, there emerged in the Greek consciousness a strong awareness of individual differences among people —a realization that has been a mixed blessing ever since.

The emphasis on individualism was no less marked within Greek society than it was in the relationships between Greek and non-Greek. In contrast to the Middle East, the centralization of political power in Greece was not necessary, nor was it then possible. There were no sprawling river valleys and flood plains waiting to be brought under human control through hydroagricultural techniques. Each polis was small and relatively inaccessible, except by sea. Kings there were, but from the very beginning their rule was circumscribed by limited resources, both human and material, and by the topography of an extremely mountainous peninsula. Even though the population expanded, there was no possibility of bringing ever-increasing amounts of land under cultivation. The only reasonable alternative to mass starvation and widespread civil disorder was, as we have noted, through the colonization of the Mediterranean littoral. The colonizing Greeks, living as they did between the barbarian tribes of Europe and the militaristic empires of Asia, formed the first advanced societies in the West to free themselves from highly controlled, despotic rule coupled with an absolute subservience to a pantheon of wrathful gods. This development, more than anything else, put its stamp upon Greek spiritual and intellectual activity, including Greek science.

Since Hellenic civilization developed independently of major coercive influences from the outside, its members were free to borrow as much, as in the case of Athens, or as little, as with Sparta, as they liked from other cultures. In the area of religion, they rejected the ossified State cults of the Middle East whose spiritual leadership was concentrated in a powerful priesthood. The early Greek deities were little more than human beings writ large, in contrast to the omnipotent beings that inspired constant fear and dread among their followers in the East. The Greeks were less concerned with gods of great power than they were with deities with whom they could bargain for favor or gain. Thus, the Hellenic gods were endowed by the great epic poet Homer and other early Greek writers with the strengths and weaknesses of mortals. While they did not die because of a special diet of nectar and ambrosia, the gods quarreled, loved, hated, ate, slept, and thirsted for power, as do mortals. Moreover, they mingled with humans and on occasion even had offspring by mortals. And though they commanded respect, and while their cloud-shrouded retreat on Mount Olympus in northern Greece was deemed sacred by all Greeks, one could never hold them in such awe or dread, as in the case of the Babylonian and Egyptian deities, for they acted so much like ordinary men and women. It is little wonder that the sin of hubris—aping the gods—is such a recurrent theme in the classical literature and drama of a people whose conduct the gods themselves so often seemed to emulate.

It is largely for this reason that the Greek intellectual could theorize about cosmology while the temple priest, with his vested interest in maintaining the political and religious status quo, could not. Babylonian science was predicated on a tradition of astronomical record-keeping for strictly religious purposes. By contrast, the natural philosophers (scientists) of ancient Greece were not tied to a profession that established clearly defined limits on the range of their intellectual inquiry. They were men who followed a diversity of professions and included among their ranks physicians, poets, and teachers. Many of them were quite well-to-do and, freed of the daily care of earning a living, took up critical speculation on the origins and nature of the universe as a leisure activity. What is most important about the Greek scientific experience, however, is that it was dependent upon original thought as opposed to religious and political considerations and inherent biases associated with such activity. Unlike the civilizations of Egypt and Mesopotamia, the Greeks developed not one but many interpretations regarding the origins and operation of terrestrial and celestial phenomena. Many, if not most, of their theories seem rather naive, even childlike, today. They truly excelled in only one field—that of geometry. Elsewhere, as has been pointed out, "their speculative theories led to curiously little effective result. Again and again, one finds in the Greeks hints of later discoveries, without these insights ever being driven home." [3] Still, there are some very interesting and forward-looking aspects to their work, from which certain key ideas and concepts eventually evolved. And despite their many disagreements and errors of fact, these thinkers were no longer prophets seeking gain at the expense of reason and critical analysis. They were philosophers of the first rank, who sought to increase human understanding by reflecting upon past experience and then relating it to their own observations of the world and the universe around them.

Homer and Hesiod

Strange as it may seem, we possess a better record as to the ideas of the universe which prevailed prior to the days of the first natural philosophers in Greece (c. 600 B.C.) than we do of the various subsequent schools of thought that developed over the next two and one-half centuries. For after the death of Homer, perhaps as early as the eighth century B.C., we do not have anything but fragments of the works of any of the major Greek philosopher–scientists up to the time of Plato and Aristotle in the fourth century B.C. To be sure, we know some of their thoughts through quotations in later works, and we possess scattered pages of their manuscripts. But these have been subjected to much editing by disciples and have undergone numerous translations into many languages including Latin, Syraic, Arabic, French, English, and German. Furthermore, we have no way of judging just how accurate any given quotation may be, or even whether it is representative of the thought as a whole of the philosopher to whom it is attributed. And, as has been the case throughout history, it is frequently impossible to separate the work of the "master" from that of his closest followers. Quite often a particular school of thought would produce an intellect more refined and perceptive than that of its founder; but due to the honor and respect accorded him by later generations of disciples turned mythographers, all wisdom and insight are considered to be his alone. Hence, Greek astronomy from its earliest beginnings until the fourth century B.C. remains clouded to contemporary scholars due to the paucity of primary source material available on the subject. There is, of course, still hope that previously unknown manuscripts will one day be discovered by researchers doing work in the archives of institutions whose collections have not as yet been fully catalogued; but unless and until such discoveries are made, we can only rely on the scattered information now available.[4]

We may take as the starting-point of the Greek conception of the universe the literary works of the great Dark Age poets Homer and Hesiod, which are in fundamental agreement with each other. The earth is pictured by them as a flat circular disk surrounded by the mighty river Oceanus. From this great waterway, which begins its flow near the Pillars of Heracles and winds to the north, east, south, and finally westward back upon itself, all other rivers take their source. Over the earth rests the vault of heaven like a hemispherical dome or great bell. Directly opposite, beneath the earth, is the region called Tartarus, which forms a concave vault in symmetry with the dome of heaven. The earth itself is partially covered by oceans, including the Mediterranean, and an even larger sea located far to the north—the same body of water Jason and the Argonauts crossed during their epic voyage in quest of the Golden Fleece. Hades, located underneath Tartarus, is supposed to be at a depth below the earth equal to the height of the highest point of heaven above the earth. No accurate dimensions of the heavens and earth are provided, except that, according to Homer, Hephaestus—cast down from the top of Olympus—falls from morning to sundown before reaching the earth, while Hesiod writes that it would take an iron anvil falling freely nine days and nights to reach the earth from heaven; and again nine days and nights to fall from the earth to the lowest regions of Tartarus. From these descriptions, apparently it was thought that Hephaestus fell 18 times more rapidly than his anvil, or, more likely, Hesiod's universe was considerably larger than Homer's.

Above the earth is the region of the invisible element ether, the proof of whose reputed existence would preoccupy astronomers from Aristotle to Newton, and even beyond. Under the vault of heaven, the sun, moon, and stars move about, rising from Oceanus in the east and slipping under its waters again in the west upon completion of their celestial flight. It is not said what becomes of the heavenly bodies between their setting and rising; they apparently do not pass under the earth because Tartarus is never illuminated by the sun's light. It has been suggested that they float on Oceanus to points where they begin their next appointed flight across the sky.[5] Such an explanation would be in keeping with the much older Egyptian belief, perhaps known to the Greeks at this time, that the sun, moon, and major stars float back over the water to their points of departure after moving across the heavens. In the *Iliad* of Homer, the Elysian fields, abode of the blessed after death, are at the end of the earth, while in the cosmology of Hesiod the dwelling place of the honored dead is subterranean. At the western end of the earth are the sources of Oceanus, and from there a branch of the river called Styx flows downward into the underground world of Hades, brother of Zeus and Poseidon.

Both Homer and Hesiod provide the reader with some knowledge of astronomical facts, although Hesiod's comprehension of celestial phenomena seems somewhat more advanced. Besides the sun and moon, the poets mention the Morning and Evening Stars, the Hyades, the Great Bear, Orion, and the Pleiades. In the *Odyssey* the constellations already have value for sailors in fixing localities and marking the time of night. Odysseus is instructed to always sail with the Great Bear on his left, while the following passage describes his resolve at the helm of his beloved Calypso:

> Keeping the raft to her course by the helm with the skill of a sailor
> Seated he steered (not for a moment did drowsiness fall on his eyelids),
> Holding the Pleiades in view and the autumn-setting Boötes,
> Holding moreover the Bear, that is called by the name of Wagon.
> (Ever she circles around and around on the watch for Orion,
> Having alone of all the stars no share in the baths of the Ocean.)

At about the same time Hesiod was writing in the *Works and Days:* "Then when the stars of the Pleiades, the daughters of Atlas, are rising Harvest begins, but the seeding takes place when they are setting."

The astronomy we find in the poetry of Homer and Hesiod resembles that employed by the much older civilizations of the ancient Middle East. It involves the use of celestial recurrences for the practical purpose of regulating daily activity. Not until a century or two later do we observe evidence of a clear break with the astronomical traditions of the past, as philosophers sought, for the first time, to truly explain celestial phenomena and their causes in a rational manner.

Ionians and Pythagoreans

The sixth century B.C. is one of those all too rare periods in man's history when a number of the world's civilizations made a collective movement toward the broader establishment of basic humanistic principles. Philosophical inquiry reached exhilarating levels in the East where Buddha, Confucius, and Lao-Tsu founded schools of thought that would forever change the course of human events. At the same time in the West, Greek culture, under the stimulus of the Ionians, had begun its remarkably vigorous and sustained recovery from the Greek Dark Ages (c. 1100–800 B.C.) that had cast such a long, impenetrable shadow over Hellas after the fall of Mycenaean civilization. The Ionians undertook the exploration of the seas, founded new colonies, and immersed themselves in the arts. They were the first to live the full and unfettered life which we today think of as being characteristically Greek. By the time we meet the earliest known Ionian philosopher, Thales of Miletus, who was born about 634 B.C. and lived to the respectable age of seventy-eight, something new and exciting in the human experience is clearly afoot on the coast of Asia Minor, something Arthur Koestler diagnosed as "Ionian fever" [6]—that Promethean quest for rational explanations behind natural occurrences.

Thales and his Ionian followers initiated a new method of attempting to explain the structure and behavior of things in the heavens by constructing what we would call theoretical models, the same method employed by contemporary mathematicians and physicists to solve problems posed by their research. They then sought confirmation for their hypotheses through analogy; that is, they compared bodies in the heavens with the more familiar objects on the earth in the belief that by constructing an acceptable model of things terrestrial an analogous model of the sky could be formulated.[7] This type of theoretical thinking can be most helpful, particularly in such fields as astronomy, but several centuries passed before major discoveries were made. From what we know of the cosmological concepts of Thales, they were little, if any, more developed than those of Homer and Hesiod.

Thales believed water to be the first principle of all things, and he conceived of the earth as a circular disk afloat on a great sea. "This is the most ancient explanation which has come down to us," wrote Aristotle, "and is attributed to Thales of Miletus. It supposes that the earth is at rest because it can float like wood upon water."[8] Thales apparently observed that all things contain at least some moisture and he therefore reasoned that water must be the primary matter of life. Like Homer, he believed in a finite universe capped by a celestial vault; on the other hand, he says nothing about the lower realms. However, since water is the first principle, we can assume that the great sea upon which the earth floats requires nothing to support it, and that like the heavens it has prescribed limits. Nor are we informed of the nature of the sun and the moon, or of what becomes of them between their setting and rising. Apparently, as in the cosmology of the Egyptians, they are borne across the waters to begin their journey anew every day.

It is in the writings of Herodotus that Thales earned his greatest, albeit highly dubious, claim to fame: We are informed by the Greek historian that in 585 B.C. the philosopher successfully predicted an eclipse of the sun which allegedly occurred during the height of a battle between the Persians and the Medes. This hardly seems probable, however, given the rather unsophisticated nature of Thalian cosmology. Yet it is known that the Ionian traveled a good deal, and perhaps he acquired such knowledge during a lengthy stay in Egypt, where astronomers had obtained accurate information about the motion of the sun and moon from the Chaldeans. In either case, the knowledge of the universe imparted by Thales is inconsequential. What makes him an important figure in the history of astronomy is that he asked questions that, so far as we know, had never before been posed: He addressed those questions not to an oracle, or a priestly hierarchy, or to the gods of Olympus—he addressed them *to nature*. This probing of the forces and process of the physical world, divorced for the first time from the strictures imposed by the inflexible tenets of an all-powerful State theology, constituted a radically new departure from the old methods of thought about the world and universe, and marks the true beginning of a new and exciting chapter in the intellectual history of Western man.

Anaximander (c. 611–546 B.C.), also a citizen of Miletus and a younger contemporary of Thales, is the second of the major Ionian philosophers. His concept of the universe involved a radical departure from that of Thales, because he introduced the idea of "infinity." Anaximander reasoned that the earth is in the very center of the universe, freely suspended without support. It maintains its position because it "is equably related to the extremes" and "has no impulse to move in one direction— either upwards or downwards or sideways." [9] He abandoned the idea that water or any other known material might be the first principle by arguing that only a substance without such common qualities as weight, color, or odor can be infinite, indestructable, and everlasting. All things are developed out of and return to this primal matter, so that an infinite number of worlds have been created only to be dissolved back into the eternal mass.[10]

Anaximander's earth, shaped like a cylindrical column with a width one-third its height, has a flat or slightly convex surface and bottom, and it floats upright surrounded by air. The heavens are spherical, enclosing the atmosphere, while celestial bodies are shaped like the solid, thick wooden wheels common to the carts and wagons of ancient times. The various celestial bodies are not what they appear to be: They are not objects in the concrete sense; rather they are enclosed rings of fire with numerous holes in their outer surface from which the light of the contained flame emanates. If one pictures a series of tires rotating about the cylindrical earth with punctures that emit bright streams of light—the sun, moon, and stars—one has a fair idea of what Anaximander's universe looks like. We are also told that the phases of the moon and the regular disappearance of other heavenly "bodies" are caused by regular and recurrent stoppages of the punctures, as are the more infrequent eclipses. And for the first time in the history of Greek astronomy, Anaximander gives some relative measurements of the heavens in relation to the earth: he estimates the circle of fire called the sun to be twenty-eight times as large as the earth, while the moon is a circle nineteen times as large. Given the fact that the earth is not smaller, but actually much larger than the moon, it is rather difficult to be charitable about the gross inaccuracy of his calculations; they are probably best forgotten, although this first attempt at measurement is very important in and of itself.

From the description of Anaximander's theoretical model of the universe, it seems obvious that he had a fertile imagination, for it is not easy to see how he arrived at his rather exotic conclusions. Koestler observes that "the machinery looks as if it had been dreamed up by a surrealist painter" while "the punctured fire-wheels are certainly closer to Picasso than to Newton." [11] Still, the use of the wheel is the first tentative attempt to construct a mechanical model of the universe, a process that would only be completed over twenty-two centuries later by Newton in a brilliant synthesis that completely revolutionized man's picture of the cosmos.

The third major philosopher of what we might call the "Ionian triumvirate" was Anaximemes (c. 585–526 B.C.), also a citizen of Miletus and an associate, if not a pupil, of Anaximander. According to Anaximenes, the earth is flat, like a table, but instead of being freely suspended, as with Anaximander, it rests upon air which is the first cause of all things, and out of which all primary materials are formed. The sun, moon, and stars are also flat and are kept from falling by the support of this same primal material. What makes this theoretical model unique, and of interest to the historian of science, is the concept of the heavens formulated by Anaximenes. He believed the stars to be attached "like nails" to a celestial vault composed of a solid but transparent crystalline material which went round the earth much "like a hat round the head." The heat of the sun is caused by the friction produced against the air by this rapid rotation. The stars, however, do not emit discernible heat because their position on the heavenly dome places them much too far away from earth. This idea of the crystalline vault, pioneered by Anaximenes, seemed so convincing that it was later adopted and modified into the well-known concept of transparent concentric spheres by Aristotle, after which it became a dominant part of both ancient and medieval cosmology. As late as the seventeenth century, Johannes Kepler spent years of agonizing frustration in an unsuccessful attempt to tie this idea into his geometric model of the universe. And Galileo, even with telescope in hand, was unable to convince certain die-hard Aristotelians that the mountains and craters of the moon were not covered by a smooth and transparent crystal globe. One wonders if Anaximenes, while in a charitable frame of mind, and had he been able to foresee what great difficulties his fanciful crystalline vault would bring to later watchers of the skies, might not have abandoned it in favor of another, less contentious, explanation.

While the intellectual fever that struck the Ionian philosophers never assumed the dimensions of a plague because of the relatively small numbers it infected, neither was this "malady of the mind" confined to the coast of Asia Minor. Like the spores of a fungus it wafted outward on the wind, across the waters of the Aegean, to infect Pythagoras, one of the major figures in the early history of science. He was born in the early sixth century B.C. on the Aegean island of Samos, but he settled between 540 and 530 at Kroton in the south of Italy, where his immediate followers and those who later thought they were continuing his work called themselves Pythagoreans, after the master. As much as, perhaps even more than in the case of any other Greek school of philosophical thought, it is impossible to separate the contributions of Pythagoras from those of his closest followers, hence no attempt will be made here to distinguish one from the other. Instead, we will examine the basic features of this philosophy and its influence upon subsequent developments in mathematics and astronomy.

Pythagoras, son of the artisan, Mnesarchos, and follower of Anaximander, was probably born between 585 and 580 B.C. From the accounts of his followers, we know that he travelled extensively in the East, as did many educated Greeks, and seems to have been much influenced in his scientific thinking by what he learned in Egypt, Babylonia, and Asia Minor. According to a story told by Aristoxenus and later repeated by Herodotus, Pythagoras was in the employ of Polycrates, the enlightened tyrant of his native island of Samos, as a diplomatic representative during his eastern travels. After he returned to the court of his king, Pythagoras became increasingly disenchanted with what he considered to be a creeping despotism not evident in the early years of Polycrates' rule. Perhaps out of disillusionment or fear for his personal safety, Pythagoras removed with his family to the less politically hostile environment of the southern Italian peninsula, where he completed his greatest work.

The major currents and philosophical ramifications of the Pythagorean school are many, but they are all based upon one major premise —*number is everything.* The Pythagoreans were the first to use numbers to express basic relationships that had been observed in natural phenomena. The best example of this is the famous Pythagorean discovery in music that a note's pitch is dependent upon the length of the string which produces it, and that the musical intervals in the scale are produced by the simple ratios: 2:1, octave; 3:2, fifth; 4:3, fourth; and so on. Not only was it thought that music is based upon numerical proportions, Pythagoras and his followers conceived of everything in nature as being governed by musical (i.e., mathematical) relationships, including the movements of the planets and the stars. They believed the celestial motions are performed with exacting regularity, just as the harmony of musical sounds depends on regular intervals. To the Pythagoreans, the cosmos is nothing less than an orderly and beautiful musical instrument, perfectly attuned by the harmonious relationships associated with numerical regularity. Each of the celestial revolutions produces a different tone "so that each of the planets and the sphere of the fixed stars emitted its own peculiar musical sound, which our ears are unable to hear because we have heard them from our birth, though afterwards [it was] asserted that Pythagoras alone of all mortals could hear them." [12] Here, for the first time in the early history of Greek cosmology, is a unified vision of the heavens. All of the components of the Pythagorean model interlock, each absolutely necessary to the proper operation of the whole. The very word "Kosmos," meaning order and harmony, as opposed to "khaos" or total disorder, is often directly attributed to Pythagoras and his early followers.

The other information that comes down to us from such ancient writers as Diogenes regarding the composition and operation of the Pythagorean universe is rather scant and highly contradictory. Pythagoras is supposed to have taught that the world is composed of the four elements earth, water, air, and fire—a central tenet of Aristotelian cosmology, but an innovation credited by most scholars to Empedocles of Agrigentum in the fifth century B.C. We are also informed that Pythagoras was the first to recognize that Phosphorus and Hesperus, the morning, and evening stars, are one and the same, and that he thought the moon to be a mirror-like body moving like the planets in celestial orbit. It would also appear that he believed the earth to be a sphere, and he maintained that the center of the universe contained a great fire around which the earth—as one of the stars—moves. Aristotle observed that the Italian school called Pythagoreans "affirm that the center [of the universe] is occupied by fire, and that the earth is one of the stars, and creates night and day as it travels in a circle about the center." [13] This is not to be confused, as has been done by some scholars, with the concept of heliocentrism first advanced by Aristarchus and later taken up by the Renaissance cleric Copernicus, for Pythagoras believed the sun also revolved around the central fire. Because no one had ever seen the central fire, it had to be supposed that the orbit of the earth lay in the plane of its equator, so the known parts of the earth, including Greece, were always turned away from the center of the orbit.

To show how deeply attached (some would say slavishly committed) they were to the concept of numerical harmony, the Pythagoreans played mathematical games on a cosmic scale whenever it suited their purpose: the classic example is their use of the number ten. To them ten was a perfect number because it is the sum of the first four numbers; i.e., $1 + 2 + 3 + 4 = 10$. Thus, they believed there must be ten bodies moving about the finite universe. However, only nine were visible to them: the earth, sun, moon, the five planets, and the sphere of the fixed stars. To reach the number ten, the Pythagorean philosopher Philolaus, a native of southern Italy and a contemporary of Socrates, invented the antichthon, or counter-earth. J. L. E. Dreyer observes: "This tenth planet is always invisible to us, because it is between us and the central fire and always keeps pace with the earth; in other words, its period of revolution is also 24 hours, and it also moves in the plane of the equator." [14] How convenient it must have been to indulge in such soaring speculations as the theoretical existence of the central fire and the antichthon without subjecting one's theories to absolute refutation by skeptical critics. Still, we must not let such enraptured detachments from reality as these, which increasingly preoccupied the members of the Pythagorean

Brotherhood, obscure its highly valuable contributions to the history of science.

Unlike their contemporaries, the Ionians, the Pythagoreans were non-materialistic in outlook. During their inquiry into the nature of the terrestrial and celestial realms, the Ionian philosophers were primarily concerned with the stuff of which the universe is made. For Thales it was water, for Anaximenes air, and for still others fire was the primary element. The Pythagoreans, on the other hand, stressed "the marshalling and defining function of numbers," and out of numbers the "orderly structure of the physical universe." [15] They truly believed (but could not discover) that the distances of the planets from the center of their orbits fits a simple mathematical formula, a quest that drove Kepler some 2000 years later to the formulation of the first planetary laws. There can be no doubt that Pythagoreanism was instrumental in establishing a close connection between mathematics and physical science which, as E. J. Dijksterhuis points out, was to become one of the two major pillars of classical science. They were, however, like the founders of the other Greek schools of natural philosophy, largely unconscious of that other major support so necessary for a deeper understanding of nature—empirical research.

Doubtless the Pythagorean passion for numbers would have led them to even greater discoveries in the science of mathematics had they not gotten sidetracked by their overzealous dedication to geometry as a universal key for unlocking all of nature's multitudinous secrets. They abandoned the gods of Olympus for a new deity—quantification—and attempted, as are an increasing number of scholars in our own era, its extension into the realms of politics, art, philosophy, and other human forms of expression, often at the expense of nonmaterialistic values. Some of their experiments, however, are most interesting, as in the Pythagorean application of music as a cure-all for what is known today as mental illness: Disturbed patients were induced to dance themselves into a frenzy to the accompaniment of pipe music and drums, after which, exhausted by the expenditure of a great amount of physical energy, they would hopefully fall into a deep curative sleep. Indeed, the Pythagoreans looked upon the body, much as they did the heavens, as a musical instrument whose various parts must be correctly attuned to assure proper balance. Koestler points out that "the metaphors borrowed from music which we still apply in medicine—'tone,' 'tonic,' 'well-tempered,' 'temperance,' are part of our Pythagorean heritage." [16] Such activities, combined with the great secrecy that surrounded the organization and membership of the Brotherhood, ultimately led to a violent reaction against the order by the uninitiated citizens of Kroton, whom the Pythagoreans had come to dominate. Toward the end of his life Pythagoras was banished, while his disciples were either slain or exiled, their quarters burned to the ground. The Brotherhood was not disbanded, however. Those dedicated to the movement spread its teachings to selected converts throughout Greece and her surrounding colonies, keeping the philosophical system alive for two centuries after the death of its founder. Further, the philosophic works of later writers dedicated to the Pythagorean teachings would continue to exert a strong influence over the intellectual activities of the Classical and Hellenistic worlds for centuries to come.

It has been observed by the modern scholar B. Farrington that "Pythagoras is the founder of European culture in the Western Mediterranean sphere." [17] He and his closest followers mark the point of departure toward a European civilization markedly different from those of India and China. The ever-growing interest of Western scholars in numbers and quantification—indeed the very belief held by Galileo, Newton, and by scientists today that mathematics holds the key to unlocking the most intimate secrets of the universe—can be traced in a direct line back to the teachings of the ancient Brotherhood. Still, it was not the mathematical philosophy of the Pythagoreans that would dominate man's vision of the cosmos for the next 2000 years. Before the scientific revolution in astronomy, based on the formulation of fundamental mathematical laws, could take place during the seventeenth century, the longest-lasting and most dominant conception of the universe in Western intellectual history— the cosmology of Plato's most brilliant pupil, Aristotle of Stagira—had to be successfully challenged.

The Stagirite

At the Kunsthistorisches Museum in Vienna one may view the famous marble bust of Aristotle, copied by the Romans from the Greek original, completed sometime during the last quarter of the fourth century B.C. By this time the Greek classical style of portraying god-like physical perfection in human form, almost completely absent of individual emotion or intellectual insight, had given way to the new realism associated with the Hellenistic style. As a result, the intense and contempletive visage of the greatest philosopher in Western history stares back at the viewer from across a historical gulf of over twenty-three centuries. The dominant physical features are bulging temples and a high, massive, yet refined forehead, the latter accentuated by a gently receding hairline. The effect, so clearly intended and masterfully executed by the unknown sculptor, is to portray brilliance in the demeanor of aristocracy. The enlarged cranium strains—indeed seems about to burst apart—in its efforts to keep the impacted genius of the brain from exploding outward. If this bust is a reasonably accurate rendering of the appearance of the philosopher, as seems most probable, it is no wonder that Aristotle's mentor Plato, a brilliant man in his own right, perhaps not without a touch of envy, chose to call his gifted pupil "the Mind"!

Aristotle (384–322 B.C.) was born in the small Greek colony of Stagirus on the Chalcidic peninsula of Macedonia in northern Greece. Nicomachus, his father, was court physician to the king of Macedonia, Amyntas III, father of Philip II and grandfather of Alexander the Great. No doubt Aristotle's childhood interests in nature and medicine were stimulated by his close association with his father's profession, which in classical times was passed down, through the apprenticeship system, from father to son. Although we have no conclusive evidence, it seems highly likely that Aristotle mastered the basic fundamentals of medicine, perhaps acting as his father's apprentice. Additional support for this theory derives from the fact that the subject was studied in Aristotle's school, the Lyceum, which he founded in 335 B.C. During his formative years, he also had the opportunity to view at first hand the machinations of the Macedonian court, whose lack of refinement and ever increasing appetite for political and military power generated in him an intense and life-long distaste for kings and princes. Coupled with his close association with medicine, it explains why he decided to follow the career of scientist and natural philosopher. Perhaps the reason why Aristotle did not become a physician like his father is that Nicomachus died while his son was still a youth, and Proxenus, a paternal relative, became his guardian. Then in 367 B.C., at the age of 17, Aristotle was sent to study at the Academy of Plato in Athens, where he remained for the next twenty years. Thus, along with Socrates and Plato, Aristotle became the third member of the great interconnected intellectual trinity of classical Athenian philosophers.

The teaching of science at Plato's Academy was deeply influenced by the Pythagorean passion for mathematics. Over the Academy entrance were written words to the effect, "Let no one enter here who knows no geometry." But despite the considerable writings of Plato that have survived, it is difficult to determine just how well Plato himself had mastered geometry. However, several students educated at the Academy made important discoveries in mathematical science, both pure and applied. The Platonists kept alive the fundamental teaching of Pythagoras and his followers, that theories about nature should rest primarily on mathematical principles, and the belief that their proof should take the form of rigorous argument, the validity of whose propositions should be made clearly apparent to any rational thinker.

But there was an even stronger strain in Platonic thought, one that counseled a withdrawal from the material world, particularly from the idea that observation and sense experience can provide us with a true knowledge of reality. Plato taught that the phenomena of the world and universe are merely imperfect and transitory reflections of ideal forms whose essence is spiritual, not corporeal. They are only limited reflections of formless matter that never completely yields to conceptualization through the senses. In the *Timaeus*, a Platonic dialogue specifically devoted to physical questions, Plato freely mingles myth with science to the point where sharp differences of opinion have developed over the structure of his cosmical system. Nor is it essential to this study to go into the various interpretations of Plato's universe, for his theoretical model exercised little, if any, lasting influence over the subsequent development of scientific theory. In fact, Plato's disdain for empirical inquiry—the very heart of Aristotle's approach to science—was a major step backward in the scientific tradition established by the early Greek cosmologists. On the other hand, even though Aristotle rejected many of the basic tenets of Platonic cosmology, by popularizing through his writings the Pythagorean doctrine of the spherical earth and the orbital motion of the planets from east to west, which he had learned at the Academy, he did ensure that Plato's scientific teachings did make a lasting contribution to the history of astronomy. Moreover, the Platonists would vie with the Aristotelians for over 1500 years to establish the supremacy of their respective philosophies. Yet in the final analysis, it is the religious quality and the poetical nature, not the scientific content, of such dialogues as the *Timaeus* that earned them a lasting and respected place in the classical literature of the world.

The legend developed that after the master's death (c. 348 B.C.), Aristotle left Athens in frustration and anger because Plato's nephew, Speusippus, rather than Aristotle, was named head of the Academy. In light of the high esteem in which Aristotle had been held by his teacher, it is tempting to believe the story is not without some basis in fact. A more likely explanation, however, is that Aristotle was ineligible to become the Academy's head because his status as a resident alien prohibited him by law from holding property, something Aristotle himself would have known from the outset. In any event, he absented himself from Athens for the next twelve years, while, in the company of a small group of friends, he traveled across the Aegean to establish philosophic circles at Assus and Mytilene in Asia Minor. It was during this period that Aristotle composed his treatise, *Politica*, in which he elaborated upon the elitist concept, borrowed from Plato's *Republic*, that the highest purpose of the polis (State) is to establish conditions in which a minority of enlightened citizens can live the good life, meaning a life based upon philosophical inquiry. Aristotle's spurning of classical Athenian democracy has earned him, as it did Plato, considerable opprobrium from contemporary scholars. Without attempting to offer a complete defense or rationalization for his motives, it should be kept in mind that due to their conquest by the Macedonians the Greeks saw their idyllic world crumble under the weight of political, economic, and moral bank-ruptcy. Classical Athens, as the center of a citizen-controlled political system, was gone forever, so that Aristotle's advocacy of rule by an en-lightened oligarchy constituted somewhat of an advance over, rather than the complete capitulation to, the despotic political climate of the times. Above all, he rejected the concept of empire embraced by the Macedonians and later practiced with such consummate skill by the Rome of the Caesars. And despite his lack of confidence in direct de-mocracy, he still preferred the polis, with its limited geographical and political boundaries and its continued emphasis on individualism, to the amorphous anonymity of citizenship in a sprawling empire under the iron-fisted control of unchecked despots. It is indeed one of the ironies of ancient history that this advocate of small political states would be-come the tutor of the mightiest conqueror of the ancient world, Alex-ander the Great.

At about the age of forty-two, Aristotle was invited by Philip II of Macedon to undertake the instruction of his son, then thirteen, at the Macedonian capital of Pella. The philosopher accepted. Aristotle looked to the greatest of Greek teachers, Homer, and chose the epic hero Achil-les as the instructional model for Alexander. It goes without saying that the young Macedonian became one of the greatest warriors of all times, and from this point of view it might appear that Aristotle succeeded with his pupil in a way few teachers dare to dream. However, most scholars agree that his influence on Alexander was somewhat negligible. Even before meeting his great teacher, Alexander had inherited from the court of his father the Macedonian predisposition for the martial arts; and he showed little evidence, before his untimely death, of putting into prac-tice the type of rulership based on the principles of philosophy so strongly advocated by his mentor. After three years at the Macedonian court, in 339 B.C. Aristotle settled at Stagirus, on the estate left him by his father, where he remained until 335 B.C., when he returned to Athens at about the age of fifty. It was against this background that Aristotle developed the cosmology which dominated Western intellectual and religious history for two millennia and was taught for the first time in the Lyceum, the rival institution of the Academy founded the same year he returned to Athens.

In discussing Aristotle's conceptual model of the universe, it becomes necessary to stress the limitations of his cosmology as well as the nu-merous errors of fact associated with them. Moreover, because his system held sway for so long, and toward the end of its reign degenerated into a kind of fanatical dogmatism in the hands of its medieval scholastic supporters as against the new and what are considered today the more satisfying discoveries of the Renaissance astronomers, it is easy for the contemporary scholar to consign this philosopher's work to that category of pseudoscientific thought considered best forgotten because of its repressive and error-producing influence. For example, Arthur Koestler, one of the most perceptive writers of the twentieth century, speaks of Aristotle's system as a "failure of nerve." [18]

Looking back over the history of astronomy and physics, it is certainly true that Aristotle's cosmology did not, in certain respects, serve the cause of scientific inquiry very well. Still, there is much in his system that deserves our attention, for certainly no one can read his *De Caelo* (*On the Heavens*) or *Metaphysics* and not be left with a strong sense of appreciation and great admiration for the depth of the philosopher's insight and the tremendous range of his intellectual powers, especially when applied to the difficult problem of determining the bases of natural causation. Furthermore, it was not by deception that his concept of the universe became the predominant Western cosmology from the decline of Hellas through the High Middle Ages. Rather it was through a powerful combination of logic, observation, and synthesis that his system gained almost total acceptance, not only by countless lesser lights, but among such men of brilliance as medieval Christendom's most gifted scholar, St. Thomas Aquinas, and its greatest poet, Dante Alighieri. If Aristotle is to blame for slowing the course of what we think of today as "scientific advancement" and "progress" for over twenty centuries, it is because he argued so well, not because his cosmic system resulted from a failure of nerve. It is important that this be kept in mind during any examination and evaluation of the Stagirite's theoretical model of the universe.

To begin with, Aristotle taught that all things in nature, then as now, are composed of a combination of basic elements. Instead of the large and ever-increasing number accepted today, however, Aristotle accepted the four elements identified by Empedocles of Agrigentum in the fifth century B.C.: earth, water, air, and fire. These elements were believed to follow certain patterns of movement, or what we would want to define as scientific laws today. But Aristotle did not conceive of natural laws based on mathematical principles. The best he could do was to attribute to his elements certain *inherent tendencies*. Because the elements earth and water are heavy, they express their ideal nature by moving downward, while air and fire, because of their lack of weight, move upward. Each of the elements has a natural center or resting place: for earth and water it is in the earth on which we live, for air and fire it is in the heavens. Aristotle's elements are not immutable, and all substances in nature, even those bearing the names of the elements, are composed of at least two of the four elements, though one or another is always predominant. In very solid and heavy bodies such as metals, for example, the earth content is very high, but water is also present. Organic objects, including the human body, as well as inorganic objects are composed of the basic elements. In *Antony and Cleopatra* William Shakespeare consciously played the part of an Aristotelian when he provided the dying queen of Egypt with the lines: "I am Fire, and Ayre, my other elements I give to baser life."

Aristotle further maintained that the four elements express themselves in the physique and temperament of the individual. Each element corresponds to one of the four body fluids (humors) identified by the Greeks, each of which governs a basic human emotion:

Element	*Body Fluid (Humor)*	*Emotion*
Earth	Black Bile	Melancholy
Water	Phlegm	Phlegmetic
Air	Yellow Bile (Choler)	Ill tempered
Fire	Blood	Sanguine

Both the ancient and medieval practitioners of medicine believed that illness was the result of an improper balance of the four humors or inclinations. To restore the proper balance, they employed techniques ranging from hot and cold baths to purgatives and surgical bleeding. We are reminded that Aristotle's concept of the humors held sway even longer than his model of the heavens, when it is recalled that, in 1799, George Washington succumbed to pneumonia complicated by general physical weakness resulting from the bleeding prescribed by his physician, and that in the twentieth century mental patients were treated by being immersed alternately in tubs of steaming hot and freezing cold water.

The elements may also be transformed from one into another, owing to the fundamental qualities inherent in each. Because it rises quickly, Aristotle attributed to fire the complementary qualities of hot and dry, the opposite of water which is cold and moist. Earth, on the other hand, is dry and cold, while its opposite, air, is moist and hot. The transformation from one element to another takes place most readily between two elements that have one quality (i.e., heat, cold, dryness, or moistness) in common, as the unbroken lines in Figure 1 illustrate. Theoretically, however, the transformation of total opposites, fire into water and air into earth, is possible, as indicated by the broken lines.

It is the striving of each of the elements to reach its natural place that keeps the world and the universe going. It is as if each element possesses an independent "will" through which it seeks to fulfill its own essence. As long as the elements never attain their natural places in an absolute sense, things will continue much as they are. Only if universal fulfillment were achieved, would the cosmos come to a complete rest. Were this to happen—if the elements no longer mixed because they had all reached their natural places—a four-sphere universe would result. At

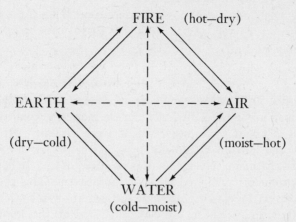

FIGURE 1.   The pattern of transformation of the Aristotelian elements.

the center would be a solid core of earth. Surrounding the earth would be an ocean covered in turn by a sphere of air. Finally, these three spheres would be encased in a circle of enduring fire, fitting like a cap over all the rest. Depending upon one's point of view, this set of conditions can either be considered the supreme example of God's perfect handiwork, a universe no longer restless but fixed forever because each of the elements has found its ideal home, or else it would be a dead universe, deprived of the animation and vitality associated with life. Aristotle apparently thought the latter because he does not envision a time when the mixing action of the elements caused by the sun shall cease. To him the universe as a whole is ungenerated and indestructible.

It was Aristotle who established the supremacy of the geocentric or earth-centered universe for centuries to come. He believes that because of its heaviness, the earth cannot be placed anywhere but at the center of the universe. This is only natural for what is heavy or has weight always falls to earth in quest of its natural place, while lightness is motion away from the center toward the heavens. He also considers the earth to be a sphere and gives two basic reasons for reaching this conclusion, one metaphysical the other empirical. In the first place, Aristotle argues that when heavy particles move uniformly from all sides toward a common center, as do the heavy elements toward earth, they must form a body whose surface is everywhere equidistant from the center—hence, a sphere. Secondly, he relies on the direct observation of lunar eclipses:

> If the earth were not spherical, eclipses of the moon would not exhibit segments of the shape which they do. As it is, in its monthly phases the moon takes on all varieties of shape—straitedged, gibbous and concave—

but in the eclipses the boundary is always convex. Thus if the eclipses are due to the interposition of the earth, the shape must be caused by its circumference, and the earth must be spherical.[19]

Neither does he consider the earth to be a very large sphere in relation to the sun and other stars, because "certain stars are seen in Egypt and the neighborhood of Cyprus, which are invisible in more northerly lands, and stars which are continually visible in the northern countries are observed to set in others." [20] In other words, were the earth a larger body, the angle of the horizon would not bend as sharply as it does so that such a relatively minor change in the position of the stellar observer would not produce this altered view of the heavens. This conclusion is essentially correct except that Aristotle did not know the stars are much farther away than had been estimated by any of the known Greek astronomers. Thus, by making celestial observations several hundred miles apart, it was easy to underestimate the earth's circumference. Aristotle tends to accept the earlier calculation of the earth's circumference done by an unknown mathematician of about 400,000 stades or 9987 miles.[21] This is about two and one-half times less than the actual circumference of about 25,000 miles. He also believes the earth is at rest, because this state corresponds best with the information collected from celestial observations that indicate it is the planets and stars, not the earth, which are in motion. An even more obvious proof of this is the total lack of the sensation of movement beneath our feet that we should all experience were the earth not at rest. And if a heavy object is dropped upon the surface of the earth it appears to fall directly upon the point above which it was dropped instead of landing at a point some distance away, as it should if the earth spins beneath it after it is released.* Aristotle, of course, was wrong, but no one had a better or more satisfying explanation for these phenomena, nor would any be available for a long time to come.

Above the terrestrial or sublunary region, where rests the motionless earth composed of the two heaviest elements and surrounded by air and fire, there is a second realm of matter that operates under a different set of natural principles. This is the eternal and unchanging celestial realm, the home of the sun, the stars, and the planets. It is composed of a fifth element, much more pure than the rest, which takes over above the upper part of the atmosphere where fire predominates. The heavens operate according to a set of physical principles radically different than those of the terrestrial region, a phenomenon directly attributable to the presence of this quintessential substance. Aristotle comments:

> It seems that the name of this first body has been passed down to the
> present time by the ancients, who thought of it in the same way as we do,

---

* A more detailed discussion of this question is undertaken in chapter 8.

for we cannot help believing that the same ideas recur to men not once or twice, but over and over again. Thus they, believing that the primary body was something different from earth and fire and air and water, gave the name *aither* to the uppermost region, choosing its title from the fact that it "runs always" and eternally.[22]

This eternal motion, resulting from the presence of ether *(aither)*, is not the rectilinear or straight line up and down movement displayed by the elements on earth. Rather it is circular or what Aristotle calls "primary motion." He, like the Pythagoreans, considers the circle to be the perfect geometric construct because it has no beginning and no end. It is eternal, forever returning upon itself in constant and unbroken rhythm, "for things only cease moving when they arrive at their proper places, and for the body whose motion is circular the place where it ends is also the place where it begins." [23] Unlike the four sublunar elements, each with their opposite in qualities (e.g., hot vs. cold) and rectilinear motion (up vs. down), ether has no contrary qualities and no contrary motion. It forever remains unalterable in form as it moves through the celestial realm in eternal rotation.

Aristotle further believes that the universe is finite in nature. "The world in its entirety is made up of the whole sum of available matter, . . . and we may conclude that there is not now a plurality of worlds, nor has there been, nor could there be. This world is one, solitary and complete." [24] This is so because every element moves naturally in a certain direction and toward a certain fixed goal. This means that all earth in the universe must move naturally towards the same center, just as the other elements must seek their true places. Since there are only four elements, they cannot exist elsewhere than in this single finite universe because a duplication of universes composed of the same elements would violate their natural tendency to seek a universal center; it would make such tendencies self-negating and destroy any possibility man might have of understanding the nature of causation. Furthermore, since the universe is bounded by the spherical ethereal realm which moves in a circle about the earth, infinity is ruled out because the completeness of the circle makes the infinite extension of matter impossible. Only the straight line has no beginning and no end, but rectilinear motion is strictly confined to the corrupt and changeable sublunar region. All matter in the universe is encased in a spherical shell with clearly circumscribed boundaries. For Aristotle the concept of infinity has only one meaning when applied to the heavens: "to be everlasting" [25] and not, as Descartes believed, to be infinite in extension.

If we step back at this point and draw a diagram of the Aristotelian universe, it will appear as follows: Situated in the center will be the spherical earth bounded by nine circular layers, one enclosed by another like encased tubes. Moving outward from the earth and its atmosphere

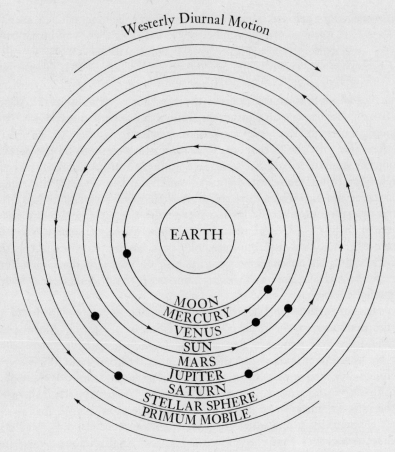

FIGURE 2.   The Aristotelian universe.

of water, air and fire we encounter the first seven crystalline spheres which carry in succession the planets: Moon, Mercury, Venus, Sun, Mars, Jupiter, and Saturn. Above them is the Stellar sphere of the fixed stars and above it the entire system is capped by a ninth sphere, the Primum Mobile or "First Mover." Each of the spheres revolves once a day about the axis of the heavens with the earth as its center.

Aristotle borrowed his concept of the homocentric spheres from the fourth-century B.C. mathematician, Eudoxus of Cnidus, who assumed that all the spheres necessary to account for the motions of the heavens are situated one inside the other in concentric pattern about the earth. Every celestial body was thought to be located not inside but on the equator of a crystalline sphere which revolves with uniform speed round the heavens, carrying with it a planet or, in the case of the Stellar sphere, the stars. The idea that the spheres are made of crystallike material explained why they are invisible to the terrestrial observer, thus making it appear that the celestial bodies move independently of any motive force. Only the outermost sphere—the Primum Mobile—carries no celestial body on its daily rotation around the heavens; instead it turns not only itself but all the other spheres as well. It is this master or primary gear that sets in motion the Stellar sphere, which in turn drives the sphere of Saturn, Saturn drives Jupiter, etc. The entire celestial model revolves around the motionless earth once every twenty-four hours.

Another interesting facet of Aristotle's concept of the universe was his belief that the closer the celestial sphere and its planet to the earth, the less perfect their composition. While it was thought that the entire celestial realm is composed of an invisible and very subtle ethereal matter, it was also observed that the surface of the moon, the closest "planet" to earth, is shadowed and seemingly flawed, much like the terrestrial surface. This is probably not true, however. Rather it results from the reflections off the crystalline lunar surface of the mountains and valleys located on the earth. Further distortions are caused by the movements of the elements water, air, and fire through the region between the earth and the moon. Perhaps, too, some of the fire and air in the upper regions of the earth's atmosphere did reach the moon in very limited quantities contributing to its visual imperfection. Since these imperfections appeared not to pertain to those planets located at a greater distance than the moon, it was concluded that their further removal from the terrestrial realm gave them an even greater perfection. Like many of Aristotle's explanations there is room for various interpretations.*

It is also worthy of mention that Aristotle did not believe that, simply because some stars are brighter than others this phenomenon could be explained on the basis of a variation in their relative positions. He concluded that all the stars are equidistant from earth; some stars simply shine more brightly than others. As for their apparent twinkling: "The planets are near, so that our vision reaches them with its powers unimpaired; but in reaching to the fixed stars it is extended too far, and the distance causes it to waver." [26] It is most interesting that one who placed supreme faith in the observation of celestial bodies in the construction of his theoretical model of the universe should have concluded, in the absence of a more satisfactory explanation, that the twinkling of the stars

---

* It is of interest to note that many centuries later the most rigid of the Aristotelians refused to entertain the possibility that the lunar surface might be corruptible and subject to change even, as was earlier discussed, after Galileo's telescope had provided the necessary visual proof. They stubbornly insisted either that the image viewed through his eyeglass was a distorted and therefore false one, or that the moon's surface was crystalline and transparent so that the observer was looking through an outer shell into the moon's craggy interior. With this latter explanation, however, there was at least a tacit acceptance that the moon was less pure in composition than had been previously believed, even though all future change on the planet was subsequently blocked by an invisible shield of crystal.

is caused by a distortion of visual perception. It is an explanation similar to that used to account for the moon's mottled surface and points out Aristotle's own awareness of the dangers inherent in relying too heavily upon sense impressions, which are often contradictory and misleading, but which nevertheless remained at the center of his investigations.

As in all self-contained or closed systems of the type constructed by Aristotle, there remain a number of questions that have never been satisfactorily resolved. Typical among them was the difficulty the Stagirite encountered in his attempt to explain the light emanating from the stars. Almost to a man, his Greek predecessors had believed the stars to be composed of fire, but not so Aristotle. The stars in his system are made of the fifth element in which they move—ether. How, then, are the stars lit if not by fire? "The heat and light which they emit are engendered as the air is chafed by their movement." [27] But this attempt at explanation raises an even more fundamental question: How can light be produced from the heat generated by the intense friction of a moving body on the surrounding air when that moving body is confined to a realm devoid of air? For air, it will be recalled, like the three other elements, is confined to the sublunar region. We are left by the master without a satisfactory answer to our question. Perhaps one was contained in the estimated 170 works published by Aristotle and subsequently lost during the gradual decline and fall of classical civilization. If so, there is little likelihood that it will ever be known to us. At best we can suggest, as a possible way of escaping the contradiction, that ether exists in a state of absolute purity only in the *outermost sphere* of the universe—the Primum Mobile—and that a light degree of contact with air takes place within the stellar sphere.

As the inheritors of a religious tradition whose roots antedate the writings of Homer and Hesiod, it is only natural that we should inquire about the possible existence of a relationship between Aristotelian cosmology and the belief in a supreme being. Moreover, it is a question which will have a direct bearing on the appropriation and modification of Aristotle's model of the universe by the theologians of medieval Christendom. By the time Aristotle founded the Lyceum, the period had long passed when the gods of the Panhellenic Olympian system could command the undivided loyalty of the Greek peoples. Beginning with Aristotle's pupil, Alexander, the dominant form of government in the Hellenistic Age was the despotism of kings, who represented themselves to their followers as at least semidivine beings. During his lifetime, Alexander was worshipped as a god in Greece and recognized as the son of a God in Egypt, where his successors, the Ptolemies, made systematic attempts at self-deification. Doubtless Aristotle was unimpressed with this new and vulgar tendency in religion, especially since he had observed no godlike qualities in his personal association with the young

Alexander. Still, Aristotle, while rejecting the gods of Olympus, as had Socrates and Plato before him, as well as the new cults of personality, was not without certain religious principles, vaguely defined though they may be, which are manifested to some degree in his cosmological system.

Aristotle writes that "God and nature create nothing that does not fulfill a purpose" [28] and that "the activity of a god is immortality, that is, eternal life. Necessarily, therefore, the divine must be in eternal motion." He reasons that "heaven is of this nature (i.e., is a divine body), that is why it has a circular body, which by nature moves forever in a circle." [29] Are the heavens, then, the embodiment of God to Aristotle? The answer must be a qualified "yes." He does not accept Plato's teachings that the planets are living gods, who move independently about the celestial realm through their own efforts. Neither does he fashion the "divine body" in the anthropomorphic form of the Hebrew God Yahweh, who continually intervenes in human affairs by causing great catastrophes such as the Noachian flood, by establishing covenants with the Hebrew prophets, or by appearing to the righteous in such guises as that of a burning bush. Rather, Aristotle's belief in divinity centers on the existence of what Dijksterhuis calls "a supreme, immaterial Prime Mover." The Prime Mover is responsible for the motion associated with the ninth sphere or Primum Mobile that controls the rotation of the entire celestial machinery. However, He does not set this sphere in motion through an overt physical act because "that would be contrary to the fact that He is *actus purus*, and thus has nothing left to realize; perfect actuality must consist in absolute inactivity." [30] Aristotle proposes that the Primum Mobile turns as a result of a craving for perfection generated in it by the Prime Mover. He borrows from Plato the idea that matter aspires after form. The universe is animated by an all-pervasive aspiration to a higher state, a greater perfection as embodied in the Prime Mover. Implicit in this belief is the concept of hierarchical organization—known to later thinkers as the Chain of Being—in which all matter, from the corrupt and mutable elements of the terrestrial sphere to the pure and immutable ethereal regions of the celestial domain, was classified in ascending order of importance. At the apex of this pyramid stood God. It was, as we shall see, out of this pagan thread that St. Thomas would fashion the Christian vestments of the medieval universe.

In looking back over the history of the first scientific revolution, modern scholars have generally credited Newton as being not only a brilliant mathematical innovator, but the first great synthesizer of fundamental scientific principles. In one sense this is a correct reading of the past, because Newton was the first to tie together the mathematical laws which he, Kepler, and Galileo had discovered into a unified vision of the cosmos. On the other hand, Aristotle was a great synthesizer of scientific thought in astronomy in his own right, as is evidenced by his melding of the four elements of Empedocles, the homocentric spheres of Eudoxus, the mathematics of Pythagoras, and the Platonic concept of form and matter into a conceptual model patterned on the pioneering work of the Ionian philosophers. It was, of course, a model radically different from the Newtonian construct. Aristotle fashioned a two-sphere universe, one celestial, the other terrestrial, and each subject to a different set of "laws." The terrestrial sphere is corrupt and mutable, the domain of on-going change. It lacks the qualities associated with the perfection of the celestial sphere: circular motion, elemental purity, immutability. It was this breach between the world of man and the world of the heavens—lasting for some 2000 years—that Newton, through the unparalleled force of his mathematical genius, was finally able to close. Equally important is the fact that Aristotle's cosmos was limited in dimension; a finite geocentric universe with clearly circumscribed boundaries. It is somewhat ironic to think that at the very time Western man was moving outward, in an attempt to overcome the geographic and cultural limitations of life in the city-state, to form empires and establish ties with cultures hitherto unknown, Aristotle's vision of the universe remained constricted to that of a sphere whose boundaries were as rigidly fixed as those of the planet on which he lived.

Yet if Aristotle's limited vision of the cosmos established an absolute antithesis between the earth and the heavens, it also brought reassurance to a frightened world by reasserting the ancient belief in order and permanence. And if the perpetuation of his scientific philosophy by later generations gave the Stagirite what Koestler calls a "millennial stranglehold on physics and astronomy," there was implicit in its general principles the modern idea of mechanization, so appealing to a Newton or a Jefferson. What is the Primum Mobile, after all, if not a subtly disguised motor, propelling in daily orbit the gearlike bands of the planetary spheres? Had the Greek intellectuals of the Hellenistic period been more favorably inclined toward experimentation—a manual undertaking they considered beneath their dignity and class standing—they might well have established in a conscious sense that which Aristotle subconsciously suggested and what Newton openly believed in—a mechanistic universe. But for Aristotle and his followers, the belief held that the only knowledge we can acquire originates in sense-perceptions. They sought to create a physical science of qualities in which the material bearers of properties are regarded as explanatory principles in and of themselves. They were little interested in the concept of multiple causation or the fundamental interrelationships between natural phenomena. Then, with the break-up of classical Western civilization in the early Christian era, the secular philosopher, so much a part of the Greek cultural heritage, became extinct. Aristotelian cosmology was forgotten in the West for several hundred years, during which time the heritage of Greco–Roman civilization was barely kept alive in the few monastic enclaves scattered along Europe's Atlantic coast and constructed atop rocky outcroppings off the western shores of Ireland. Only 1000 years later, in the thirteenth century, was Aristotle's thought revived and cleansed of its pagan associations under the auspices of the medieval Church, an institution far more concerned with the flight of angels than with the operation of machines, either man-made or celestial.

The Forgotten Heliocentrist

When the astronomers of antiquity chose to call the seven heavenly bodies located between the earth and the stars "planets," they ascribed to them certain qualities of motion that were to baffle celestial observers until the time of Johannes Kepler. For planet means wanderer, an object whose movements are irregular and subject to change without notice. With the exception of the sun and moon, the most disturbing of all the characteristics associated with the planets were those of occasionally standing still, then reversing course and going backward for a time— phenomena known to astronomers as "stations" and "retrogressions." The Babylonians, who had deified the major celestial bodies, attributed this behavior to the will of capricious gods; but the Greek astronomers, who showed much more interest in the discovery of the natural causes behind celestial occurrences, were unwilling to accept such facile explanations. The Greek natural philosopher who wanted to construct a convincing theoretical model of the universe had to somehow account for these irregular meanderings in a more scientific manner. The search for an acceptable solution to this problem had become the major preoccupation of Hellenic astronomy by the fourth century B.C. And with the general acceptance of Plato's dictum that all celestial bodies must move in perfect circles, the task of Greek astronomers was narrowed even further; it now became necessary to find a coherent system whereby all planetary movements, no matter how irregular, were the result of one or a combination of uniform, circular motions.

As we have seen, Aristotle attempted to explain planetary motion by borrowing from Eudoxus the concept of homocentric spheres in which the planets move round the earth in a concentric series of circular orbits. However, Eudoxus had assigned to each of the planets not one but several individual spheres, in order to account for all of the different movements ascribed to any given body. His system is described by Dreyer thus:

> Every celestial body was supposed to be situated on the equator of a sphere which revolves with uniform speed round its two poles. In order to explain the stations and arcs of retrogression of the planets, as well as

their motion in latitude, Eudoxus assumed that the poles of a planetary sphere are not immovable but are carried by a larger sphere, concentric to the first one, which rotates with a different speed round two poles different from those of the first one. As this was not sufficient to represent the phenomena, Eudoxus placed the poles of the second sphere on a third, concentric to and larger than the two first ones and moving round separate poles with speed peculiar to itself.[31]

Eudoxus found it possible to represent the motion of the sun and moon by assigning three concentric spheres to each of these bodies, while for the more complex motions of the planets Mercury, Venus, Mars, Jupiter, and Saturn four spheres were needed. The fixed stars required only one sphere to account for their diurnal (daily) movement around the heavens because they appeared to change not at all. Thus the total number of concentric spheres in the Eudoxian system was twenty-seven. By having each sphere rotate at the appropriate axial tilt and speed the astronomer was able to approximate—in a *very general* manner—the various movements of the planets. Callippus of Cyzicus (c. 370–300 B.C.), for technical reasons that are not essential to this discussion, later added seven more spheres to those of Eudoxus to bring the total number to thirty-four. This is where matters stood when Aristotle decided to incorporate the homocentric system into his model of the universe.

In my discussion of the Aristotelian system, I stressed the broad outlines and major implications, both metaphysical and scientific, of the philosopher's universe, without going into great astronomical detail. It is, however, necessary to backtrack for a moment, if one is to gain a better appreciation for the extremes to which Aristotle was driven in order to protect the integrity of his theoretical model. This is especially true as regards his concept of the crystalline spheres.

The nine spheres of Aristotle, ranging from the lunar sphere to the Primum Mobile, were alone remembered during the Middle Ages when they became a central part of the medieval view of the universe. Yet Aristotle, like Eudoxus and Callippus, believed in nests of spheres-within-spheres as the best method of accounting for the peculiarities of planetary motion. However, he disagreed with them on one very important point. To Eudoxus and Callippus the rotating homocentric spheres were simply geometric figures without physical character; in other words, they provided a convenient method of constructing a model of the universe in purely geometric terms, but one they knew existed on paper only. To Aristotle the spheres were truly material objects, crystalline shells that surrounded and carried one another along. This presented him with a major problem not previously encountered by his nonmaterialistic fellow philosophers; namely, that the concentric spheres must be mechanically connected, while, at the same time, the movements unique to each of them must not be passed along to its neighbor. Perhaps as much out of desperation as of logic, he ingeniously inserted an additional number of spheres between the innermost sphere of each planet and before the outermost sphere of the planet immediately beneath it. These were the so-called "neutralizing spheres," which turned in the opposite direction of the working spheres. They not only tied the successive nests of spheres together but also prevented the motion of Saturn, for example, from being transmitted to Mars, and so on down to the lunar sphere. In this manner the nest of each sphere could be started from scratch so to speak, its movements made independent of those of its neighbors. By the time Aristotle had finished adding these *cordons sanitaires*, the number of spheres had reached a grand total of fifty-five.

Besides the problem of wildly escalating numbers, another major difficulty remained. Since the motion of the Primum Mobile was apparently neutralized vis-à-vis the planets by their individual protective spheres, what drove the celestial bodies through the Aristotelian heavens? Aristotle's answer: Fifty-five spirits or "unmoved souls," one for each crystalline sphere, were necessary to keep the system operating. Obviously, things had gotten completely out of hand at this point. It is no wonder that everyone from his immediate followers to the medieval scholastics of the great European universities of the Middle Ages chose to disregard this aspect of Aristotelian cosmology by paring down the homocentric spheres to the more manageable number, nine! Meanwhile, the ultimate solution to the problem of eccentric motion would have to rest in other hands.

At about the same time as Aristotle was designing a model of the universe that would successfully challenge all rival systems until the Renaissance, the last of the Pythagorean line of astronomers, Aristarchus of Samos (c. 310–250 B.C.), was developing a radically divergent, yet scientifically more accurate vision of the cosmos. For Aristarchus proclaimed that the sun, not the earth, is the center of the universe. Aristarchus is frequently referred to by ancient writers as "the mathematician," to distinguish him from other notable contemporaries of the same name. Yet due to his advanced knowledge in the field of geometry, it was a title that he might well have earned regardless of this peculiar set of circumstances. In his only extant work, *On the Size and Distances of the Sun and the Moon*, Aristarchus demonstrates both an ability as an original thinker as well as a talent for the meticulous observation of celestial phenomena. So accurate were his mathematical calculations that he improved the length of the solar year by adding $1/1623$ to the accepted figure of $365\frac{1}{4}$ days. Through the method of triangulation, he calculated that the sun was nineteen times as far from the earth as is the moon, and also that it was nineteen times as large. Modern measurements, aided by the telescope, show that Aristarchus' ratio was too small by a factor of more than twenty, but his application of mathematics to the problem of determining cosmological dimensions is indicative of the detail with which the universe was being investigated by scientists of his time. Moreover, his method of calculating the distance of the sun and the moon from the earth was followed by astronomers until the time of Copernicus.

Unfortunately, the work containing his most revolutionary ideas has not been preserved. Yet even though the historical evidence is second-hand, there can be little if any doubt that Aristarchus was the first to broach the heliocentric (sun-centered) hypothesis of the universe. Our best secondary source of information is the greatest mathematician of antiquity, Archimedes, who in his little treatise titled *The Sand Reckoner* states:

> His [Aristarchus'] hypotheses are that the fixed stars and the sun remain unmoved, that the earth revolves about the sun in the circumference of a circle, the sun lying in the middle of the orbit, and that the sphere of the fixed stars, situated about the same center as the sun, is so great that the circle in which he supposes the earth to revolve bears such a proportion to this distance of the fixed stars as the center of the sphere bears to its surface.[32]

Plutarch makes an equally important reference to Aristarchus in his work *On the Face of the Moon* when one of his characters remarks:

> Do not, my good fellow, enter an action against me for impiety in the style of Cleanthes, who thought it was the duty of Greeks to indict Aristarchus of Samos on the charge of impiety for putting in motion the Hearth of the Universe, this being the effect of his attempt to save the phenomena by supposing the heavens to remain at rest and the earth to revolve in an oblique circle, while it rotates at the same time, about its own axis.[33]

These passages, in addition to several others, offer quite conclusive evidence that Aristarchus of Samos had indeed carried the development started by Pythagoras some three centuries earlier, and continued by Philolaus and Heraclides, to its logical conclusion: the construction of the heliocentric universe in which the earth rotates on its axis once every twenty-four hours during its annual movement round the sun.

The most obvious advantage of this system was that it made obsolete the fifty-five spheres of Aristotle. No longer would the stations and retrogressions of the planets have to be masked by a metaphysical veil of ethereal crystal. Since the earth is not at the center of the planetary orbits, we see things celestial first from one side and then the other so to speak. Because the earth passes all of the other planets except Mercury and Venus (they pass earth) in its orbit about the sun, their motion appears irregular and at times even backwards or retrograde, although they are actually moving steadily round the same center. The stars are so far away that our ever changing viewpoint from earth makes no difference, for they appear to remain stationary. This was basically the same view advocated by Copernicus, albeit in much greater detail, some 1800 years later.

From the vantage point of historical perspective, this all seems so simple that one is at first amazed to learn that the work of this Greek Copernicus gained few, if any, converts, that the master had no known disciples. This major, indeed revolutionary hypothesis, over three centuries in the making, was disregarded for nearly two millennia. As Koestler points out, "This paradox would be easier to understand if Aristarchus had been a crank, or a dilettante whose ideas were not taken seriously. But his treatise *On the Sizes and Distances of the Sun and Moon* became a classic of antiquity, and shows him as one of the foremost astronomers of his time." [34] The question then remains: why did the Aristarchian system fail?

The explanation for its failure does not lie, as the above quote from Plutarch suggests, in a fear of religious persecution. Cleanthes brought no charge of impiety against Aristarchus, nor is there any suggestion in the historical record that any other adherent of an unorthodox cosmological system was either persecuted or indicted for his beliefs during the Hellenistic Age. The simplest and most plausible explanation is that the accumulated weight of scientific evidence available at the time was in opposition to the heliocentric view. Although Aristarchus put forth an appealing hypothesis on the movements of the planets and the stars, it was simply that, a hypothesis and nothing more. While his speculations happened to be correct, he did not have evidence enough to gain the support of even a small number of adherents. The geocentrists, on the other hand, could offer convincing support for their system from a number of quarters.

For example, when viewed from the earth, the relative size of the fixed stars, as opposed to the constantly shifting planets, always remains the same at all times during the year. This was accepted as clear proof that the stars never vary in their distance from earth, and that the earth is at rest in relation to them. For if the earth indeed changes positions, as Aristarchus claimed it did, it would undoubtedly produce visible changes in both the brightness of the stars and in the patterns of their constellations. Thus, the acceptance of the heliocentric doctrine required for its proper operation a universe immense in its dimensions. For if the earth is in motion and the pattern of the stars remains unchanged, it has to be supposed that the diameter of the earth's annual orbit round the sun is miniscule in comparison to the distance of the stars from earth. In other words, the effects of planetary motion on the relative position of distant objects such as the stars would be cancelled out only across a vastness of space undreamed of by the astronomers of the classical world. To illustrate this point, consider the following example: Suppose you are standing in the middle of a room looking directly at a picture hung on one of the walls. Take five steps in any direction and look at the picture again. Regardless of the direction in which you move, whether forward, backward, or sideways, your perspective of the picture changes quite markedly. Then go outside and pick out an object, such as a large tree or building, on the distant horizon. Take those same few steps in any direction again. This time your perspective remains virtually the same, for even though the point of observation is changed, your movement will be rendered visually negligible by the much greater intervening distance between you, the observer, and the object under observation. This is the very same thing that happens when viewing the stars from earth.* While they were bold, visionary thinkers, willing to consider original speculations, the Greek astronomers were not capable of making the gargantuan leap from a closed or finite world to what would have amounted to a virtually infinite universe. They collectively dismissed the prospect as untenable.

Yet a consideration far more important to the ancient critics of the heliocentric theory than the apparent absence of stellar parallax was the simple fact that the sun-centered universe defied common sense, even as it does today.

> Indeed, one's mind at first boggles at the thought of the whole material globe, together with . . . our houses, and possessions, not to mention ships, animals, and trees, hurtling every twenty-four hours over a distance of some 2,000,000 miles, and all the while spinning round eastwards at up to 1000 miles per hour.[35]

These same difficulties plagued Copernicus, Tycho Brahe, and Kepler in the sixteenth century. By then, however, the grip of classical science had been sufficiently relaxed by a major shift in philosophic outlook and by mounting empirical evidence which offered repeated indications that Aristarchus had indeed been correct after all. Even so, the balance on the scale of scientific thought was reversed ever so slowly and painfully. Given our knowledge of these subsequent developments, it would go against reason to blame the contemporaries of Aristarchus for rejecting what was to them little more than a speculative hunch. The belief in the geocentric universe was destined by the contemporary state of scientific development to prevail, as was the belief in perfectly circular planetary motion. Yet while few, like Aristarchus, questioned the proposition that

---

* Naked-eye observation of "stellar parallax"—meaning the apparent change in the position of the stars caused by a change in the observational position resulting from the orbital motion of the earth—is not possible. Only with the perfection of the telescope centuries later could stellar parallax be conclusively confirmed and used as proof of the Aristarchian–Copernican hypothesis. For a more detailed explanation of this phenomenon see Thomas S. Kuhn, *The Copernican Revolution* (Boston: 1957), pp. 157–160.

the earth occupies the central position in the cosmos, it took a man of genius, living toward the end of the Hellenistic Age, to maintain the integrity of Plato's and Aristotle's circular doctrine.

Circles on Circles

Aristotle's cosmology had become so dominant by the second century B.C. that his theoretical model of the universe had gained acceptance by a majority of Greek natural philosophers. There remained, however, one major stumbling block to a total embracement of the Stagirite's system—the yet unsolved problem of the homocentric spheres. As we have seen, Aristotle had been partially successful in explaining the stations and retrogressions of the planets by adding several spheres to those previously invented by Eudoxus. Still, the theory of the concentric spheres kept each planet at the same distance from the earth at all times during its revolution. This did not and could not account for the great changes in the brightness of the planets, particularly Mars and Venus, and made the idea of invariant distances appear contrary to both common sense and to sense perception. At this point, the first major crisis arose in the history of astronomy, and there developed three significant responses to it. First, it could simply be ignored. While the anomaly of changing planetary brightness was of major concern, no other natural philosopher had presented a more convincing theoretical model of the universe than had Aristotle. Why not, then, simply give him the benefit of the doubt? After all, no system was without certain imperfections and this one, whose advantages greatly outweighed its disadvantages, when judged on the basis of Greek scientific development up to this time, was clearly superior to its rivals. The second alternative invited modification of the Aristotelian system in such a manner that its essential framework could be salvaged, although in somewhat altered form, to better fit the empirical data. If this could be accomplished, the teachings of the master would not only prevail, they would be considerably strengthened. Furthermore, given the genius and natural curiosity of the Greek natural philosophers, it did not seem likely that they would long remain satisfied with no better explanation of the problem than that offered by Aristotle. The third, and most radical course, involved scrapping Aristotelian cosmology altogether and substituting in its place a totally new theoretical model. Depending upon the personal inclinations and training of the various individuals involved, all three courses were more or less pursued simultaneously. Some ignored the problem altogether by focusing their attention on what were considered by them to be more pressing matters; others tinkered with the model in a variety of ways, usually with little success; while still others, like the revolutionary Aristarchus, came up with sweeping alternatives. In the end, only one of the three positions prevailed; but because of the ingenious manner in which the crisis was handled, Aristotelian cosmology, while modified, not only survived, it determined Western man's interpretation of the fabric of the heavens until the Renaissance.

The man who provided Aristotle's geocentric system with a new and highly durable lease on life was Ptolemy (Claudius Ptolemaeus, c. A.D. 100–170), a celebrated astronomer, mathematician, and geographer who resided not in Greece but in Alexandria, the namesake city of Aristotle's most famous pupil. The reader who is little interested in biographical detail will be spared at this point, because virtually nothing is known of Ptolemy's life, including the dates of his birth and death. We can only fix within broad limits his most active period in astronomy by the dates, which he himself mentions, of celestial observations taken between A.D. 127 and 150. Beyond this, there is much speculation but little or no concrete evidence.

Fortunately, although there are few facts on the life of the man, his important works have survived largely intact, a fate not accorded to many of the treatises written by other Greek scientists whose personal history we know much better than Ptolemy's. He titled his major astronomical work *The Mathematical Collection*, which subsequently became known as *The Great Astronomer*, a somewhat more appropriate appellation. This title was not destined to survive either because, after the Muslim conquest of Egypt, Arab astronomers of the ninth century applied the Greek word for great, *Megiste*, to the work and prefixed the definite article *al* to it. The book has been known ever since as the *Almagest*.[36]

Rather than ignore, as had some, the differences between observed facts and Aristotle's theoretical speculations on the course of planetary movement, or present a new and revolutionary theory on the subject, as had others, Ptolemy decided to attempt a solution to the problem within the framework of Aristotelian cosmology. He makes clear from the outset his objective of taking the middle road by carefully enunciating the arguments in favor of Aristotle's physics in the first of the thirteen books into which the *Almagest* is divided. Ptolemy accepts all of Aristotle's major propositions including the concept of the geocentric universe and the belief that, due to its central location, the earth must be immovable. He contends that if the earth indeed moved, as some earlier philosophers including Aristarchus had said, then many of the phenomena experienced and observed by man would be radically changed. He trots out the now familiar, but crucial, arguments that since all bodies fall to earth, it must be at the center of the universe; and the earth must be fixed in position or otherwise an object thrown into the air should not fall back to the same place as it is observed to do. Just as important, he believes the earth to be a sphere and that the heavens move spherically around it: "If, for example, one should assume the movement of the stars to be in a straight line to infinity, as some have opined, how could it be explained that each star would be observed daily moving from the same starting point?" [37] It is at this point, in the opening pages of his lengthy treatise, that Ptolemy tips his hand; he refuses to abandon the circle—the geometrical construct so sacred to the Greeks from Pythagoras to Plato and from Plato to Aristarchus. "The circle among plane figures offers the easiest path of motion, and the sphere among solids . . . the circle is the greatest of plane figures, and the heavens are greater than any other body." [38] These principles established, Ptolemy takes on the formidable task of finding a rational explanation for the irregular motions of the planets within the context of the sphere.

The roots of the system so painstakingly developed by Ptolemy can be traced back to the great second-century B.C. Greek astronomer, Hipparchus. His contributions to the science of the skies included, among other things, the discovery of the moon's parallax, the development of a system of mapping and classifying the stars according to their magnitudes, and most important, from Ptolemy's point of view, the use of the epicycle. The original purpose of the epicycle—or circle on a circle— was to explain the retrograde motion of the planets. It worked in the following manner: a large circle, the deferent, D, (Figure 3a), was drawn on a piece of paper to represent the path of the planet, P, as it circles the earth. Next, a smaller circle, the epicycle, E, was drawn with its center on the circumference on the deferent. During the time when the planet is at the arc on the epicycle farthest away from earth, it appears to be at

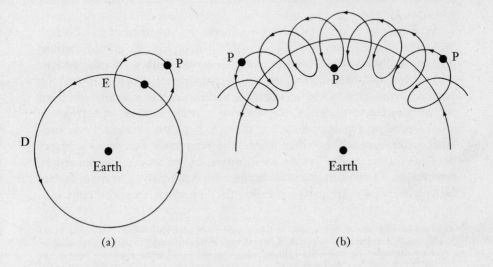

FIGURE 3.   (a) An epicycle; (b) a typical path through space.

its dimmest and when at the arc nearest earth, it is most brilliant. In this manner, the concept of circular motion was preserved, although the planet moves in a series of loops (Figure 3b) around the rim of the deferent, much as the enclosed cages of the modern "Ferris Wheel" move in loops around the big wheel, which in turn rotates about its axis.

The epicycle had a further advantage, for it could be used to explain retrograde motion in a considerably more convincing manner than Aristotle's nests-of-spheres. Picture, for example, a bicycle pedal as it turns round the sprocket. As the pedal is pressed downward from its highest point, it also moves forward. Then, after reaching the midpoint between the top and the bottom of the sprocket it begins to move backward and continues to do so until it arrives at the midpoint of its upward arc, when once again it moves forward. Imagine the planet, P, in Figure 3(a) is a bicycle pedal. While the pedal moves on the arc of the epicycle outside the deferent it appears to the observer on earth to advance, when moving along the arc inside the deferent it seems to move in a retrograde or backward manner. At the two opposite points on the circumference of the deferent, where it is intersected by the epicycle, the planet appears to stand still before once again changing positions and reversing the direction of its motion. By utilizing the system of one epicycle and one deferent the rates of revolution of the two spheres may be adjusted to fit the observations for any planet. The larger the epicycle, the greater the variation in the planet's movement. Hipparchus employed the system to plot the minor epicycles of the sun and moon only; Ptolemy was much more ambitious in his use of the epicycle and extended the system to include all seven planets.* The stellar sphere, owing to its vast distance from earth, needed no epicycles because its movement appeared regular and unchanging.

Ptolemy's system would have been much less complicated and his task made considerably easier if the irregularities in the planetary motions previously discussed had been the only anomalies with which the ancient astronomers had to deal; but there was yet another problem. Were the planetary motion pictured in Figure 3 continued into a second or third trip around the earth, we would discover that the new set of retrograde loops begins at a point somewhat "short" or to the west of those outlined in the diagram. In other words, the retrograde motion of a planet does not occur at the very same position of the zodiac on successive revolutions. In an effort to explain this added anomaly, a second device, also pioneered by Hipparchus, called the "eccentric" was brought into

---

* It should be pointed out that Hipparchus used epicycles to explain the rather limited but significant irregularities in the positions of the sun and moon based upon very careful celestial observations, not because these two "planets" appeared to change direction or stand still. Their epicycles were so small that the course reversal attributed to them was very rapid and therefore invisible to the terrestrial observer.

play. The eccentric is a displaced circle that carries the planet round
the earth, but not in such a manner that it is equidistant from earth at
all times. In Figure 4(a) the distance between the earth, E, and the
center of the eccentric, A, accounts for the irregularity associated with
the motion of the planet, P. To explain even more complicated motion,
Ptolemy placed epicycles on epicycles (Figure 4b) and epicycles on
eccentrics (Figure 4c). In this cumbersome manner the elliptical orbits
of the planets could all be defined within the context of the *perfect
circle*. By changing the size, position, and speed of his "cosmic wheels"
Ptolemy was able to create a large, indeed infinite variety, of circles,
arcs, and ovals. It was even discovered that if the system were tinkered
with enough, rectilinear figures could be produced.

Had Ptolemy's interest in recording highly accurate celestial observations been less keen, it is doubtful that his geometry of the heavens would
have reached such an advanced stage. But even with the perfection of
Hipparchus' epicycles and deferents, and their extension to all the then
known planets, Ptolemy still found it difficult to reconstruct on paper
the intricate celestial maneuvers dictated by the results of his observations. For reasons of mathematical economy, he introduced yet another
geometric figure called the "equant," the third and last device used to
account for those aspects of planetary motion yet unexplained.* In the
equant the center of the planet's motion coincided with the center of
the earth, as in Aristotle's crystalline spheres. However, in this instance
the deferent's rate of rotation is required to be uniform not with respect

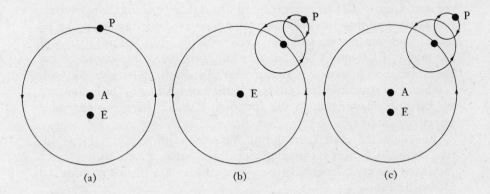

FIGURE 4.   (a) An eccentric; (b) an epicycle on an epicycle on a deferent;
(c) an epicycle on an epicycle on an eccentric.

* An equant is simply a substitute for a minor epicycle, and produces the very same
observational result.

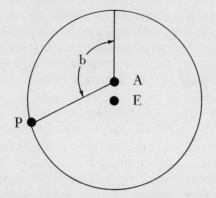

FIGURE 5. An equant. In this case the planet, P, moves on the earth-centered circle but at an irregular rate determined by the condition that the angle, b, vary uniformly with time.

to its geometric center, the earth, but with an equant point, A (Figure 5).

By the time Ptolemy had perfected his system, as many as forty wheels in various combinations were required to move the celestial machinery through the heavens—only fifteen less than the fifty-five homocentric spheres employed by Aristotle. Many versions of what became known during the Middle Ages as the "Ptolemaic system" were subsequently developed, and any number of epicycles could be employed so long as they satisfactorily explained away the particular planetary anomaly under observation. In fact, so interchangeable were the celestial gears composed of deferents, epicycles, eccentrics, and equants, one type could be and frequently was freely substituted for another, something Ptolemy himself was quick to realize but apparently was not troubled by in the least. Clearly, he was not interested in a rigid or uncomplicated system but rather one that worked, no matter how exotic it might appear on paper. Such was not the case, however, with the great Spanish patron of astronomy, Alphonso X of Castile, who, upon being introduced to the Ptolemaic system several centuries after the death of its inventor, was overheard to remark in frustration: "If the Lord Almighty had consulted me before embarking upon the Creation, I should have recommended something simpler."

Four hundred years later, during the seventeenth century, John Milton expressed a poet's contempt for this system which had not yet yielded to the new Copernican astronomy. In Book VIII of *Paradise Lost* Milton declaims:

> From man or angel the great Architect
> Did wisely to conceal and not divulge
> His secrets to be scann'd by them who ought
> Rather admire; or, if they list to try
> Conjecture, he his fabric of the heavens

Hath left to their disputes, perhaps to move
His laughter at their quaint opinions wide
Hereafter; when they come to model heaven
And calculate the stars, how they will wield
The mighty frame; how build, unbuild, contrive
To save appearances; how gird the sphere
With centric and eccentric scribbled o'er,
Cycle and epicycle, orb in orb.

The sentiments expressed by Alphonso and Milton are shared, no doubt, by many modern students of the history of astronomy. Yet a complete grasp of the many intricacies of Ptolemy's complex mathematical system is not essential to a basic understanding of his important contribution to the new science of the heavens. Doubtless there are few experts in the fields of mathematics, astronomy, or the history of science today, who have diligently waded through each of the nearly 500 pages of complicated diagrams, charts, and formulas contained in the *Almagest*.

That Ptolemy was a conservative, even a reactionary in certain respects, is undeniable. He took his arguments that the earth must be the center of the universe right out of Aristotle: Nowhere in his work is there even the slightest inkling that the heliocentric theory ever seriously crossed his mind. He did not even concede the possibility of the earth's rotation. Furthermore, Ptolemy's approach to astronomy was strictly quantitative, whereas Copernicus was as much concerned with the question of aesthetics as he was with the mathematical character of the universe. In fact, one of the things that set Copernicus to searching for a new approach to astronomy was his reticence to accept, without qualification, the ugly celestial machinery bequeathed by Ptolemy to Western civilization. Koestler observes that, "There is something profoundly distasteful about Ptolemy's universe; it is the work of a pedant with much patience and little originality." [39] Significantly, Ptolemy completed his geometric calculations without so much as developing one new idea of major theoretical value. His successors simply followed his lead by adding epicycle to epicycle and eccentric to eccentric. Nor did he provide a set of geometrical constructions capable of accounting for all the motions of all the planets at the same time, as Eudoxus and Aristarchus had done. Instead, each planet had its own peculiar machinery which was related to but not integrally connected to a universal system.

The word "astronomer" originally meant one whose theories on the planets and stars must do justice to the observed phenomena, or "appearances." This meaning underwent a gradual and subtle change until, by Ptolemy's time, it signified one who simply "saved appearances", so long as he could formulate a hypothesis which resolved the anomaly under investigation. Ptolemy was indeed a master at saving appearances, for he resolved, to the satisfaction of most, the irregular motions of the planets by ingeniously combining circles until the proper orbit had been defined, that is, an orbit that corresponded to the visual record of its constantly shifting position.

Obviously, the epicycles and deferents which Ptolemy substituted for the homocentric spheres of Eudoxus and Aristotle did not fit comfortably into the latter's theory of the crystalline spheres, and whether or not Ptolemy even believed in such spheres is not known; they receive no mention in his writings. Perhaps his silence on the subject speaks more loudly than words. To have challenged Aristotle on this point would have invited confrontation on other critical issues, an undertaking which the Alexandrian apparently considered both unnecessary and distasteful. He much preferred to limit the scope of his inquiry to the field of geometrics. Evidence of this surfaces in the opening page of the preface to the *Almagest* which contains a revealing statement worthy of quotation. Ptolemy refers to the three theoretical genera devised by Aristotle to explain natural phenomena, the physical, the mathematical, and the theological:

> And therefore meditating that the other two genera of the theoretical [meaning the theological and physical] would be expounded in terms of conjecture rather than in terms of scientific understanding: the theological because it is in no way phenomenal and attainable, and the physical because its matter is unstable and obscure, so that for this reason philosophers could never hope to agree on them; and meditating that only the mathematical, if approached enquiringly, would give its practitioners certain and trustworthy knowledge with demonstration both arithmetic and geometric resulting from indisputable procedures, we were led to cultivate most particularly as far as lay in our power this theoretical discipline.[40]

Quite clearly Ptolemy, seemingly in anticipation of Galileo some 1500 years later, wanted mathematics to rule supreme in the scientific realm. By consigning Aristotle's nonmathematical universe to the physical and theological realms, he cleverly avoided an open breach with the great philosopher. Had Aristotle, or Ptolemy for that matter, been capable of formulating a mathematical physics, the course of ancient science would have taken a much different turn. As it was, physics was divorced from mathematics until the seventeenth century, a situation that enabled Ptolemy to construct an abstract geometric universe, not by openly attacking Aristotelian physics, but by largely ignoring it. Theirs was a relationship based upon expediency and convenience, not one of compatibility and mutual support. Nevertheless, in its own right the Ptolemaic system is an amazing achievement and has no parallel in the history of science until the work of Kepler and Newton. Although his complex system never quite reconciled the differences between theory and observation, Ptolemy and his successors were able to predict, within very narrow limits, the planetary positions at any given time.

The rise of Alexandrian astronomy in the second century saw the separate tides of Greek and Babylonian astronomy flow together. For the first time as a group, Greek astronomers took pleasure in employing numerical precision to explain the observed motions of the planets and stars. This activity was stimulated to a great degree by the growth of judicial astrology which was just then coming into its own; the great accuracy of Ptolemy's planetary position predictions gave added impetus to the art. Astrology was, in fact, considered as important as tide-predictions and navigation in an era when the need to know and prepare for one's fate proved much more relevant than one's physical power to alter the environment. The work of Ptolemy thus took on immense value because of its utility.

During this period, rational scientific inquiry was gradually squeezed out by a growing preoccupation with questions of a technological kind on the one hand, and those dealing with religion on the other. With Ptolemy, the pioneering Greek quest in astronomy reached both its climax and its virtual end. It had been an exhilarating and multifaceted odyssey, during which many of the principal scientific theories of the universe had been postulated. Given this rich background, one might have wished that such a great intellectual episode had ended on an even higher road than that traversed by Ptolemy's cosmic wheels. Yet it is dizzying to think just how very far the Greeks carried the torch of scientific inquiry in the relatively short time it had been entrusted to their care. It was in their novel way of thinking about the world that man was first able to relate the visible changes in Nature to the permanent principles underlying them. And it is something to have invented and developed the whole concept of "scientific theory" at all. And what enlightened people would not be proud of having originated the very idea of philosophical inquiry? Though many of their concepts were at first vague and not a little naive, they were later clarified with the precision associated with scientific activity. Many of these same concepts, albeit in modified form, still preoccupy scientists today. This seems even more remarkable when we consider that in all of classical Greek history only a small number of men—perhaps as few as four or five hundred—undertook this Promethean quest.

Then, for reasons still not completely understood, the intellectual brilliance so necessary to the continuance of the journey went into a long and unbroken hibernation. Western man would not see the likes of it again for the better part of fifteen centuries, not until that flashing aurora borealis of scientific genius erupted in the seventeenth century. In the meantime, the world of the ancients passed away, blighted beyond recognition by the hoar-frost of cultural decay and deluged by a flood tide of barbarian invasions.

3. The Medieval World and Universe

The Demise of Classical Civilization

WITH THE DECLINE OF CLASSICAL GREEK CIVILIZATION, Latin culture fell heir to the legacy of Hellenic science. But for the most part, the Romans, who were more interested in practical considerations as opposed to questions of theory, ignored their considerable inheritance and contributed little, if anything, to the science of astronomy. From the publication of the *Almagest* to the dawn of the Copernican era—a period nearly double that from Thales to Ptolemy—no astronomical discoveries of major importance were made. Then, during the latter part of the third century A.D., there began to appear on the Roman horizon disturbing signs of cultural decline and moral decay. Rome became increasingly unwilling and ultimately unable to solve its basic social conflicts born of festering class antagonism, slavery, and monarchical imperialism. Thus, the Germanic conquerors who swept in from the north and east did not destroy a vital and vigorous culture, rather they extinguished an already moribund political and social system honeycombed from within by an advanced case of dry rot. Gradually but inexorably, the weakening empire moved ever deeper into the shadows until it was finally swallowed up by the long dark night of the Middle Ages.

While it is possible for twentieth-century humans to grasp the significance of this series of events intellectually, their emotional impact, for the most part, is beyond recall. After all, how can the modern student of history truly experience the ruin and devastation of a great civilization that had lasted for almost 1000 years? And how can the modern mind come to grips with the savagery and violence, so devoid of sympathy, that overtook Western Europe during the fifth century and lasted for a dozen generations? Still, if one would understand the medieval world and the long and painful experience involved in its gradual resurgence and ultimate recovery from the longest and most bitter winter in the memory of Western man, it is essentially on an emotional level that one must come to grips with Rome's demise.

For reasons that have never been clearly established, in the year 370 the Huns, a nomadic pastoralist people, launched a major invasion into Europe via the Caspian steppe, an avenue of easy access to the continental heartland. During the next seventy years they controlled an enormous "empire" stretching from the Volga river in the east to the Pyrenees mountains in the west. The records left us by their Latin chroniclers are uniformly hostile, as would be expected of refined and educated men, who suddenly found themselves at the mercy of an illiterate and uncultured barbarian horde. Yet their accounts are in agreement on so many fundamental points that there is little reason to question their basic accuracy. The very physical description of the Huns proved sufficient in and of itself to strike terror into the hearts of their enemies. They are pictured as resembling beasts more than men, with ugly faces highlighted by flat noses, slanting eyes, and large ears. Swarthy of skin and squat of body, they were bowlegged almost to a man, supposedly because male children were taught to ride a horse even before they could walk. Nor was their reputation enhanced by the fact that ratskin caps adorned their shaggy heads. As if their unpleasant physical appearance was not enough to earn them the enmity of the citizens of Rome, their repulsiveness was heightened by the filthiness of their personal habits. These included, among others, a total lack of regard for even the most fundamental rudiments of hygiene and a diet, when at war, that consisted of raw meat kept under the saddle supplemented with a drink composed mainly of fermented mare's milk called *kumitz*.

However, as warriors the Huns were unsurpassed and employed military tactics hitherto unknown in Europe. They were amazingly accurate mounted archers, and for this reason they generally shunned direct attack in favor of encircling the enemy's position on horse back while shooting at him on the run, a highly effective technique employed by the plains Indians of North America many centuries later. When in retreat, they were capable of turning completely around in the saddle at a full gallop to unleash a deadly barrage of arrows into the ranks of their adversaries, who were usually caught off guard.

Unlike many of the Germanic tribes from whom they exacted tribute, these Asiatics had no permanent homes and no kings. They lived in individual groups under the leadership of chieftains or *primates*, the name given them by the Latin historian Ammianus Marcellinus, who wrote in the late fourth century. The *primates* attained their dominant positions through a combination of military skill, physical prowess, and personal magnetism. Only rarely, as under the leadership of Attila (c. 432), were these fiercely independent warrior bands brought under centralized control, and even then the potential for internecine rivalry remained high.

Almost four centuries before the Hunnic irruption into southeastern

Europe, the Romans began to take notice of certain other peoples moving south behind the Celts into the country between the Elbe and Rhine rivers. They called them the *Germani*, a term that defies precise definition, but from the Roman point of view designated the major barbarian powers in Western Europe. Julius Caesar wrote the earliest surviving first-hand account of the Germans during his long military campaign in Gaul. In his *Commentaries* (55 B.C.) the future dictator of Rome provides a valuable description of the political and social organization of the Germans. An even more comprehensive record was left by the Roman scholar and historian Tacitus. In his treatise *Germania* (c. A.D. 100) he paints a somewhat idyllic picture of German domestic life, but one that has been verified in its more important aspects by later historical accounts.

While Caesar stressed that grazing combined with fishing and hunting constituted the backbone of the Germanic economy, Tacitus more astutely emphasized their dependence upon tillage farming. He relates that none of the tribes occupied cities but lived in small villages in which each man's home is separated from the residence of his neighbors by a large open space. Much of the tribal work was left to women, slaves, and males deemed unfit for combat, while those men considered physically suited for warfare spent much of their time eating, sleeping, and hunting. Intertribal quarrels and interfamily feuds were commonplace because the basic social unit in the Germanic tribe was the clan, or kinship group. Several clans made up a tribe, although there appears to have been no central leadership except in times of war. This barbarian society was led by warriors banded together through an institution which Tacitus named the *comitatus*. By promising his companions (*comites*) a life of adventure and plunder, in addition to food and military equipment, a leader or chieftan attracted a following that pledged him its loyalty and military support. Out of this simple arrangement evolved the more complex lord–vassal relationship, the predominant social and political system of the Middle Ages.

While the Germanic tribes were not always engaged in warfare, they were in a state of constant preparation for it. Under the circumstances, it was only natural that religious life be focused on their gods of war. The most prominent of these, Thor and Wodan (or Wotan), were worshipped in ceremonies that included human sacrifice of enemy prisoners. The Germans also practiced the wholesale destruction of captured booty, the remains of which were either cast into ponds and lakes or burned in peat bogs—a form of conspicuous consumption anticipatory of that endemic to modern technological society with its foul-smelling city dumps and sprawling suburban landfills. In the case of the "barbarians," however, the act of destruction was infused with a deep sense of religious purpose long since forgotten by modern man.

Such, then, were the main heirs of classical learning. Idyllic though their style of life might have seemed to Tacitus, the Germans could only remain so from the Roman point of view while they could be held in check by the military defenses of the empire. Of course, no one at this time, least of all Tacitus, ever dreamed that in the not too distant future these inhabitants of rude mud and wattle hovels would breach the Roman frontier in a hundred places to ravage the empire and its capital with impunity. By the beginning of the Christian era, several different tribes of Germanic peoples had slowly filtered across the frontier to occupy the most northerly and least populated regions of the empire. Some became civil servants while many others were settled by the government on the great agricultural estates, the *latifundium*. Julius Caesar and later Roman military leaders were so impressed by their skill as warriors that they hired the Germans by the thousands to augment the Roman legions, which suffered from a chronic shortage of able-bodied men. In fact, during the three and one-half centuries that separated Caesar's brief reign from the accession to the emperorship by Constantine in 312, German soldiers had become the majority in the Roman army. This, it was recognized, constituted a potentially dangerous situation, but what else could the Romans do? They needed this reserve of manpower to guard against the additional thousands of outsiders who were tempted by the luxuries and wealth of the weakening empire.

While armed Germanic expeditions into Italy had been attempted on several occasions, the first truly disastrous incursions did not take place until the late fourth century. Beginning in 370, the Huns made their dramatic appearance on the European scene by crossing the Volga and Don rivers, after which they dealt a swift and stunning defeat to the Ostrogoths. The impetus of their movement continued unchecked as they built up an extensive nomadic empire controlled from a central base located on the Hungarian plain. The Germanic tribes were no match for the Asian invaders and either had to submit to Hunnic overlordship, with its humiliating demands of obedience and forced tribute, or flee westward in search of protection behind the Roman frontier.

One of the largest Germanic tribes, the Visigoths, chose the latter alternative and sought refuge south of the Danube in Roman territory. Because of its weakened condition, the Roman government had no choice but to admit the Visigoths en masse in 376, a fateful decision that would mark the beginning of the end of classical Western civilization. Both peoples remained suspicious and distrustful of each other, and in 378 the Visigoths, frustrated by the oppressive conditions imposed upon them by their imperial governors, rose in revolt. The army sent to meet this threat was decimated at Adrianople: the road to Rome now lay open to the barbarians. In 410 the ancient world seemed momentarily to stand still: under their leader Alaric, the Visigoths captured and plundered the great city. Rome of the Caesars—the eternal city—had unbelievably fallen victim to the pagan hordes. Where relatively crude barbarians had once been ruled by a highly civilized people, the tables were now turned.

Yet despite their aggressive militarism, the Germans had not come to Rome either to rule or destroy it; instead, they wanted to share and enjoy the wealth and luxuries unknown in their more primitive societies. Even the non-Germanic Huns, who were totally illiterate and unreservedly hostile to anything they did not understand, did not undertake the destruction of the great buildings scattered all over the Roman world. Likewise, the thought of maintaining them never entered their heads. Nor did the Germans harbor any idea of substituting a new culture for the one they overran. While many a petty German chieftain took possession of a Roman estate and sought to play the role of Roman lord and master, their success was only superficial at best. On the one hand, they possessed no deep appreciation for what Latin civilization was, or how it functioned. On the other, the German leaders found it difficult to sustain the loyalty of frustrated followers by assuming the trappings of a foreign culture. Even this feeble pretense at preserving the outward appearance of a once vigorous civilization would soon wear thin. Within a matter of a few generations the educated layman virtually disappeared from Western Europe, while the Latin language split into dozens of different dialects. In some areas, especially along the Rhine and upper Danube, the few remaining Latins could preserve neither their language nor their customs and traditions. The heretofore brilliant, often blinding light of classical culture was gradually reduced to a pitiful flicker. It would not burn brightly again for upwards of six centuries. That it did not die out completely in this darkest of dark ages is a supreme tribute to those who no longer dedicated their existence to secular society and all that it involves. The future of Western culture, including its science, now rested in the hands of those committed to the city of God.

The Heavenly City

The study of the Middle Ages, which until rather recently languished in our institutions of higher education, has assumed a new scholarly importance since the end of World War II. Up until then, the citizens of Western society tended to assume that the forward march of scientific and technological progress would proceed unimpaired into the foreseeable future. Thus, how could medieval man, inheritor of the decimated culture of his ancestors, yet ignorant of so much of what they had known, who had to renew the quest for a civilized existence, possibly teach anything to those living in the second half of the twentieth century? Yet ever since the atomic bombing of two Japanese cities in 1945, followed by the development of even more sophisticated thermonuclear weaponry, the contemporary consciousness has been gradually permeated by the discomforting thought that we could all be on the threshold of a new Dark Age. We live, like our ancestors of the late fourth and early fifth centuries, in an age of anxiety and dread. When viewed from this perspective, the study of medieval Europe takes on a renewed and very special interest.

Few who survived the sack of Rome by the Visigoths were able to view the events surrounding this disaster in a manner even remotely approaching objectivity. As commonly happens during a crisis of major proportions, the victims feel compelled to blame someone—anyone—for their suffering and misfortune. In this instance, they turned on an old scapegoat—the Christians. It had only been a century since the Roman Emperor Constantine, in 313, bestowed upon Christianity a status equal to that enjoyed by the traditional pagan cults of Rome, and only 30 years since, in 380, Theodosius I commanded all his subjects to become orthodox Christians. So while the Church had gained a fairly broad base of popular support, there were large numbers of individuals whose conversion to the new religion was only nominal and many others who rejected its teachings altogether. Romans of the old school had never developed a deep interest in religions as such, and they were inclined to see a connection between Christianity's rise and the rapid decline of Roman political power. Many upper class Romans, in particular, were highly suspicious of such Christian teachings as pacifism, otherworldliness, disinterest in public affairs, and an intolerance for the old gods. Still, the followers of Christ had survived the contempt, ostracism, and outright bestiality of their detractors in the past. In fact, the courage shown by their martyred brothers and sisters had only strengthened their resolve to persevere. They eventually succeeded in reversing the roles of the pagan religions and their own by taking the offensive. When Christianity became the official state religion their struggle and suffering were seemingly vindicated. To the faithful, this triumph proved beyond any doubt what they had proclaimed all along—that Christianity is the only true religion. Now faced with the new and extremely serious charge that they, not the Germanic barbarians, were the true despoilers of the empire, certain influential members of the Church hierarchy once again felt compelled to answer the accusations leveled at them by their detractors. Foremost among them was Augustine, Bishop of Hippo in the Roman province of Africa from 396 to 430.

Aurelius Augustinus (Augustine) was born in 354 on the Algerian coast. His father, Patricius, remained a pagan until late in life while his mother, Monica, was a devoted Christian, who introduced her son to the teachings of Christ. Perhaps because of his father's opposition to the

Christian faith Augustine was not baptized in infancy, and continued to
be torn by conflicting impulses on religion throughout the greater part
of his adult life. As a young man, he openly rejected his mother's religion because he found it too simplistic and unphilosophical, an opinion
formed during his student days in Carthage, where he came under the
spell of philosophical inquiry after reading a popular treatise of Cicero's,
the now lost *Hortensius*. Throughout his early adult life he passed from
one religious system to another, unable to derive lasting spiritual satisfaction from any. He was first attracted to the teachings of Manichaeanism
which claimed to be the true Christianity. But he soon discovered its
leaders were intellectually second rate and incapable of providing satisfactory answers to the complicated questions he put to them. His zeal for
spiritual enlightenment unabated, Augustine next turned to the writings
of the Neo-Platonists whose mysticism held a fascination for him that
continued throughout the remainder of his life. Augustine was later to
combine elements of this philosophy with the teachings of Christianity.
Of primary interest to him was the Neo-Platonic concept that the inward experience is superior to the outward; that to attain good, one must
"return into" oneself.

Meanwhile, Augustine formed a liaison with a woman of low birth
by whom he had a son. This relationship was later written about in his
famous autobiographical work *Confessions*, in which the future saint tells
the story of his restless youth and of his conversion to Christianity. It
is a remarkably candid and revealing essay whose analysis of the emotional side of the Christian experience in the face of sin remains un-
equalled. What is even more important is his attack upon his former
scholarly pursuits as impediments to a true understanding of God. It was
Augustine who provided the first strong intimation of the direction in
which educated men of the Middle Ages would channel their talents
and energy:

> But what value did I gain from my reading as long as I thought that
> you, Lord God who are the Truth, were a bright, unbounded body and I
> a small piece broken from it? What utter distortion of the truth! Yet this
> was my belief; and I do not now blush to acknowledge, my God, the
> mercies you have shown to me, nor to call you to my aid, just as in those
> days I did not blush to declare my blasphemies aloud and snarl at you like
> a dog. What, then, was the value to me of my intelligence, which could
> take these subjects in stride, and all those books, with their tangled prob-
> lems, which I unravelled without the help of any human tutor, when in the
> doctrine of your love I was lost in the most hideous error and the vilest
> sacrilege? And was it so great a drawback to your faithful children that
> they were slower than I to understand such things? For they did not
> forsake you, but grew like fledglings in the safe nest of your Church,

nourishing the wings of charity on the food of the faith that would save them.[1]

Secular affairs and classical scholarship would no longer constitute the primary considerations underlying human existence. They had only opened the way to moral decay and vainglorious pride, when self-contempt and escape from the temptations of the flesh were called for.

In 387 Augustine was baptised a Christian in Rome by Ambrose, one of the greatest Church leaders of the time. At one stroke the proselyte abandoned his chosen career as a teacher together with plans to marry. Shortly thereafter, he left Italy for Africa along with a small group of friends. For the next four years he occupied himself primarily with writings in which he leveled attacks against enemies of the faith. Then, during a visit to Hippo in 391, Augustine was reluctantly persuaded to accept ordination into the priesthood as assistant to the aging bishop, Valerius. Upon the latter's death five years later, Augustine entered the episcopate in which he served for the remainder of his life.

The multiple strands of Augustine's brilliant career are far too complex to be examined in detail in an introductory essay on the history of science. Some might even question the value of discussing his work at all. Yet there is one crucial aspect of Augustinian theology, already hinted at in the passage quoted from the *Confessions*, that would have a direct and enduring effect not only upon the religious philosophy of medieval Christendom, but also on the entire intellectual climate and temper of the age. This in turn would determine the future course of development in Western science until the Renaissance.

Three years after the storming of Rome by the Visigoths, Augustine began his masterpiece *The City of God (De Civitate Dei)*. When finished in 426, the voluminous work contained some twenty-two books and upwards of 1000 pages. Despite the many changes in Roman Catholic doctrine since its publication some 1500 years ago, it was destined to become one of the most influential works of either medieval or modern theology. The first third of the book is devoted to a detailed refutation of the pagan charge that the capture and sack of Rome by the Germans was a direct result of supernatural vengeance wrought against its citizens by the deities of its ancient ancestral religion. Indeed, the great theologian proceeds to argue that the greed, corruption, and depravity of Rome's own Latin population, not the wrath of forsaken gods, paved the way for the barbarian conquest. The Romans, in fact, are little better than their German enemies, for both are guided by materialistic considerations. He attempts to show that the pagan gods never brought their followers any lasting benefits, and that the difficulties in which the Romans find themselves are essentially of their own making. Whether or not this largely propagandistic part of the book won any new adherents to Augustine's position is a matter of conjecture. However, after having presented the case against Christian guilt, he moves on to a subject much dearer to his heart, and the major underlying reason for his writing the book.

In his mind's eye Augustine envisions two cities, one heavenly (*civitas dei*), the other earthly (*civitas terrena*). He did not, however, employ the word *civitas* in the political sense, because the organization and functions of the state as such are not matters of primary interest to the cleric. Rather they are the symbolic embodiment of the two major spiritual forces—good and evil—that have competed for man's allegiance since his creation by God. The great struggle of the universe is not between Church and State, but between two opposing ways of life. Neither good nor evil is perfectly represented in any worldly institution, the Roman Catholic Church included; they are everywhere intermingled beyond separation. The earthly city is best characterized by "the love of self extending to the contempt of God," the heavenly city by "the love of God extending to the contempt of self."

Since the heavenly city is eternal, it existed before the creation of man and constitutes the everlasting abode of God and the angels. Man, a sinful creature by nature, is incapable of gaining acceptance into this kingdom without the saving grace of the Creator. Although man possesses the ability to choose between good and evil, it is God who provides him with the motive or desire to select one path over another. Because He has known all things for all time, God realizes that some men will lead holy lives, while others will refuse to cooperate with His plan. Augustine was the first major Christian theologian to teach the doctrine of predestination, a concept that would be taken up again in various ages by the Albigenses, Calvinists, and Jansenists, among others. Even before creation, according to Augustine, God had fixed forever the number of those to be saved as He had already chosen the saintly and elect, who would forever dwell in the heavenly city, while abandoning the rest of mankind to eternal damnation. The battle between the elect and the fallen was joined almost immediately following the creation as is evidenced by the fratricidal combat between the first sons of the first family ever to walk the face of the earth:

> Of these first two parents of the human race, then, Cain was the first-born, and he belonged to the city of men; after him was born Abel, who belonged to the city of God. For as in the individual truth of the apostle's statement is discerned, 'that is not first which is spiritual, but that which is natural, and afterward that which is spiritual,' * whence it comes to pass that each man, being derived from a condemned stock, is first of all born of

* Corinthians, XV, 46.

Adam evil and carnal, and becomes good and spiritual only afterwards, when he is grafted into Christ by regeneration: so was it in the human race as a whole. When these two cities began to run their course by a series of deaths and births, the citizen of this world was the first-born, and after him the stranger in this world, the citizen of the city of God, pre-destined by grace, elected by grace, by grace a stranger below, and by grace a citizen above.[2]

In contrast to the heavenly city, the earthly city is occupied by the fallen angels and the impious. There, those who "live after the flesh" are destined to eternal punishment at the hands of Satan. This realm is the very incarnation of sin and depravity, the antithesis of Christian love, and of no lasting value whatsoever. Augustine conceived of human history as the unfolding of God's divine will; for all that has happened, or will ever happen, has been planned down to the most minute detail. All men can know or can ever hope to know is to be found in the Scriptures.

Even though the doctrine of predestination was later rejected by the Roman Catholic Church, the influence of St. Augustine proved enor-mous. Augustinian mysticism, drawing heavily from Neo-Platonic thinkers, left an indelible mark on the thought of succeeding generations of theologians, who were the heirs to the scientific tradition of classical antiquity. Where intellectual curiosity and reason had once been allowed free play, "faith" and "orthodoxy" became the new bywords of the Christian era. This is not to say that Augustine or the later Fathers of the Church were diametrically opposed to the acquisition of knowledge, or to all classical learning. Rather, their thinking denotes a permanent shift in both the emotional and intellectual temper of the age. For despite a record of remarkable growth, Christianity and its supporters remained on the defensive. The Fathers thought it essential that the pagan learning of the Greeks and Romans be depreciated. The Germanic in-vasions were not yet at an end, and even after they ceased the remainder of the civilized world, like an endangered tortoise, contracted inside a protective shell of social indifference and impenetrable antiintellectualism. Science, as a part of secular learning, seemed useless at best, except when it proved essential for daily life. At worst, it was dangerously dis-tracting. And although the Church's attitude toward science was not uniform throughout the medieval period, between 400 and 900 its position, on balance, was antiscientific. Astrology, for example, which had deep scientific roots and enjoyed widespread popularity among all classes in Rome, was condemned by Augustine both because of its pagan associations and the belief that God alone knows what the future holds for mankind. Thus, the Church acted much like a giant mirror, reflecting the negativism, alienation, and despair that permeated the whole of early medieval society. The agonizing fall of classical civilization gave rise to the Christian expectation of evil, pain, sin, and death. Unlike the openness and optimism associated with the Hellenic world, the early medieval world became the very symbol—if not the incarnation—of negativism and despair. Lewis Mumford writes:

> The whole of life for the Christian derived from his methods of encountering negations. Whereas in all the older civilizations, man had freely sacrificed to their gods, with Christianity its god had taken human form and had accepted sacrifice in order to redeem sinful man and free him from the anxiety and guilt that issued forth from his condition.[3]

Science thrives best in a free and open climate where, as in ancient Greece, the intellect is unfettered by the demands of social conformity and religious orthodoxy; where men possess confidence not only in their way of life but in themselves. A certain degree of self-pride, what the Greeks called *hubris,* must be present if men are to reach for the stars. During the Middle Ages society suffered, as much as anything else, from an acute and prolonged crisis of confidence. Both the individual and the larger society of which he was a part became acutely introverted, lost as it were in the dark reaches of inner space. The dedicated Christian acknowledged and to a considerable extent embraced the unpleasant realities of his time. Instead of seeking security in the company of many men united under the banner of the State, he sought the spiritual solace, sacrifice, and regimentation that came from withdrawal behind the cloistered walls of the monastery in the company of a small group of similarly minded individuals. Or, if this form of self-denial did not quench his spiritual thirst, he could elect, as many did, the even more radical alternative of lonely retreat into the unknown wilderness.

Since, during the violent persecutions of the faithful, it was the general directive of Christians to save themselves if they could without renouncing their faith, the original monk was usually to be found in a cave or rudely constructed hut where, sustained by the offerings of others, he could spend all of his time in the search for perfect holiness. The growth of asceticism in the West was deeply influenced by the example of such Near Eastern religions as Manichaeanism and Gnosticism. The climate of eastern countries, like Egypt, where these cults flourished, made life in the out-of-doors possible the year round: thus, it seems hardly coincidental that early Christian monasticism also took root there. As the opportunity for martyrdom in the gladitorial arenas of the State diminished, finally to end altogether by decree of the Emperor Gallienus in 260, the incidence of morbid self-torture, taking a wide variety of forms, rose drastically. In its more radical stages, asceticism was characterized by contemporary critics as a way of life "worse than the existence of a pig." Hermits did indeed graze in the fields like animals, roll without clothing in thorn bushes, swim naked in snake-infested swamps, and seek release from the sexual drive, which in their view epitomized the very essence of carnality and evil, through self-castration or other forms of mutilation. The most famous of the ascetics, St. Simeon Stylites (d. 459), began his life of self-denial "as a rooted vegetable in a garden" after which he undertook the construction of his famous pillar. There he lived for over thirty years, sixty feet up in the air, atop a platform just wide enough to lie down upon. All of his meager material needs were supplied by loyal followers, who daily filled a basket lowered to the ground by means of a rope. Every afternoon St. Simeon received pilgrims and gave advice on spiritual matters while maggots fed on his self-inflicted wounds, which he intentionally kept open for that purpose. When a worm chanced to fall off he would replace it with the admonition, "Eat what God has given you!"

Madness? In some cases the answer is doubtless, yes; in the majority of instances, no. Rather it is more an extreme example of the spirit of a dismal and humorless age carried to its ultimate if, from the modern point of view, illogical conclusion, of the attempts of a relatively small number of individuals to take upon themselves the supreme task of atoning for the sins of a fallen civilization. While Rome had totally abandoned its better self to the materialism and decadence associated with the late classical age, the ascetics turned their backs on nearly every impulse that makes us human. Neither society had sought the middle ground; both the citizens of Rome and the zealots of medieval Christendom eschewed the Hellenic quest for the golden mean.

Hermitic asceticism represents the most radical phase of the earthly search for access to the gates of the heavenly city. While certainly not typical of the activities of a majority of the Christian community at this time, neither could the Church Fathers, who viewed its more extreme forms as a perversion of Christ's basic teachings, suppress its continued practice. After all, how would one go about punishing individuals who donned spiked collars, hair shirts, and had themselves flogged to the point of insensibility? Excommunication also remained out of the question because much of the Church membership stood in awe of these exploits. The mania, like the periodic plagues that swept across medieval Europe, would simply have to run its course. Meanwhile, St. Augustine and the other Christian Fathers gave their wholehearted support to a more reasoned approach undertaken by those who sought isolation from the temptations of the flesh and the cruel barbarism of the age, while still carrying on a useful and productive existence.

Like the ascetic movement of which it was an outgrowth, monasticism had its origins in the Middle East. By the fourth century it had already begun to attract many intellectually sensitive persons for whom it offered the opportunity, in an otherwise turbulent world, to cultivate both the soul and the mind. Two of the most prominent and influential organizers of the early monastic movement were the fourth century Egyptian saints, Antony and Pachomius. Under the Antonian system, which prevailed in lower Egypt, there was no prescription for community life. Each monk had his own quarters where he could carry on whatever practices he liked; only on Saturdays and Sundays during religious services did he meet regularly with his fellow ecclesiastics. In contrast, the monastery founded a decade later by St. Pachomius became the first to operate under rules governing community activity. These were elaborated upon a few years later by St. Basil (329–379), who prepared a set of monastic regulations which subsequently became the model in the Greek East. Not until about 529 did St. Benedict of Nursia formulate his famous rule of poverty, obedience, labor, and religious devotion that has guided the development of Latin monasticism in the West ever since. As the movement grew, it eventually led to a division in the ranks of the clergy. The monks, living as they did by definite rule (or *regula*), became known as the regular clergy while the priests, bishops, and archbishops, who carried out their duties in the midst of the world (or *saeculum*), became known as the secular clergy.

With the old trunk of Roman civilization severed, the emerging monasteries became the new shoots of the severly damaged but still surviving root system—the principal refuge of early medieval culture and learning. Although the pattern was somewhat altered during the High Middle Ages, these institutions were not essentially dynamic centers of new thought, but lonely and often inaccessible enclaves where the disparate pieces of a fractured culture were preserved until, centuries hence, they could be assimilated by an intellectually resurgent Europe. Conservative though the monasteries were, however, their influence upon the society of the Middle Ages would be difficult to exaggerate.

Prior to the fall of Rome, Western civilization was, as it is today, basically a city civilization. It was to its urban centers that those interested in a better education and a broader range of opportunities were drawn. Now, however, the great Latin cities fell prey to widespread depopulation, economic decline, and physical decay. In their place the monastery became the new polis, a self-contained community where likeminded people formed a permanent common bond in an effort to achieve a Christian life on earth. These new "cities" were communal in nature, based upon a denial of private property, secular power, and class consciousness; within their boundaries work—both physical and mental—was ennobled. Although the Benedictine rule imposed specific obligations upon each individual, it was rarely severe to the point of austerity. The monks were sustained by a more than adequate diet of wholesome fruits and vegetable supplemented by a liberal daily ration of wine. Meat did not have a place in the diet, except in old age. Each monk received good clothing, although bathing took place only irregularly unless they were ill. In each monastery, the abbot had absolute authority, even to the point of inflicting corporal punishment for gross disobedience, an extreme measure that was only occasionally invoked. The abbot's responsibility was considerable because he answered to God not only for his own acts but for those of his subordinates as well. While his decisions were final, before resolving any weighty matter he was required to call all the monks together for consultation, during which their advice was freely solicited. And when a vacancy occurred in the abbacy all the monks participated equally in the election of a man known for his humility, wisdom, and virtue, even if in order of seniority "he be the last in the community." In a very real sense, the Benedictine monastery was the realization of man's ancient dream of creating a society of equals dedicated to the best life possible.

The monks were generally the best farmers in Europe, and, as anyone with a casual acquaintance of the Middle Ages knows, they reclaimed waste lands, drained swamps, and made many discoveries leading to soil improvement and greater agricultural productivity. It was also in the monastery that the Latin language was employed in daily conversation, through which means it survived many of the horrible mutilations it suffered elsewhere in the provinces of the old empire. Here, too, Greek medicine was practiced with great success. But the most important activity encompassed the preservation of classical scholarship, for each brother toiled patiently and long in the cause of learning and truth.

No place took to the Christian message more enthusiastically than did Ireland. Monasteries sprang up in every quarter of the island, and on the rocky outcroppings rising above the sea off the western coast. From tiny Skellig Michael to massive Cashel, work in the monastery's scriptorium went on continuously for centuries as the monks copied texts from deteriorating papyrus onto more durable parchment either for their students' instruction, or to be sent in the leather bags carried by missionaries in their little skin-covered boats, or on their backs in quest of the *pagani*, the people of the countryside. In an essay titled "Old Ireland, Her Scribes and Scholars," historian of religion Robert E. McNally paints a vivid picture of work in the scriptorium:

> The scribe worked a large portion of the day hunched over a slanting desk with his bare feet on the cold, stone floor. With goose quill in his first three fingers, he spent the long hours tracing the so-called Irish letters in jet ink on well polished white parchment or vellum. The equipment for executing these precious works of art was simple indeed—a supply of well sharpened quills, pigments for producing red, yellow, blue and green, a sharp blade or pumice stone for correcting the text, sand for drying ink and a keen knife for sharpening the pens. From every point of view scribal

work was difficult. It was ordinarily done in silence, frequently in hunger, invariably in discomfort. The scriptorium was poorly lighted; and on wintry days, the cold freely penetrated and benumbed the body of the scribe.[4]

There was, of course, occasional relief from the tedium involved in the copying of manuscripts, as is evidenced by notations made in the margins of works being copied: "It is time for dinner." or "Wonderful is the robin there singing to us." and "Our cat has escaped from us." However just as common were complaints born of occupational strain: "Alas—my hand!" followed a few lines later by "O my breast, Holy Virgin!" Still, the very fact that such simple complaints could be made serves as additional testimony that within the monastery walls order and serenity prevailed, whatever the confusions of the outer world.

The connection between man and science was not altogether lost within the cloister. Physical labor, while a required part of every monk's day, did not occupy him from sunrise to sunset as it did slave or serf. The five-hour work day was typical, with more than an ample amount of time left over for spiritual contemplation and intellectual activity. Virtually every facet of monastic life was regulated either by the hourglass or sundial; later by the clock. And "from the monastery, this time-keeping habit spread back to the marketplace, where in the classic era it had perhaps originated; so that from the fourth century on a whole town would time its activities." [5] In order to establish a basis for accurate time-keeping, the monasteries had to keep alive at least a limited knowledge of the constellations and planetary movements. This information also served as the basis for fixing with exactness the dates of major religious observances such as Easter. The science of computation *(computus)* in the form of calendric reckoning thus became an important, if less influential, part of the new religion, just as it had in the older Axial religions of the ancient Near and Middle East.

An equally, if not more, important practical application of scientific principles to daily life in the monastery was the Benedictine development and perfection of various types of mechanical devices, even though the introduction of labor-saving machinery owed little, at least in the beginning, to the desire for greater freedom from physical toil. St. Benedict's insistence from the outset that manual labor constitutes a fundamental human obligation never ceased to apply. On the other hand, physical work was no longer identified in the monastic mind with endless, mindless drudgery: certainly there could be no harm in balancing off back-breaking toil with devices that only served to enhance the productivity of God's heavenly city on earth. For this reason the regulations of the Cistercians, a contemplative order founded in France by reformist Benedictines, favored constructing their monasteries close to rivers capable of supplying adequate amounts of waterpower to propel machines. The importance of this undertaking is seen in the following contemporary description of Clairvaux Abbey, in Migne, at the time of St. Bernard in the twelfth century:

> The river enters the abbey as much as the well acting as a check allows. It gushes first into the corn-mill, where it is very actively employed in grinding the grain under the weight of the wheels and in shaking the fine sieve which separates flour from bran. Thence it flows into the next building, and fills the boiler in which it is heated to prepare beer for the monks' drinking, should the vine's fruitfulness not reward the vintner's labor. But the river has not yet finished its work, for it is now drawn into the fulling-machines following the corn-mill. In the mill it has prepared the brothers' food and its duty is now to serve in making their clothing. Thus it raises and alternately lowers the heavy hammers and mallets of the fulling-machines. Now the river enters the tannery where it devotes much care and labor to preparing the necessary materials for the monks' footwear; then it divides into many small branches and, in its busy course, passes through various departments, seeking everywhere for those who require its services for any purpose whatsoever, whether for cooking, rotating, crushing, watering, washing, or grinding. At last, to earn full thanks and to leave nothing undone, it carries away the refuse and leaves all clean.[6]

Judging from this portrait, it is difficult to imagine more diverse uses or ingenious controls of water and mechanical power on such a limited scale.

The use of constellations to compute time and dates, accompanied by the gradual introduction of machinery into the monastery, comprises almost the total content of the thin rivulet of science and applied technology that ran through the early centuries of the Middle Ages. Certainly, as regards astronomy, no advances were made; in fact, just the opposite occurred. The serious study of the heavens was in retreat after the fourth century as the writings of the most influential Fathers of the Church so graphically illustrate. J. L. E. Dreyer observes that when the modern scholar turns over the pages of some of the Christian Fathers, it might be imagined that he is reading the opinions of the Baylonian priests written down thousands of years before the Christian era, because "the ideas are exactly the same, the only difference being that the old Babylonian priest had no way of knowing better, and would not have rejected truth when shown to result from astronomical observations."[7] A stern judgment indeed, but an accurate one nonetheless. Grégoire of Tours, a Latin scholar of the sixth century, most aptly characterized the contemporary state of learning in a single sentence: "The mind has lost its cutting edge, we hardly know the Ancients." Hellenic science, despite the advanced degree of its intellectual merit, was still considered the product of a pagan culture, an association that made it strongly distasteful to the early leaders of the Church who preferred to let it languish.

A typical representative of the patristic school of thought is St. Lactantius, whose seven books on *Divine Institutions* were written during the first quarter of the fourth century. Lactantius was the first and perhaps the strongest opponent of the Greek belief in the rotundity of the earth, the truth of which he challenged with complete success. In the third book of his writings, called "On the False Wisdom of the Philosophers," the saint presents a series of fallacious and naive arguments against the spherical doctrine, and the existence of the antipodes. Lactantius contends that were the earth a sphere, people would walk with their feet above their heads, rain and snow would fall upwards, while the heavens would hang lower than the earth. He also denies the argument of the natural philosophers that heavy bodies are attracted toward the center of the earth. He would go on to point out more errors in pagan science, but, as he explains, his book is almost finished and the problem has received sufficient attention. (With this thought we would certainly agree.)

A somewhat more tempered position was that taken by Basil the Great, who, in 360, published a long treatise on the six days of creation. Evidently acquainted with the works of Aristotle, St. Basil did not chaff against the opinions of the learned Greeks as did Lactantius, although he did not accept them either. He believes in the Old Testament notion of the earth as a disk over which the sky is stretched like a tent, while both are surrounded by water. The idea originated in Genesis 1:6,7:

> And God said let there be a firmament in the midst of the waters and let it separate the waters from the waters. And God made the firmament and separated the waters which were under the firmament from the waters which were above the firmament.

Basil explains that God's purpose in selecting this particular arrangement was to protect the earth from the all-consuming celestial fire. And even though he was aware of and unable to refute the Greek belief in the annual motion of the sun, Basil refused to accept the concept of the spherical heavens.

The idea of the earth being a disk covered by water also received the enthusiastic endorsement of the famous sixth–century monk and geographer Cosmas, in his well-known treatise *Topographica Christiana*. Before taking his monastic vows Cosmas had been a merchant and seaman well acquainted with North Africa and India. This knowledge earned him the surname *Indicopleustus*, meaning the Indian traveller. Even though he did not hold an influential position in the Church hierarchy, his acceptance of and expansion upon the celestial system outlined by Basil became the prevalent opinion for the next several centuries.

Cosmas considered the Hellenic belief that the earth is the center of the heavens as totally absurd because the earth is so heavy that it can rest only at the bottom of the universe. He also branded the idea of planetary motion caused by invisible epicycles as equally erroneous. Why, asks Cosmas, must the planets be animated? And even if they are, then why are the sun and moon without their own epicycles? "Is it that they are not worthy on account of their inferiority?" or "Was it from the scarcity of suitable material the Creator could not construct vehicles for them?" He concludes with the execration: "On your own head let the blasphemy of such a thought recoil." [8]

And so it went. Other attacks, most of them less virulent, followed, but with essentially the same result. The Fathers of the Church condemned Greek astronomy almost to a man, yet no one took upon himself the weighty task of working out a detailed system to replace the unacceptable cosmological teachings of the pagans. In fact, of all the early Christian Fathers St. Augustine, founder of the pessimistic doctrine of predestination, must be considered the most enlightened when it came to Greek science. Unlike Lactantius, Augustine did not treat the scientific scholarship of the ancients with ignorant contempt. While it is true that he upheld the Scriptures in all important matters, he also seemed willing to yield to ancient science whenever it did not conflict with the Bible. This attitude is reflective of the saint's deep commitment to classical scholarship as a young man, and his life-long struggle to reconcile Neo-Platonic philosophy with Christian doctrine, a task ultimately consummated in his great work *The City of God*. Throughout his life, the dual forces of secular learning and absolute faith pulled him first one way, then the other. To his credit, he refused to adopt the polemical position taken by the other Fathers towards pagan science, which from their perspective constituted a direct threat to Christian orthodoxy. For a man who lived in an age of bigotry, fear, and expanding ignorance, and who died even as the Vandals laid seige to his beloved city of Hippo, St. Augustine's attempt to keep at least a halfway open mind on the subject of ancient science is deserving of our admiration.

Christian Time and History

A new age, a changed consciousness, an altered conception of time and history—or so the historical record would indicate. For just as the sense of primitive time developed by Paleolithic man had given way to the ancient belief in cyclical recurrence, so, too, would the latter view yield to a new interpretation of time based upon Christian ideology.

Both Plato and Aristotle had entertained the idea that once every few thousand years, the sun, moon, planets, and stars return to the same relative positions they had held in the long forgotten past. The celestial bodies than proceed to pass through the same series of configurations as they had done countless times before. It was believed that this cyclical process was paralleled by an exact repetition of worldly events. Aristotle conjectured that if this were in fact true, "he himself was living *before* the Fall of Troy quite as much as *after* it; since, when the wheel of fortune had turned through another cycle, the Trojan War would be reenacted and Troy would fall again." [9] The Romans, no less than the Greeks and Babylonians, also clung to various notions of cyclical recurrence in time: some conceived of an endless spiral of degeneration within each cycle, while others subscribed to a theory of infinite and meaningless undulations.

The triumph of Christianity led to a radical departure from the classical view of time and its understanding of the nature of historical development. Christianity took over from Judaism the belief that the historical process has definite meaning beyond that of cyclical determinism. For both Jews and Christians, history is comprised of a series of *unique* events, none of which have either previously occurred, or can be exactly repeated. As far as human understanding goes, the beginning point of history is the creation of man and the universe as described in the book of Genesis. The Creation is followed by an indefinite future within historical time. Historical events are not believed important in and of themselves, however. The Old Testament is unconcerned with the acts of men or nations, unless they have a direct relationship to God. Drawing up a record of secular events that have no bearing upon religious faith for use by future generations is of no value. The Scriptures alone provide an absolute guide to proper human conduct. There is nothing to be learned unless man's actions are interpreted against this background.

Accordingly, the pagan belief in the purposeless undulations of time was branded as blasphemy by the Christian Fathers. How could it possibly be reconciled with God's promise of salvation and eternal happiness in the heavenly city? Would not the saintly be cast out of Paradise and the damned released from Hell only to relive their lives in historical time *ad infinitum?* The absurdity of this thought moved St. Augustine to the point of anger:

> These things are declared to be false by the loud testimony of religion and truth; for religion truthfully promises a true blessedness; of which we shall be eternally assured, and which cannot be interrupted by any disaster. Let us therefore keep to the straight path, which is Christ, and, with Him as our guide and Savior, let us turn away in heart and mind from the unreal and futile cycles of the godless. [10]

"Christ," after all, had "died once for man's sins; surely after rising again He will die no more."

The Incarnation of Christ constitutes the central event in Christian time-reckoning and becomes a dividing line in history. The years before Christ's birth, B.C., continuously decrease, while those after, A.D., continuously increase toward an end—the Second Coming—whose exact date in the future remains unknown. From the Christian perspective the key to understanding secular history is man's sin and God's saving grace. The period between the death of Christ and his promised reappearance is the decisive time of testing and probation of mankind. During this period man experiences a supreme tension of conflicting wills; whether, on the one hand, to accept God's call, or to abandon it on the other. Eternal damnation is the reward for those who deny their Creator's love, while timeless salvation awaits the righteous. Thus, the Christian interpretation of history stands or falls on the acceptance or rejection of Christ as the Son of God. Existence in historical time is important only because it provides the opportunity for eternal happiness and perfect grace beyond time. "I am intent upon this one purpose," St. Augustine wrote in the *Confessions*, "not distracted by other aims, and with this goal in view I press on, eager for the prize, God's heavenly summons." [11]

To the Christian, then, no secular historian can possibly discover that Jesus is the Son of God. One must accept it as an article of faith, sufficient unto itself, for all time. For the individual who chooses to do so, temporal time loses all meaning because, as Augustine observes: "Then I shall listen to the sound of your praises and gaze at your beauty ever present, never future, never past." [12] From this perspective, history is the dramatic unfolding of the struggle between two societies—one secular and constantly changing, the other eternal—and of the ultimate triumph of the City of God.

This apparent lack of concern for the nature of secular events would seem to represent another major step away from any rational attempt to rekindle the now dying fires of scientific inquiry. Science, after all, is concerned only with the natural world, not the spiritual kingdom of God. During the next several centuries this attitude would indeed impede the rebirth of a viable scientific tradition. Yet while we may deplore the Church's often fanatical persecution of secular thought, which is certainly one of the least palatable aspects of the medieval age, we must also recognize that Christian doctrine—including that on time and history—nurtured the seeds of a delayed renascence in science. Only in the West, where Christianity took root most deeply, did the view develop that the historical process is unique and infused with a special purpose. The idea that human progress is possible, both in materialistic and social terms, is a direct outgrowth of the linear concept of time first propounded in Jewish thought and later borrowed by the Christian Fathers. Little did they know, of course, that their attack on the ancient belief in cyclical recurrence would eventually backfire—that Western man's historical optimism, a belief that has done much both to invigorate and endanger his culture, would one day make science the new god. Theirs was a search for a religious philosophy capable of assuaging the debilitating effects of cultural breakdown by providing an ethic for those alienated from a society gone berserk. Only from the perspective of the twentieth century have we come to realize that it also created the potential which has permitted the human condition to drift toward the establishment of a permanently irreligious, totally secularized culture. But so long as man's outlook remained pessimistic, centered on the otherworldly, the Church had little to fear. Only with the growth of the new humanism during the late Middle Ages, and its accompanying belief that man can indeed create a better life for himslf *within* historical time, were the conditions right for the rise of a renewed interest in science and the eventual secularization of that which had once been considered sacred.

A Light From the East

While astronomy and the other sciences languished in the intellectually hostile environment of Europe during the first half of the Middle Ages, a new power, inspired by the teachings of the spiritual leader Mohammed, was beginning to assert itself in the seventh century. Islamic civilization first took root in the roughly quadrangular peninsula of Arabia, which to the north borders on the rich and well populated lands of Mesopotamia and Syria. Shortly after the death of the great prophet in 632, the Arabs broke out of their desert strongholds by launching a *jihad*, or Moslem holy war, against their infidel neighbors. They quickly dominated the surrounding civilized countries of Egypt and Syria, which, until then, had been parts of the Byzantine Empire. Mesopotamia fell to the Arabs a short time later followed by North Africa, Spain, and parts of India. In less than a generation the Moslems had created a world empire whose proud warriors, enterprising merchants, and brilliant scholars wrought a major economic and cultural renaissance in the Middle East.

The tremendous energy that impelled the political expansion of the Arabs was accompanied by an intense interest in assimilating what the more ancient and richer societies of the subjugated countries had to offer. The conditions for intellectual freedom in the area of scientific inquiry were especially good, owing to the religious tolerance of the Islamic faith. Although stimulated by religious zeal, the Moslem conquest was essentially political in nature. Its leadership opposed the indiscriminate slaughter of unbelievers as well as their compulsory conversion to Islam. A major reason for this attitude is that, according to early Islamic law, only those subjects who had not become members of the faith were liable to taxation: in fact, too many converts later became a source of embarrassment to the ruling class. Furthermore, unlike the Bible of the Christians, the Koran presented no graphic picture of the world and universe to compete with the teachings of the ancient natural philosophers. Thus, the eventual acceptance of Aristotelian cosmology by Arab scientists failed to produce a conflict between science and religion anything like the one that dominated the intellectual life of late medieval Europe.

The new Moslem Empire inherited the ancient manuscripts and scientific traditions that Christendom had either lost or chosen to ignore. This legacy, combined with valuable scientific knowledge carried westward by Moslem scholars from India, became the basis for what is commonly called Arab science. In fact, the use of the term no more implies that Arab science truly arose in Arabia than that it is attributable to the Arabs alone. Its chief centers of development were located elsewhere—in Damascus, Baghdad, Toledo, and Cordova. Neither were the scholars who contributed to Arab science from one locality; they belonged to many different nations and not all were members of the Moslem religion. The one thing they had in common, however, was their use of the Arabic language as a medium of communication. Arabic was to assume the same importance for Islamic cultures as Latin had for early Christian Europe.

The first activity of those citizens of Islam concerned with science was one of assimilation. The knowledge that was coming in from Greek, Latin, and Indian sources could only be universalized by translating the original texts into Arabic, after which new information was then added. From about 750 onward, this task was mainly carried out under the auspices of the greatest patrons of Arab science, the Caliphs of Baghdad. Their splendid new city was constructed a few miles up the Tigris river from the ruins of Babylon, the ancient city near where the science of astronomy first began. Their lead in subsidizing the work of translation was followed by sultans of smaller countries like Egypt and by wealthy gentlemen, who had both the economic resources and leisure time to cultivate their artistic and scholarly interests. This high regard shown by the wealthy patrons of the new science appears to have resulted less from altruistic impulses, however, than from a deep concern for the security of their own lives and political fortunes. Through astronomy the destiny of the individual could hopefully be ascertained, while his health was sustained by an increasing knowledge of ancient Greek medicine. Like the kings and pharaohs of old, the ruling elite of Islamic society coveted the knowledge imparted to it by their astronomical and medical experts, who they believed held the key to a long and successful life.

Yet Arab science was much more than the province of wealthy dilettantes. Under the Baghdad Caliphite of the eighth century, scientific studies began to exert an increasingly stronger influence over the whole of Moslem culture. Primary knowledge of Greek science was garnered from the Nestorian Christians, who having taken refuge in Persia from the persecutions of the Byzantine Church, founded scholarly institutions in which astronomy was intensively studied. From India came a literature of astronomical and mathematical writings called "Siddhantas," a collection of the works of several scientific scholars. In them one encounters an essentially Aristotelian world picture—originally carried to the East in the wake of the Alexandrian conquest—complete with a geocentric earth and the epicyclic orbits of the planets, albeit in less complicated form than those outlined by Ptolemy. And early in the ninth century, the latter's own work on astronomy would itself be translated into Arabic and renamed *Almagest*, the Arabic word for "the very great composition."

In addition to astrology, there were other inducements for the Moslems to pay special attention to astronomy. The science aided the faithful in determining the direction in which to turn during their daily prayers and in selecting the appropriate dates for religious observances. Those who made their living by braving the deserts in caravans often journeyed at night to escape the unmerciful heat of the sun. They increasingly utilized the excellent tables of stellar and planetary movements, so carefully constructed by Arab astronomers, to establish their positions when out of touch with civilization for weeks, sometimes months on end. Under these treacherous conditions the slightest error in calculation could and frequently did lead to a prolonged and agonizing death. To further assist those guiding the caravans, Arab astronomers greatly improved the scientific instruments left them by the Greeks. The most notable of these, the astrolabe, gave accurate measurements of the altitude of the sun and other celestial bodies. With the revival of science in the West, the astrolabe and other Arab observational devices were used by both the scientists of the stars and the sea-going navigators of early modern Europe to explore and chart the waters of the still largely mysterious planet, earth.

The golden age of science in Baghdad lasted little beyond the tenth century. In the meantime, the Western countries under Moslem control kept the intellectual activity of the Arab world alive. In the Spanish cities of Seville, Cordova, and Toledo the planetary theories of Ptolemy continued to be taught, while scholars in the neighboring European countries displayed a renewed interest in the lost treasures of ancient science. Even as the political power of the Arabs in the peninsula waned, as a result of an invigorated European civilization, the Christian king, Alfonso X of Castille, followed the example of the Caliphs and summoned astronomers of all faiths to his court to prepare the famous Alfonsine Tables published in 1252. These periodic charts were employed by both Arab and Christian astronomers during the next three centuries to calculate the positions of the planets and major stars. With Alphonso the study of astronomy disappeared from southern Spain but, as we shall presently see, not from Western Europe for a second time.

Even though the scientific tradition of the Arab world declined in almost direct relation to Moslem political power, it constitutes a vital chapter in the history of science. For if their knowledge had not been reassimilated into the scholarly tradition of Europe during the twelfth and thirteenth centuries, the study of physics and astronomy might well have moved even further eastward, into Asia. While the Arabs made original and fundamental advances in mathematics, optics, and chemistry, the special significance of their work in astronomy lay not in the areas of innovation or bold theoretical analysis. It cannot be assumed that to be a scholar always implies being an original investigator. Islamic civilization is important to the history of science primarily because it preserved and disseminated the records of the ancients for later use by European scholars. The Arabs were quite content to work within the framework of the Aristotelian–Ptolemaic tradition, and, because of this, no significant progress in astronomy took place. But innovation and advancement were not what were called for at this time. Christendom first recovered its ancient heritage from Moslem culture through Latin translations of Arabic editions of the Greek originals. Thus, while the impulse towards continual progress was lacking in the Arab mind, the preservative functions of Arab scholarship formed the main bridge across which classical science reached Western Europe. Islamic civilization quite literally offered the Western world a rare second chance to recover the wealth of knowledge and wisdom of its fallen civilizations. Fortunately, this opportunity was not ignored as it previously had been after the decline and fall of the Roman Empire.

Revival

During the Renaissance, the practice first developed of dividing the history of the Western world into three major time periods: ancient, medieval, and modern. And although this proved a most convenient form of designation, it also produced a conceptual rigidity in the thought of the average layman about the past that historians have been attempting to dispel ever since. It is still a widely held belief that the human condition has improved only during two great historical periods: the time of the ancient Greeks and Romans, and our own age of mechanical development and industrialization. A major reason for the continued acceptance of this attitude lies in the modern notion that from the fall of Rome to the beginning of the Renaissance in the fourteenth century, no major cultural changes transpired. Nothing, of course, could be farther from the truth. The long European winter of the Dark Ages did not readily yield to the light and warmth of the Renaissance spring; the thaw required several centuries and the combined energies of countless thousands of unknown men and women to complete. It began at least as far back as the tenth century and culminated in what we now call the Twelfth Century Renaissance.

During the late Middle Ages, there was a gradual increase in the tempo of all major aspects of European culture accompanied by a significant change in the Western outlook on life. The simple passage of time had done much to reduce the tension, anxiety, and dread that had so deeply marked the first centuries A.D. The writings of St. Augustine were apocalyptic and highly anticipatory of the imminent destruction of secular society, of which the great African Doctor believed Rome's fall clearly signalled the beginning. Neither St. Augustine nor any of the other early Christian Fathers had foreseen the present order of things enduring for more than 1000 years. Yet mankind had somehow survived, civilization was still weak but alive, and the idea increasingly recurred to thinking men that perhaps they should consider making the best of life on earth—that even if the individual was not the arbiter of his own destiny, he could at least influence his future to some degree. Gradually, the people of Latin Christendom began to discard their heavy winter garments fashioned out of the uncomfortable fibers of fear, repentance, and otherworldliness, and exchanged them for the less irritating and restrictive attire of the individual intent upon not merely existing but living in a world and environment molded to his own needs and desires. Deeply encased in this outlook, and soon to take root, lay the seeds of the modern mind.

The causes underlying this refreshing change in temper were numerous and varied. For the first time since the conversion of Europe to Christianity, the majority of the population enjoyed a degree of relative political security. Arab power was on the decline, Viking militarism had been dissipated, while France and England had entered into the germinative stages of modern nationalism. Great intellectual changes were also afoot. During the latter part of the eleventh century students from throughout Europe were beginning to assemble in steadily growing numbers to listen to readings and commentaries by masters on ancient texts. A century later these gatherings became so large that formal regulations and charters were drafted officially establishing the first universities.

In these vigorous new institutions, all the ingredients of a rapidly developing civilization were brought together for the first time, thus laying the basis for a creative and original tradition of European scholarship.

Europe's cultural renaissance would not have been possible, however, had it not been for the economic revival of the later medieval period. As usual when economic factors are concerned, it is difficult to state precisely what was the cause and what was the result; but we can be certain of some fundamental relationships. An increase in agricultural production was absolutely necessitated by the growth of new trading centers whose inhabitants were dependent upon large quantities of imported food. These towns and cities arose in turn to meet the demands of a burgeoning commercial revival. Luxury goods that could only be imported from the East were increasingly sought by the wealthy as the economic and political independence of the feudal manors gradually gave way to expanding cultural contacts. It was this interaction of political, cultural, and economic factors that prompted the rediscovery of ancient learning, of which the reclamation of science was an integral part.

While it cannot be said that Aristotle was not known in Western Europe before the thirteenth century, very few of his works, and none of those primarily concerned with cosmology, had as yet been translated from Arabic and Greek into Latin. As a consequence, Aristotelian natural philosophy was understood only in the most fragmentary manner and only then as it appeared in the secondary encyclopedic works compiled by such Latin scholars as Boethius and Chalcidius. The major corpus of the Stagirite's work on physics, cosmology, and metaphysics became known to Europeans only after about 1250 through the contacts developed between Latin scholars and their counterparts in the Arab world.

Several major Arab thinkers, including the brilliant scholars Averroës and Avicenna, had studied and commented upon Aristotle's physical system; when knowledge of their work reached Europe via Spain and Italy, it claimed the full attention of many Western intellectuals. The main center of transmission in Spain was through the Moorish city of Toledo. Even though it had been conquered by the Christians in 1085, centuries before the fall of Granada and Cordova, Toledo remained the seat of oriental science. There, a college of translators was founded whose work became of utmost importance to the development of Western science. The most productive of the many gifted Latin translators who settled in this great intellectual center was one Gerard of Cremona. Gerard and the scholars under his direction seem to have turned out Latin translations of Arab manuscripts with virtual assembly-line regularity: Gerard himself has been credited with completing 92 such translations. In this same period southern Italy and the Kingdom of the Two Sicilies constituted a Mediterranean crossroads where the Latin

West came into contact with both Byzantine and Arab culture. It was in Solarno, on the Gulf of Naples, rather than in Spain, that the earliest center of European science arose. As early as the ninth century, this city possessed a famous medical school whose curriculum drew directly upon what remained of Greek science garnered from Byzantine and Arab sources.

Still, it is difficult to realize just how ignorant the learned men of Western Europe were in comparison to Arab scholars of the time. According to Alfred North Whitehead, "in the year 1500 Europe knew less than Archimedes who died in the year 212 B.C." [13] Thus the thirteenth-century process of absorbing and assimilating Arab scientific knowledge proved much more difficult than appearances might at first indicate. Theirs was not simply a matter of translating familiar subject-matter into the Latin language; doubtless they would have been most pleased had this been the case. Quite frequently the language itself proved inadequate to the task at hand, for Latin contained few of the technical terms needed to properly express this new and foreign world of ideas. European scholars had to take on the additional responsibility of creating the terminology through which these new concepts could be rendered into intelligible form. Ironically, this placed them on the same footing with the Arab scientists who had faced a similar task centuries before as they struggled to make the contents of Greek scientific writing understandable in Arabic.

Further complications arose from the fact that Arab science had not been communicated to Europe in unadulterated form. Despite their unquestioned acceptance of most of the master's teachings, the Arabs themselves had never read Aristotle's works in the original Greek. They became available to Islamic culture only in Syraic translations acquired during the Moslem conquest. By the time Aristotle's writings had once again become known to Europeans, they had been translated at least three times: from Greek into Syraic, Syraic into Arabic, and Arabic into Latin. Add to this the consideration that Aristotle himself wrote tersely, often without adequate explanation, and one begins to gain an appreciation of the serious problems confronting European thinkers. No wonder the resulting picture of the Aristotelian universe contained some horrendous distortions. Moreover, medieval scholars had not yet mastered Greek in sufficient numbers to translate the few original manuscripts discovered in the archives of the great European monasteries. Until they did, a fair number of misconceptions and corruptions continued to prevail. And even after a knowledge of Greek became widespread, there was much in the work of Aristotle and his contemporaries that remained forever beyond recovery. Lastly, many of the Latin translators had no interest in science as a discipline in and of itself. Consequently their work frequently fell considerably below the high standards of accuracy and detail demanded by rigorous scientific thinking.

This confusion did harbor one significant advantage. Once the medieval view of the universe based on Aristotle's works had been established, virtually any and all contradictions in the Stagirite's cosmology could be attributed either to improper translation or to a lack of solid evidence to the contrary. For compared to their own understanding, the scope of Aristotle's personal knowledge and insight had been so broad and penetrating that it was only natural for European scholars to hesitate before concluding that the master might have erred. After all, these were the descendants of medieval thinkers, who for generations had harbored an acute sense of their own intellectual and cultural inferiority. They possessed (or so they thought) no worthwhile historical legacy of their own. The Schoolmen tended "to telescope the astronomical doctrines of Eudoxus, Aristotle, and Ptolemy, and to squeeze out all disagreements and inconsistencies." As they looked back over a millennium "they could not reconstruct the historical development of Greek thought in perspective." [14] Nor, it should be pointed out, did they wish to do so. After centuries of chaos, this attitude was based on an overwhelming desire for intellectual conformity and cultural respectability through the imposition of a single mode of thought. It received great encouragement from Europe's most powerful institution, sacred or secular, the Roman Catholic Church.

Earlier in this chapter the observation was made that neither the Church's attitude nor its practices regarding science were uniform throughout the Middle Ages, but that until the tenth century its position was predominantly antiscientific. Then, for a number of reasons yet to be discussed, toward the end of the first millennium A.D., this hostility gradually cooled and to a considerable extent altogether disappeared. From the latter part of the tenth to the middle of the sixteenth century, the Church played a dominant role in the medieval revival of ancient science. This reversal of positions might at first appear totally out of character for an institution that had once been diametrically opposed to the influence of pagan classical scholarship. Yet it should also be remembered that, despite its position of dominance in medieval society, the Church was as subject to the new currents of intellectual change sweeping across Western Europe as any of the political or social institutions of the day.

As we have observed, the religion of the early Middle Ages had been fatalistic, anxiety-ridden, and opposed to most things secular as a compromise with Satan, the Prince of Darkness. Because he bore the stain of original sin, man was believed to be inherently wicked, incapable of accomplishing good works without direct benefit of God's grace. God

Himself was the sternest of taskmasters, selecting for reasons unknown to mortals those men and women who would enter the heavenly city and those who were to experience eternal damnation in Hell. During the High Middle Ages this pessimistic philosophy was significantly tempered by the growing intellectual confidence of a resurgent European culture. In the monastic institutions, particularly among the teaching orders, men were given the economic resources and official encouragement not only to preserve but to build up a body of knowledge and pass it on to future generations in unbroken succession. Even had this been possible a few centuries earlier (which it was not) the practice would have doubtless been strongly discouraged because of the belief in the imminent destruction of the secular city.

By the thirteenth century, some very different religious conceptions had replaced those of the early Christian Fathers. The Church, like the human community to which it ministered, had also survived the rigors of almost complete cultural breakdown; and now that social and material conditions were markedly improved, the Last Judgment hardly seemed at hand. The world no longer appeared to be quite the solitary, nasty, and brutish place that the Germanic barbarians had made it in the wake of Rome's wrenching demise. Since God had thus far spared the world His ultimate wrath, perhaps He regarded man worthy of greater consideration than the Church had previously accorded him. Perhaps man could cooperate with God in achieving the salvation of his soul by following the commandments as laid out in the Scriptures and by actively partaking of the services of the Church.

In a very real sense, Christianity in the thirteenth century worshipped an entirely new God, a deity whose rage and wrath had been conquered by the qualities of divine justice and mercy. Two brief examples must serve to illustrate the influence of the new humanism in religion: the theory of the priesthood and the theory of the sacraments. Neither of these concepts was new to the Church, for priests had existed for centuries as had certain sacraments such as communion. Yet the exact functions of the priest and the precise nature of the sacraments had never been clearly formulated. The theory now came to be accepted that the priest, because of his ordination by a bishop, who had been confirmed by the Pope in Rome, became the inheritor of a portion of that authority which Christ had conferred upon His apostle Peter. The latter had been appointed Christ's vicar on earth, a position inherited by each successive Pope. This meant, in effect, that by virtue of his ordination, even the lowliest village priest possessed the power to cooperate with God in performing certain miracles and in relieving repentant sinners from various temporal consequences of their sinful acts.

These miracles were the sacraments, which by the twelfth century were accepted (as they are today) as seven in number: baptism, confirmation, penance, the Eucharist (or communion), marriage, ordination, and extreme unction (or the last rites). The Roman Catholic Church defines a sacrament as an instrumentality through which divine grace is communicated to men. To partake of the sacraments administered by the priest was to offer evidence of man's willingness to work with God, through the offices of the Church, to attain eternal salvation. The establishment of these two basic doctrines led almost immediately to both a rise in the power of the clergy and a general strengthening of Church authority. During the later Middle Ages the Church undertook a systematic attempt to extend its moral influence over all of its lay members, an effort that for upwards of three centuries enjoyed widespread success.

No longer was the Church on the defensive as it had been in St. Augustine's day when the hierarchy considered it necessary to maximize the attention given to the problems of Christian theology. The main areas of continental Europe had been converted to Roman Catholicism by the time the West was reestablishing commercial and intellectual ties with Byzantium and the Islamic civilization of Spain, Syria, and North Africa. At long last, the spiritual and intellectual authority of the Church was unrivalled; pagan and secular learning were not only permissible again but desirable, provided, of course, they were pursued within the boundaries established by clerical authority. By now, the ridiculous commentaries on ancient science by such patristic writers as Lactantius had not only become anachronistic but an embarrassment. The "nature of things" had about reached the point at which they had been left by Aristotle some 1500 years earlier: the earth was round again, and the two-sphere universe was once more taken pretty much for granted. The way now lay open for the right person to accomplish a final reconciliation between Aristotelian cosmology and Christian theology. That person was St. Thomas Aquinas.

The Thomistic Synthesis

If the non-Christian natural philosophy of Aristotle was to become generally acceptable to European intellectuals, and particularly to members of the Roman Catholic hierarchy, the proper avenue of access ran through the gates of the great medieval universities and directly into the lecture halls presided over by the Schoolmen of the twelfth, thirteenth, and fourteenth centuries. Unfortunately, from the modern point of view, scholasticism is quite likely to be regarded as devoid of practical sense, an intricate and tiresome quibbling over metaphysical concepts of little practical importance. Yet in the broadest sense, the term encompasses virtually the entire educational system of the Middle Ages, and the most famous of the scholastics deserve to be ranked among the greatest Western scholars of any age: Peter Abelard, Roger Bacon, John Duns Scotus, William of Ockham, Albertus Magnus, and Thomas Aquinas.

The primary consideration underlying all scholastic inquiry concerned the relationship between the Mind of God and the minds of individual men. Since the scholastic believed the physical world to be an organism whose existence was sustained by Divine Will, it seemed both logical and rational to measure the value and truth of human knowledge by tracing it back, so far as possible, to the presumed source of all intelligence, God. It was only when medieval scholars permitted their academic discussions to challenge the established beliefs of the Church that they experienced serious difficulty. For the first time in centuries rational inquiry began to reassert itself, as is evidenced in the curriculum of the universities, whose main subjects included mathematics, logic, science, medicine, Roman and canon law, theology, and the "new" and exciting subject, Aristotelian philosophy.

It should not be assumed that the study of Aristotle was accepted immediately or without reservation. The Stagirite's works contained many observations that seemed alien to Christian theology and thus engendered fears, doubts, and deep suspicions in several influential quarters. During the early thirteenth century, the ecclesiastical authorities at the University of Paris banned lectures on Aristotelian natural philosophy and metaphysics. As might have been expected, however, this action only heightened scholarly curiosity; the very search for a method of censoring the ancient works led to an even more detailed study of their content. Elsewhere, the action taken by the authorities was less severe, and Aristotle's works eventually became an indispensable part of the Latin philosophical and scientific tradition—and of scholasticism itself.

One of the individuals most responsible for interweaving Aristotle's teachings with the larger fabric of Latin culture was the Dominican priest and university scholar Albertus Magnus (1206–1280). Albertus was by birth a Swabian, who entered the Dominican Order at an early age. After teaching in a number of German schools, he took his doctorate at the University of Paris, where he began a distinguished career as a teacher. It is accepted almost without question that Albertus became the most learned man of his generation; his lasting fame rests mainly on his monumental treatises—thirty-eight quarto volumes in the last published edition—written over a period of several decades. The far-reaching titles of his many works are sufficient evidence in and of themselves to prove that Albertus was a staunch disciple of Aristotle. His interests in metaphysics and natural science are nearly as wide ranging as those of the master, and large sections of the Dominican's work are devoted to Aristotelian philosophy interpreted in the light of Christian doctrine.

Even though his defense of Aristotle's work subjected Albertus to frequent and sometimes virulent attack from fellow Christian thinkers, his acceptance as an orthodox theologian is attested to by his enrollment among the saints.

It was during his tenure at the University of Paris that the suspicion arose that the study of Aristotle might lead to heresy, causing the authorities to temporarily ban lectures on the Greek philosopher and his works. Albertus cleverly defended his scholarly pursuits against this action by coming forth with the following statement: "I expound, I do not endorse Aristotle." Whether or not this utterance can be considered a falsehood depends upon how literally one wishes to interpret it. Certainly Albertus was no simple copyist of the Stagirite; time after time he expresses opinions contrary to Aristotle's observations. On the other hand, the thrust of a major portion of his voluminous writings is undeniably pro-Aristotelian. In the end, a compromise was reached whereby the teaching of "philosophy" in the faculty of arts permitted the discussion of Aristotelian tenets, while the teaching of "truth" in the faculty of theology did not. This tentative resolution of the issue implied the existence of two "truths"—one of a philosophical nature and the other the revealed truth of the Christian religion. Whenever the two systems conflicted, the revelation of the Scriptures was to take precedence. This same principle was invoked some three centuries later when, after the death of Copernicus, and before his treatise *On the Revolutions of the Heavenly Spheres* was placed on the Index of Prohibited Books, the Church permitted the teaching of the heliocentric system as theory, but not as fact. Nonetheless, the proponents of Aristotelian natural science and cosmology had won a major victory, the exploitation of which was carried much farther than even Albertus might have dreamed, and by his own greatest pupil. Indeed to some it might seem that the most significant accomplishment of St. Albertus was the training of Thomas Aquinas (1225–1274). To adopt such a narrow view, however, would be to deprecate the work of a great scholar without whose influential writings Aquinas, whom many consider the most outstanding Christian thinker of all time, would have been lost. For like all great scholars, St. Thomas built his formidable intellectual structure atop the foundation laid by his predecessors.

Thomas owed his surname to the fact that he was born near the Italian town of Aquino, at Roccasecca, on the road from Rome to Naples. His father controlled a modest feudal estate, the location of which must have served as a constant source of anxiety to the family, for it lay on a boundary disputed by the Pope and Emperor Frederick ii. Perhaps because of a parental desire to spare their precocious child the adverse consequences that might follow from the ongoing civil strife of the region, Thomas was placed as an oblate (prospective monk), while still a young boy, in the nearby monastery of Monte Casino. Perhaps, too, the parents dreamed that the obviously talented child would one day become abbot of the great institution, a position that would certainly work to the family's advantage. But after spending nine years in the famous sanctuary, Thomas returned home, in 1229, when the Emperor ordered the monks expelled for their overly zealous obedience to the Pope. Shortly thereafter, while still in his early teens, he was sent to the University of Naples where he spent the next six years. The University provided a stimulating intellectual environment that proved highly receptive to the scientific and philosophical works of classical civilization then being translated from Arabic into Latin. Against this background young Aquinas joined a new religious order, founded only some thirty years earlier, the Friars Preachers or Dominicans. It was a fateful decision, for it virtually guaranteed that his great intellectual energy was not to be confined to the pastoral and apologetic work which preoccupied the more firmly established orders. The Dominicans also substituted the more democratic organization of the mendicant friars in place of the traditional form of authoritarian government and regimentation normally associated with monastic life. They were by origin a learned order and openly embraced intellectual pursuits while abandoning the manual labor deemed basic to the monastic tradition. The Dominicans, like the members of similar orders, quite literally sustained themselves by begging their daily bread and also paid more attention to foreign missions than to charitable work at home. These practices, among others, kept the enrollments relatively low. By entering into a more active life of preaching and teaching, Thomas took the liberating path beyond the narrow boundaries of the feudal world into which he was born and the monastic influence under which he was raised.

For Aquinas, liberation from the feudal world also entailed the literal necessity of physical escape from the domination of his parents, a protracted and trying ordeal. His Dominican superiors had quickly recognized the young man's scholarly potential and assigned him to the University of Paris for advanced study. But on the road to France he was abducted by agents in the employ of his father. Both men proved equally stubborn in their convictions, and a year passed before Thomas was finally freed. In 1245, at the age of nineteen, he at last entered the convent of Saint-Jaques, the university center of the Dominicans at Paris, where he began study for a brilliant career as a scholar–theologian under the direction of Albertus Magnus.

Thomas arrived at the university during a period of major intellectual ferment. As has been observed, Church authorities were in a quandary over what course to follow regarding the rapidly growing interest in Arabic–Aristotelian science. The works of Averroës, Spain's most gifted

Arabic philosopher and interpreter of Aristotle, were just then becoming known to the Parisian masters. The Moslem scholar's dualistic concept of faith and reason had generated particular interest and sparked considerable debate not only at Paris, but among Latin scholars throughout Europe. Averroës argued, in effect, that there are two truths—one based on religious faith, the other on reason. However, by refusing to subordinate reason to faith, the Spanish-born philosopher had already anticipated and rejected in advance the compromise worked out at Paris decades later. To Averroës religion and rational knowledge not only do conflict, they must inevitably be contradictory. They represent two distinct types of knowledge, many of whose basic principles are diametrically opposed and will forever remain irreconcilable. As a thinking but imperfect creature, man must accept such inconsistencies and deal with them the best he can, which does not mean the subordination of one truth to another. Thomas found it impossible to accept these conclusions, and the young Dominican rose in spirited but respectful protest against his pro-Averroist colleagues, an act that launched his life-long quest for a method whereby the philosophy of Aristotle could be synthesized with the doctrines of the Christian Church.

To begin with, St. Thomas accepted the basic structure and virtually all of the physical characteristics of the Aristotelian universe. These included such basic concepts as the four elements, a centrally located, corrupt, immovable earth, and an unchanging, finite, celestial realm containing ethereal bodies that move through the heavens in perfectly circular orbits.* St. Thomas also employed, in altered and more detailed form, a hierarchical concept borrowed from Plato and Aristotle called the Chain of Being, the existence of which, from the Middle Ages down to the late 1700s, most educated men were to accept without question. According to this idea, Nature can be best understood by comparing it to a giant ladder or flight of stairs. At the bottom one encounters insensate matter in the form of simple minerals, in addition to other more complex inorganic substances. Somewhat further up the first flight, life forms begin to appear, with each arranged in ascending order of complexity and function; finally, atop the first landing, stands man, the pinnacle of terrestrial creation. From man, the second flight of stairs reaches up into the ethereal realm populated by purely spiritual creatures, the angels. Directly above them, at the apex of all creation, stands God.

It takes little imagination to see how this concept supported the Aristotelian belief in the two-sphere universe. Everything subject to change,

---

* In the Thomistic system, the epicycles of Ptolemy were not so much forgotten as abandoned as excess baggage. Such was not the case among certain late medieval and Renaissance scholars, however. Copernicus, for example, employed several Ptolemaic principles—including epicycles—in the construction of the heliocentric system.

including all creatures of a corporeal nature, are consigned to the earth, while the incorruptible spiritual bodies abide in the heavens. It was also believed that no new species are ever created, nor are any destroyed; individual members of the various terrestrial species die (the celestial angels are of course immutable), but their kind will always survive, the chain is never broken. Primary matter sustains the growth of vegetative life; vegetables rise upward to support the sensitive or animal species; and sensitive life to the intellectual as embodied in man who, of all terrestrial creatures, most perfectly mirrors the Creator. This concept has never been more aptly described than by the eighteenth-century English poet Alexander Pope in his *Essay on Man:*

> Vast chain of being! which from God began,
> Natures aethereal, human, angel, man,
> Beast, bird, fish, insect, what no eye can see,
> No glass can reach; from Infinite to thee,
> From thee to nothing.—On superior pow'rs
> Were we to press, inferior might on ours;
> Or in the full creation leave a void,
> Where, one step broken, the great scale's destroy'd;
> From Nature's link whatever link you strike,
> Tenth, or ten thousandth, breaks the chain alike.

As with the other physical qualities of the Aristotelian universe, St. Thomas wholeheartedly embraced this great law of continuity. But on the subject of the Supreme Deity, who occupies the apex of the scale, the Dominican parts company with the Stagirite. As has been previously observed, Aristotle's concept of divinty was centered around the idea of a supreme, immovable Prime Mover whose inactivity is responsible for the motion of the heavens. Since He has nothing left to realize, He has no need to act. The planets move as a result of their craving after perfection, a yearning passively generated in them by the Prime Mover. St. Thomas, on the other hand, assigned to God a much more active part in the ongoing process of universal motion. The Creator did not mold the universe in all its richness, diversity, and splendor, only to cast it off into space to ride the winds of fortune and shape its existence as best it could. In the Thomistic universe, God plays a major role in the activity of all things, both finite and eternal, while at the same time He sustains their being. Harmony and order are writ across the face of the universe, and this would not be possible save for God's own involvement in physical occurrences. If, indeed, the universe described by St. Thomas were to be summarized in a single word, we should have to call it "theocentric." While Aristotle taught that only a single class of beings— the spirits which propel the planets—yearn after divine goodness, St. Thomas asserts that all of Nature's creatures together, from the meanest species to the highest, reflect in their own individual manner the good-ness of their Creator. All species, interlocked in their activities, cooperate with each other as they strive to satisfy their common appetite for God. And although man is the highest form of terrestrial life, he is dependent upon the mineral elements to nurture the plants and animals that sustain his existence. Here order is manifested in its highest form. Nature's scale thus becomes a veritable flood tide of advancement, a harmonious, rhythmic cascade which should thrill even the most skeptical of ob-servers.

According to St. Thomas, one of the major manifestations of God's continuing involvement in the benevolent governance of the universe is His ordering and directing of spiritual creatures. The saint wrote ex-tensively on the proposition that all corporeal bodies are governed by God through the angels, the movers of the heavenly spheres. Once again he accepted Aristotelian physical proposition that the heavenly bodies are incorruptible and subject to no change except that of local motion. However St. Thomas also allowed that their circular movements are due to intelligent guidance combined with Aristotle's physical concept of natural form. The various angels constitute a natural hierarchy combined of three major orders: Seraphim, Cherubim, and Thrones. It is the Cherubim, whose name signifies "fullness of knowledge," that assume responsibility for the planetary movements. The angels themselves lack corporeal substance, being composed of pure intellect. And since only a corporeal body can move a corporeal body by physical force, all matter is confined to the sublunar realm. Hence, angelic movers exercise their powers over the heavenly spheres by a command of will. The Baroque artist's conception of a winged, angelic cherub, pressing its body against a planetary globe, while graphic, is highly metaphorical.

Aquinas enlarged the province of angelic rule to include an immediate governance of terrestrial things, a consideration wholly alien to Aris-totle. In his treatise *Summa Contra Gentiles* he discusses the direct re-lationship between angelic will, planetary movement, and sublunar change:

> The heavenly bodies move because they are moved; so that the end of their movement is to attain a divine likeness. Likewise, the heavenly bodies, although more noble than the lower bodies, nevertheless intend by means of their movements the generation of the latter bodies, and to bring to actuality the forms of things generated.[15]

The very conception, birth, and eventual demise of individual men, as with all other living creatures, is subject to celestial movements willed by God through the angels. This extension of angelic powers to the sublunar region marks the gap that separates the Aristotelian spirits from the great Angelic Virtues of Christian cosmology, despite obvious elements of continuity. It is little wonder that the late medieval proponents of astrology closely allied themselves with the Thomistic philosophy of the angels to support their contention that the heavens do indeed control the destiny of man.

Man, as has been intimated, also occupies a special place in the Thomistic universe; in one sense all things were created for man because God has directed all things to contribute to man's benefit. The glory of man is that he stands at the pinnacle of creation. He unites the lower orders in his material makeup and at the same time exhibits the only intellectual development and spiritual force on earth. He is the only creature in the universe in which the physical and intellectual functions are combined. Through exercising his rational powers man can draw the hierarchy of nature to himself in a unified form; he can understand it and use it to his advantage. Thus, man is a microcosm, the miniature embodiment of nature, the macrocosm. He becomes nature's high-priest, acting as its representative in its dealings with God. Not even St. Thomas himself could have truly realized the future implications that would follow from this line of reasoning. During the seventeenth century the idea of man as the master of an orderly universe was consciously merged with the linear concept of time, also derived from Christian thought, thus giving birth to the underlying rationale of the scientific revolution.

For all the opposition expressed by the Christian Fathers toward ancient science, St. Thomas had discovered in Aristotelian cosmology a system of immense value to the Church. Aristotle's world view harmonized beautifully with the Christian concept of Divine Providence, for together they made sense out of the world inherited by a generation of men striving to revitalize a civilization devastated by centuries of barbarism. The very pagan learning that patristic writers had once found highly distasteful, indeed heretical, was now embraced with a fervor and tenacity rarely surpassed in the history of Christianity: the old dogmatism was thus replaced by the new. Furthermore, the Thomistic–Aristotelian synthesis marked the effective culmination of a trend that had its genesis in the early Christian era—the complete subordination of science to religion. From the historical point of view, the Church hierarchy took the same extreme position that the priests of Babylon and Egypt had assumed over 3000 years earlier; that is, it closed its mind to any scientific observations which challenged, or even appeared to challenge, orthodox theology. Cosmological ideas had once again become an important part of religion, but to question any one of them was to attack the entire authority of established religion, or so it was thought. In the long run, such intransigence in the light of new scientific discoveries struck a body blow to the tremendous prestige enjoyed by the Church, an injury from which it never fully recovered. But for the moment, the whole corpus of ancient science and medieval theology were brought together and combined into a single, harmonious picture of the cosmos.

While it was Aquinas and like-minded scholars of the thirteenth century who certified the compatibility of Christian belief with ancient learning, the very detail and complexity of their works tended to obscure the broad outlines of the over-all picture of the new Christian universe just then emerging. The cosmography of St. Thomas needed a popular spokesman, and it found a brilliant one, surprisingly enough, in a poet. It is normally a risky business to draw conclusions about the state of scientific knowledge of any age by referring to poetical works, but in the case of Dante Alighieri it is a legitimate undertaking. As a pupil of Brunetto Latini, an avid encyclopedist, Dante studied and fully mastered the strucure of the universe as outlined by Aristotle with the commentary of Thomas Aquinas. In his masterpiece the *Divine Comedy*, Dante brings the new world view sharply into focus by undertaking an imaginary journey of cosmic dimensions.

He began, of all places, in Hell, a conical cavity reaching to the earth's center. Around this abode of eternally damned souls one encounters the infamous seven concentric circles, each of gradually decreasing diameter, where punishment is meted out to the fallen. The lower one moves into the bowels of Hell, the more horrible the pain and suffering. At its very center dwells Lucifer, Prince of Darkness, whose underground universe has obviously been constructed along Aristotelian lines. With his Latin guide Virgil, Dante is ready to pass through the bottom of the abyss, but at the last instant the poet pauses to catch a final momentary glimpse of the Infernal Kingdom and observes Satan, frozen upside down in a block of ice. Dante is at first perplexed, but Virgil explains that Satan appears to be upside down because they have begun their climb to the other side of the earth on which they emerge after passing through Purgatory, a large hill emerging out of a great ocean directly opposite the holy city of Jerusalem, the navel of the dry land.

The remaining part of the poet's odyssey transpires in the realm of the celestial spheres which, after passing through the sublunar region of fire and air, are discovered to be ten in number. The first, of course, is that of the moon followed in order by Mercury, Venus, the sun, Mars, Jupiter, and Saturn. Dante observes that the farther the planet from the terrestrial realm, the more pure its composition. He also encounters various levels of spirits or angels who are attached to each of the planetary spheres according to their rank, all of which serves to illustrate the ever increasing glory to be found in the universe. As in Aristotelian cosmology, the eighth sphere is that of the fixed stars, the ninth is the Primum Mobile. The velocity of circular motion displayed by the latter sphere is hardly believable owing to its desire to be merged with the

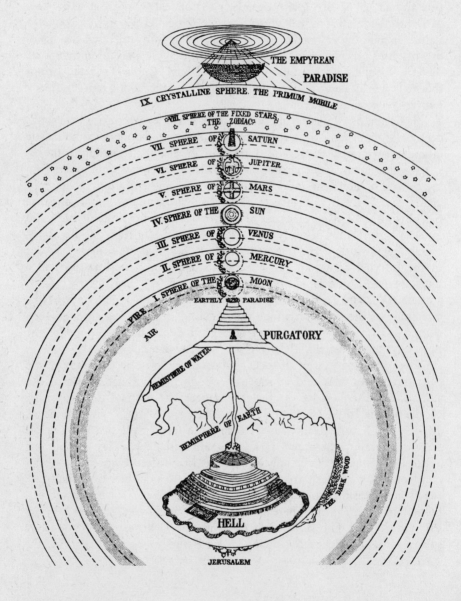

stationary and most divine of the ten celestial spheres, the Empyrean Heavens. There, surrounded by the Heavenly Host, dwells the God of the Holy Church:

> O splendor of God through which I saw
>> the lofty triumph of the true kingdom,
>> give me the power to tell *how* I saw it!
> There is a light up there which makes the Creator
>> visible to that Creature who,
>> only in seeing Him, has its peace.
> And it extends in the form of a circle
>> the circumference of which
>> would be too large a belt for the sun.
> It is composed wholly of rays
>> reflected on the top of the Primum Mobile
>> which gets life and power from it.

Dante's grandiose vision serves as a fitting conclusion to the thousand-year quest for the recovery of intellectual respectability in the West. When he died in 1321, the earth once again occupied the center of the universe. And even though much of Greek science, philosophy, and poetry were still only imperfectly known to European thinkers, Western civilization had at long last reached the point where it could begin to build upon its ancient foundations.

The medieval picture of the universe just described obviously contains more of Aristotle than of Christianity. The Church had added its angels and the concept of Paradise, but all the other essential elements came from Greek science. Hence, it was basically the Greek view of the universe that would be challenged and eventually overthrown by the astronomers of the sixteenth and seventeenth centuries. Nonetheless, the Thomistic synthesis was a brilliant intellectual accomplishment, as was Dante's poetical rendering of its scientific, religious, and mystical content. Yet even as the Church settled upon its particular version of "scientific truth," the first faint stirrings of discontent could be discerned among the ranks of skeptical intellectuals. As yet, of course, no one could have possibly imagined what profound changes this seething restlessness would ultimately bring, or how deeply the final product of this growing discontent would influence the course of modern thought and history.

FIGURE 6.   Dante's scheme of the universe. Slightly modified from Michelangelo Caetani, duca di Sermoneta, *La materia della Divina Commedia di Dante Allighieri diachiarta in VI tavole*, Monte Cassino, 1855. Drawing taken from: *Studies in the History and Method of Science*, vol. I, ed. by Charles Singer (Oxford University Press, 1917), p. 31.

The Rebirth of Astrology

Before we move on to a discussion and analysis of the various contributions made by the individual astronomers who pioneered the new era in science, something must be said of astrology, to which repeated references will be made in subsequent chapters of this book. We have already noted that the mainstream of astrology had its source in the Fertile Crescent of the Middle East, where man first laid down his hunting weapons and took up the plow and scythe to create an agricultural society. The Babylonian priesthood made the discoveries which prepared the way for the development of astrology, and its members were responsible for its ever growing influence upon Middle Eastern religion and government. After the decline of Mesopotamian civilization, knowledge of astrology spread eastward to China where it was used, as before, by the emperors and their supporters to predict the future and to sustain their political power. It later surfaced in Greece where, during the classical period, astrology was quite widely practiced among the agrarian population in conjunction with the mystery religions of oriental origin.

The Romans, despite their reputation as a hard-headed and practical race, were also a highly superstitious people. This tendency, no doubt acquired from their early religious beliefs and the strong influence on Latin civilization of Greek culture, provided a highly receptive environment for the professional astrologer. The emperors heaped great wealth upon those whose predictions were fulfilled and banished or put to death those practitioners whose forecasts did not come to pass. Roman astrology also enjoyed widespread appeal among the masses who sought advice on a regular basis from the large number of astrologers resident in the empire's many cities.

The decline and eventual collapse of the Western Roman Empire also meant the end of European astrology for the next several centuries. Not only was the complex knowledge required to carry on the art lost, along with much of the rest of classical learning, but the conversion of Rome to Christianity had created an environment quite hostile to the astrologer. His profession was condemned by St. Augustine and other patristic writers on the grounds that God alone knows what the future holds for man, and that those who believe otherwise are either charlatans or sorcerers in league with the forces of evil. But even as astrology languished in Europe, it was kept alive across the Mediterranean Sea in the great Moslem centers of learning, Baghdad and Alexandria. Arab astrology became widespread during the eighth century and, as in earlier cultures, was inseparable from astronomy. In fact, only after astrology had been reintroduced to Europe, through the same channels that classical science and philosophy were recovered, did a semantic differentiation between astrology and astronomy manifest itself, an occurrence that seems to have begun late in the fourteenth century.[16]

Throughout its long history, astrology was never more popular in the West than during the late Middle Ages and Renaissance, the uncompromising rejection of its basic principles by the early Christian Fathers notwithstanding. The recovery of Aristotelian cosmology played a major part in the favorable change of attitude undergone by both religious and secular leaders. In the first place, Aristotle had supplied the physical mechanism—the frictional drive of the celestial spheres—through which heavenly bodies were considered capable of producing changes in the terrestrial realm. And few educated men of the Renaissance could deny the purely physical impact of the planets on the earth as evidenced by such processes as the raising and lowering of the ocean's tides and the changing of the seasons. Even St. Augustine could not help but be impressed by the evident interdependence of natural terrestrial phenomena and the positions and movements of various celestial bodies. He did not believe, however, that this cause and effect relationship extended to the acts of man. Still, the door had been opened, at least in principle, to the recognition of astrology as a reliable predictive tool.

It was in the writing of St. Thomas and like-minded scholars of the thirteenth century that astrological theory received its strongest official support. The saint steadfastly maintained that all bodies are governed by God through the angels, and that this relationship involves at least indirect control over the sublunar region as well: "Accordingly the movements of heavenly bodies, in so far as they cause motion, are directed to generation and corruption in the world beneath them. And it is not unfitting that the movements of heavenly bodies conduce to the generation of these lower things." [17] Moreover, the Christian belief in God and the angels carried an implicit assumption that Satan and his demonic servants also exist. The populace accepted without question the everyday presence of occult forces and the power of magic, as did many of Europe's most famous scholars. Newton, himself, spent at least as much time in the pursuit of the occult as he did in formulating his mathematical laws, a side to his character that we still hear far too little about today. In the study of history one must be on one's guard against seeing only the obvious and missing the significant, and the significant thing here is that the supernatural was believed in by everyone. So little was known of nature that the entire creation still appeared as an unexplained miracle, a universe of mystery and wonder in which anything was possible.

Astrology also exercised a strong influence over the conduct of certain professional activities. The physician, for example, employed astrology on an ever increasing basis to cure the ill. At the School of Medicine at the University of Bologna the saying ran: "A doctor without astrology is like an eye that cannot see." The medieval alchemist also conducted his secret experiments in quest of the Philosopher's Stone—the secret of transmuting lead into gold—in strict accordance with the signs of the zodiac. The most common alchemical processes and the signs under which they were completed are listed below:

♈	Aries	Calcination	♎	Libra	Sublimation
♉	Taurus	Congelation	♏	Scorpio	Separation
♊	Gemini	Fixation	♐	Sagittarius	Ceration
♋	Cancer	Solution	♑	Capricornus	Fermentation
♌	Leo	Digestion	♒	Aquarius	Multiplication
♍	Virgo	Distillation	♓	Pisces	Projection

Countless scores of misguided dreamers squandered both their health and material goods in the hope of becoming wealthy beyond imagination or of discovering the secret of eternal life. All, of course, were unsuccessful, but their efforts helped lay the groundwork for the science of modern chemistry.

By the fifteenth century, kings, popes, generals, and aristocrats, in addition to the lower classes, all made use of the astrologer's services. But even more significant is the fact that, although they were on the very frontier of scientific discovery, most of the major astronomers also employed astrology at one time or another. Until well after the death of Copernicus the product of astronomical research had few other socially significant applications. Astrology provided the primary motive for wrestling with the problem of planetary movements. Tycho Brahe, the greatest observational astronomer of the Renaissance, spent much of his time casting horoscopes for the wealthy, including those of the members of the Danish royal family. Brahe's famous pupil, Johannes Kepler, the giver of the first scientific laws of planetary motion, believed in astrology until his dying day, although he berated the crude practices used by many of the charlatans associated with the profession. Even the scientifically minded Galileo, a contemporary of Kepler and Brahe, occasionally made astrological predictions, although with little apparent success. Only in the wake of the Newtonian synthesis, when astronomy and astrology were split beyond repair, did the professional astronomer gain public stature in his own right. Astronomy prevailed because its claims, as in any true science, could be tested and proved while those of astrology could not.

4. Nicolas Copernicus: Conservative and Revolutionary

A New World of New Ideas

THE BEGINNING OF THE SIXTEENTH CENTURY marks a great watershed in the intellectual, spiritual, and political history of Western man. The medieval world had begun to crumble and with it the fortunes of the universal empire controlled by the papacy of Rome. Kings, aided by the financial support of the burgher class, established the first national monarchies, as secular power superseded ecclesiastical power for the first time in 1000 years. Within a few decades, a recalcitrant monk, Martin Luther, would do further serious injury by splitting off from the Church, to crush forever the little remaining hope that a universal monarchy of European Christendom would ever endure.

For longer than men could remember, the clergy had constituted the intellectual leadership of European society, just as it had controlled its spiritual and political affairs. Now this distinction too became forfeit as the laity began to gain influence in the universities and to undermine spiritual authority through the development of new judicial doctrine. Even the peasantry began to feel the power of royal law, as administered by a professional bureaucracy in the employ of princes and kings. During this transmutation of Europe, leadership on the Continent shifted from Italy in the south to the northern states of France, England, and Holland.

The early 1500s have frequently been referred to as The Age of Exploration, a characterization that covers a number of separate yet related events. From the shores of Western Europe the great voyagers launched their perilous ocean-going expeditions across the forbidding Atlantic; and, to the amazement of their conutrymen, they discovered new worlds and wild landscapes that extended far beyond the imagination of Europe's most gifted thinkers. The secure little medieval domain, with its limited plant and animal life fully accounted for in the Bible by God's saving grace and Noah's sturdy ark, found itself theologically

111

incapable of coming to grips with the mariners' discoveries. For although the voyagers went in quest of gold, glory, and adventure, they returned with observations and biological specimens which tempted the curiosity of scholars, even as they defied explanation according to traditional religious doctrine. The great certitude which the Church could formerly guarantee vanished; the old basis of life was shattered.

These unstable and turbulent earthly conditions were also reflected in man's thinking about the stars. If the study of history has shown anything, it is that stereotypes are most readily discarded during periods of general intellectual and social ferment. Europe during the Renaissance and Reformation was in just such a period. The time was ripe for astronomical innovation; the old cosmology had proven as inadequate in its explanation of new celestial phenomena as it had in its attempt to account for the discovery of unmapped terrestrial continents. Thus began a truly exhilarating adventure for both the man of action and the contemplative savant. For the former, the earth lay open to exploitation, both political and economic, while for the latter, the skies invited systematic investigation, as never before. Centuries later these divergent efforts were destined to merge into a single movement, the primary aim of which was to appropriate the gifts of nature for human control. By the early twentieth century, the New World opened up by the explorer, adventurer, and professional administrator had been tamed by the scientist, engineer, and inventor. The beautifully self-contained and egocentric world of medieval Christian theology lost all credibility in light of the new reality founded upon mathematical formulas and universal laws. Everything that had once appeared large and mighty shrank to the smallest significance in a universe where the earth no longer occupied the center of all existence. Concepts such as "eternity" and "infinity" which had once had only religious significance now related directly to time and space.

How could it have been otherwise, one is tempted to ask, once it was shown that man's home is a small planet moving about a middle-sized star which is itself part of a galaxy composed of some hundred thousand million stars; and that our sun's galaxy, the Milky Way, is only one of a system of galaxies, scores of which are within our telescopic range. No wonder, then, that the old world view buckled and collapsed under the weight of its own insignificance. Man found himself in need of a new religion to cope with his new knowledge, whose limits still appear to many to be as boundless as the universe itself. He discovered it in the abstract symbols, rational systems, mathematical laws, and objective measurements of his new passion, Science, which seemed to eclipse and consign to the shadows the God of Adam, Moses, and Abraham.

Ironically, it fell to one of the Church's truly dedicated servants to conceive of the revolutionary celestial framework in which the seeds of the new power system and religion were to eventually germinate and take deep root. The man was Nicolas Copernicus, Canon of Frauenburg, who, like Joshua of the Old Testament, issued a command to the earth's nearest star that clearly signaled the dawn of the Modern Age: "Sun, stand thou still!"

Young Man of the Renaissance

When confronted with the task of describing and analyzing the life of a great man or woman, the historian almost instinctively looks to the childhood years for clues to subsequent developments. This has been particularly true ever since Freud's pioneering work in the field of psychoanalysis. In the case of Nicolas Copernicus, however, precious little information about his formative years is available; in fact, less is known of his childhood and adolescence than that of any of the other giants of early modern science. His mother, Barbara, might well have died before Nicolas, the youngest of four children, was able to form a permanent impression of her, and his father, Nicolas Kopernik Sr., passed away when young Nicolas was only ten. He, his brother Andreas, and two sisters, Barbara and Katherine, of whom very little is known, became wards of their maternal uncle Lucas Watzelrode (Waczenrode), an irascible and difficult taskmaster, but one who protected his own and for whom Nicolas seems to have developed great respect and a degree of affection, it not love.

The family name Kopernik, probably derived from the word "copper," has been traced by modern scholars at least as far back as the thirteenth century, to a village in Upper Silesia, then a part of eastern Germany, but now part of western Poland.* Since people frequently used to take their names from the places where they lived, it was not at all unusual that Kopernik should have become the family surname. It was probably carried eastward by Copernicus' ancestors during the fourteenth century, when a considerable number of Silesians immigrated into Poland at the request of its king. The Germans had already founded a number of prosperous trading centers, including Danzig, Torun, and Cracow, and were prized by the Polish government for their enterprising spirit and openmindedness.

Nicolas Copernicus was born on February 19, 1473 in the fortress town of Torun (Thorn), one of the bridge-heads of the Vistula in Prussian Poland. His father, a wholesale dealer in copper, had moved to the bustling commercial center from Cracow sometime in the late 1450s.

---

* *Koper* is also the Polish word for the plant dill, and it is possible that the family surname derived from this term.

Nicolas Kopernik Sr. became a respected member of the merchant class and before his untimely death was appointed a member of the Torun magistracy. Kopernik married Barbara Watzelrode, a member of a distinguished German family whose Silesian forebears had settled in Torun several generations before her husband's arrival. Yet despite the fact that the Watzelrodes served as counselors and migistrates, playing an important, if not dominant, role in local commercial and political affairs, all records pertaining to Barbara, the mother of Copernicus, have been lost, if, indeed, any ever existed. And of Copernicus' sisters, we know only that the eldest, Barbara, became a nun and eventually Abbess of Kulm, while the younger sister, Katherine, was wed to a merchant of Cracow. On the other hand, Nicolas' older brother Andreas, of whom we shall hear more later, spent several years as the companion of his brother both during their student days in Italy and after their return to Poland, when they assumed their official duties as Canons of the Roman Catholic Church.

While the death of a father is hardly cause for celebration, particularly when he leaves young children and is a respected member of the community, one cannot help but think that the course of modern history may have flowed in a significantly different direction had not Nicolas Kopernik Sr. died in 1483. For had his father lived, it seems more than likely that Copernicus would have trained to enter the commercial world, a circumstance that doubtless would have deprived modern science of one of its truly indispensable thinkers.

Lucas Watzelrode was twenty-six years older than his nephew when he became ten-year-old Nicolas' guardian. He had established a reputation as an eminent scholar, having studied at the Universities of Cracow, Leipzig, and Prague. Later, he crossed the Alps into Italy to study at the University of Bologna, where he won high honors and graduated as a Doctor of Canon Law in 1473. In 1489, at the age of 42, six years after adopting the orphaned Kopernik family, Lucas was advanced to the vacant Bishopric of Ermland or Varmia. This particular appointment was unlike many of those granted during the Middle Ages, which were frequently little more than lifetime sinecures bestowed upon the well-connected offspring of aristocratic families.

Ermland, bordered on the north by the Baltic Sea, was virtually surrounded on its other three sides by lands under the control of the Teutonic Knights, a Germanic Order of militant monks. Patterning themselves along lines of the Knights Templars, who fought in Palestine to recover the Holy City of Jerusalem from the Moslems, the Teutonic Knights concentrated their attention on converting by conquest the Slavic peoples who had settled in the eastern Germanic region of Prussia. The Knights founded Torun, birthplace of Copernicus, as a military outpost that marked the boundary between their territory and that of

Poland. At first welcomed by the Poles, the rapacity of the Order eventually led to a rebellion of their Prussian subjects. During the course of the conflict the Polish king, who recognized the error of having sanctioned the development of a foreign military order so close to his own borders, supported the rebels. The Knights were eventually defeated, driven westward, and forced to abandon over half of their territory. Thus, only ten years before Copernicus was born, Torun came under Polish rule.

It was against this background that Lucas Watzelrode, Bishop of the Roman Catholic Church, assumed the unenviable task of securing Ermland's independence in the face of much superior German and Polish forces. Until his death at the age of sixty-four, he put up a relentless fight against the Knights. During his lifetime Lucas was hated with a burning passion by his Germanic enemies, who referred to him as none other than "the Devil in human shape." Yet his efforts proved so effective that they prepared the way for the eventual dissolution of the outmoded Order, even though the Bishop did not live to see it himself. When he died, while in apparent good health, rumors were generated to the effect that he had been poisoned on instructions from the Knights themselves. While probably not true, the rumor itself is indicative of the hatred between the two rivals; the Knights must have long rejoiced at the passing of their most tenacious and uncompromising adversary.

From all accounts, it is apparent that Lucas Watzelrode took the responsibilities of guardianship as seriously as he did those of his high office. He was neither a kindly nor affectionate man, but he provided for his charges as well, if not better, than could have been expected of their real father. Perhaps, too, he saw in his youngest nephew the prospect of greatness, for he bestowed upon him every one of the considerable advantages within his power to grant. It was not that Copernicus' brothers and sisters were slighted, they were not; Andreas, for example, received the same appointments and educational opportunities as his brother. Still, these seemed always to have come after Nicolas had been provided for, a significant deviation from the accepted social practice of attending to the interests of the eldest male of the family first. Lucas Watzelrode obviously saw in Copernicus special qualities which, in the Bishop's eyes, set him apart from his siblings; apparently the Bishop could not help but try to assist their development in every way possible.

Although one can speculate a good deal about the childhood and adolescence of Nicolas Copernicus, the first major event in his adult life of which we have specific knowledge occurred in the winter of 1491–92, when he was eighteen. Copernicus was matriculated in the Faculty of Arts at the University of Cracow, a respected center of higher learning known mainly for its excellent instruction in the fields of mathematics and astronomy. In the only surviving document pertaining to his three years of study at the University, we are informed that "Nicolas, the son of Nicolas of Torun" has paid his fee in full. The document also records the entry of some seventy other students including brother Andreas, although for reasons unknown the elder Kopernik paid only part of the required matriculation fee.

While the University of Cracow has received far less attention from contemporary scholars than many of Europe's other leading universities, the distinguished institution attracted eminent humanists and students from throughout Europe, including many from as far away as Sweden, Italy, and France. Cracow itself was a city bolstered by the considerable financial resources of a prosperous burgher class. Certainly it was deserving of its reputation as one of the main centers of European commerce and science. Located as it was near Nicolas' home town of Torun, it must have provided a secure but stimulating environment for the somewhat withdrawn yet intellectually eager young Copernicus. After all, it was his father's former home before he settled in Torun and the residence of his elder sister Katherine, with whom he and Andreas may have roomed. Uncle Lucas also had numerous influential connections which doubtless helped smooth the transition from adolescence to university life.

One of the greatest teachers at Cracow in those days was Albert Brudzewski, the Polish astronomer and mathematician, who lectured extensively on the subject of Aristotelian cosmology. Although it is not known for certain whether Copernicus attended Brudzewski's formal lectures, he probably took private lessons from the master. From what we know of his teacher's work, however, there is little reason to believe that Brudzewski may have consciously imparted to his gifted pupil the suggestion that the earth is in motion and circles the sun. Brudzewski was obviously as staunch a private supporter of the Aristotelian–Ptolemaic system as he was a public advocate. Whether or not, by focusing Copernicus' attention on the geocentric construction of the heavens, he may have inadvertently lighted the spark that would one day engulf the scientific world in revolution is a question no one can answer.

While at Cracow, Copernicus pursued numerous other interests commonly associated with the traditional concept of the Renaissance man. For leisure, he practiced drawing and completed a self-portrait with the use of a mirror. Unfortunately, it perished long ago along with most of the correspondence and other important papers pertaining to his multifaceted career. During this period, Copernicus also began his famous collection of books on mathematics and astronomy, most of which remained a part of his personal library until his death in 1543. Many of these works, which contain jottings and calculations in the astronomer's hand, have been preserved, and Copernican scholars have learned a good deal about the evolution of Copernicus' thinking on astronomy from them. It was at Cracow, too, that Copernicus acquired a superb mastery of the Latin language. Because there were many foreign students in attendance, Latin remained the universal tongue; and although we cannot be certain, it seems quite probable that it was at this point in his academic pursuits that Nicolas Kopernik, after the fashion of his day, chose to Latinize his family name. He became the now familiar Copernicus.

It appears that Copernicus did not take a degree at Cracow; perhaps he never intended to do so. For various reasons, a significantly smaller percentage of university students earned the Bachelor's degree then than now; in fact, of the seventy or so young men who enrolled in the University with Copernicus, only about a fifth took the B.A. Andreas appears not to have been among them either. Now almost twenty-two, Nicolas left Cracow in 1494 for Heilsberg (Lidzbark), the Episcopal administrative center for the Diocese of Ermland. He probably did so at the request of his ever watchful uncle, who now saw a golden opportunity to further the career of his favorite nephew.

A position in the Canonry of Frauenburg Cathedral was soon to become vacant, and the Bishop was most anxious to have his nephew appointed at the appropriate time. A major complication arose, however. It did not concern, as one might think, the rather tender age and limited training of the candidate. In 1489, Giovanni di Medici, the future Pope Leo x, had been appointed a cardinal at the age of fourteen by virtue of the influence wielded by his father, Lorenzo the Magnificent. By any standard of comparison this supreme example of nepotism made a twenty-two-year-old canon appear somewhat ancient. Rather, the problem arose because the right of making appointments to vacant clerical offices belonged to the Bishop of Ermland only during the even months of the year. During all uneven months the appointive power rested with the Pope. When Copernicus was summoned to Torun by the Bishop the canon whom the young man was to replace, Matthias de Launau, was terminally ill but had not yet succumbed. As luck would have it, Launau's end came early in September—an uneven month—and the position was denied Copernicus when the Pope appointed his own candidate. Even though it was accepted practice among the powerful, both within and outside of the Church, to advance their relatives and close friends as quickly as possible, there is something unsettling about Lucas' premature summons and the intervening vigil predicated on the unspoken hope that death would come at the "proper" time, meaning, of course, in an even month. In correspondence that has since vanished Copernicus is said to have lamented his misfortune to his friends.

Yet time was on the side of the youthful Copernicus and within a short time a second vacancy occurred—this time in the more favorable month of August. Copernicus was duly appointed a Canon of Frauenburg, a secure lifetime position with a healthy prebend (income). As has been remarked by one of Copernicus' biographers, "Nowadays we should think it wrong for a man to become a priest or a minister for the sake of making a living. But in those days very few people thought anything about it." [1] Copernicus never advanced beyond the basic vows required of his office and thus did not partake of the sacrament of ordination. In fact, of the sixteen Canons of Frauenburg resident at the Cathedral after Copernicus completed his formal education in Italy, only one or two were ordained priests. The rest, like Copernicus, were professional bureaucrats and administrators. Doubtless Copernicus was a devoted Catholic, but he seems not to have stood in petrified awe of the Church. When, at the age of seventy, he was finally persuaded to publish his revolutionary book, *De revolutionibus (On the Revolutions of Heavenly Spheres)* he displayed both tact and courage by dedicating the great astronomical treatise to Pope Paul III. In the dedication he requested, in effect, that he be given the freedom of intellectual inquiry to follow the truth wherever it might lead. Perhaps the long years of intimate association with the mundane duties connected with his uncle's office and that of his own gradually dulled his once youthful exuberance for Mother Church. At any rate, it was not, as many have suggested, out of fear of any action the Church might take against him for the public espousal of his unorthodox views that caused the Canon to withhold publication of his book. The cause for delay sprang from another source to be discussed in more detail later.

Shortly after he was awarded a formal appointment to the vacancy in the Frauenburg Chapter, Copernicus, following in the now familiar footsteps of his uncle Lucas, crossed the Alps into Italy for the purpose of continuing his academic studies. Except for a single brief visit he would not be present at Frauenburg for the next fifteen years, but he drew his prebend as regularly as if he had never ventured outside the Cathedral walls. Copernicus enrolled in the University of Bologna, the institution from which Bishop Watzelrode had been awarded a doctorate in the 1460s. By this time the Renaissance was in full flower, and, true to its ideal of the universal man, the young Copernicus studied a wide range of subjects: medicine, astronomy, Greek, mathematics, philosophy, and Roman and canon law.

During the Middle Ages, and continuing into the Renaissance, the individual who visited a strange land frequently found himself at a considerable disadvantage. Like the metics (resident-foreigners) of ancient Athens, he enjoyed none of the political rights and privileges of the local citizenry. Thus university students commonly banded together for mutual protection and the fraternity of one another's company. During the winter semester of 1496, Copernicus joined the *natio Germanorum* or Association of German Students. Once again the young man followed the path of his uncle, who, a generation earlier, had joined the "nation" while a student at Bologna. While his major subject of study was Canon Law, Nicolas, as at the University of Cracow, showed a keen interest in mathematics and astronomy. At Bologna he had the good fortune of becoming acquainted with Domenico Maria Novara (1454–1504), one of Renaissance Europe's most acclaimed professors of Astronomy. Although technically Novara's pupil, it would be more accurate to describe Copernicus as the astronomer's technical assistant. It appears likely that Copernicus roomed in Novara's home, a common practice that enabled many a poorly paid professor to supplement the limited income derived from teaching. Galileo, too, lodged several students in his home while he was a professor at the University of Padua. This arrangement also made it convenient for Copernicus to assist his teacher in making his regular, nightly astronomical observations. Proof enough of Novara's special influence on Copernicus is the fact that the future astronomer's earliest recorded observation was made in Bologna on March 9, 1497. Nicolas watched the moon approach Aldebaran, a double star in the constellation Taurus; he noted that occultation occurred at eleven P.M.

Two years after his arrival at Bologna Nicolas' brother Andreas also enrolled in the University. By this time a third vacancy in the Frauenburg Canonry had occurred and Andreas received the appointment. During their years of absence from Ermland the income from their clerical offices was administered by their uncle, and toward the end of 1499 the brothers found themselves in the embarrassing position of having run out of funds, though whether because of profligacy or simply poor planning we do not know.

Fortunately the Bishop's secretary, who was on a mission to Rome, stopped in Bologna to pay the brothers Kopernik a visit. They informed him of their plight, and the sympathetic secretary put them in touch with Ermland's representative at the papal court. In a short time they were once again solvent and presumably chastened by their uncomfortable experience. Lucas Watzelrode was not a man to countenance foolishness, if foolishness there was, and from this time on we never again hear of Copernicus erring in economic matters. In fact, the future astronomer became known in later life as somewhat of a tightwad whose only material weakness seems to have been a penchant for purchasing expensive books on astronomy and mathematics.

The conduct of Andreas was another matter, however, and during the remainder of his tragically foreshortened life his fortunes, both financial and otherwise, went from bad to worse. When Andreas returned to Ermland after completing his studies he was infected with

---

* Occultation is the passage of a celestial body across a line between an observer and another celestial body. In this instance, it occurred when the moon moved between the earth and Aldebaran.

either leprosy or syphilis, both incurable diseases until the twentieth century. Despite every effort to check the spread of the disease, the best that could be done was a postponement of the inevitable. Andreas, his body terribly disfigured by the ravages of the affliction, died sometime before 1520 in self-imposed exile.

After spending three and a half years at Bologna, Copernicus left the University to visit Rome in 1500, proclaimed by the Church as a Jubilee Year. He had not taken a degree, but neither did he plan to terminate his academic studies at this point. Nicolas and Andreas afterward returned home in 1501 and applied at once for a second leave of absence. Their request was granted, but only after Nicolas agreed to a *quid pro quo*. In return for permission to resume his studies in Italy the young man agreed to undertake the study of medicine. Such an arrangement was not at all uncommon, for medical training was frequently provided for the laity by religious chapters. Moreover, the availability of a physician was of direct benefit to clergy as well. No such request was made of Andreas who, after crossing the Alps with Nicolas, set off for Rome. Nicolas did not return to Bologna but journeyed instead to Padua, whose reputation as a center for medical studies eclipsed that of any of its sister institutions. Though Copernicus spent the next two years at Padua, he did not intend to take a medical degree; rather, he endeavored to become conversant with the most popular methods of treatment of the day. This knowledge would later be used to aid both his terminally ill brother and Bishop Watzelrode, in addition to the common citizens in and around Frauenburg.

By the spring of 1503 Copernicus' extended leave from Frauenburg was shortly to expire, and the young Canon had not yet obtained an official diploma, as was required by the statutes of his Chapter. Strange as it may seem, he chose not to take his diploma at Padua, but instead he switched to the lesser known University of Ferrara, where, in 1503, he was awarded the diploma of Doctor of Canon Law. Upon careful examination of this puzzling behavior, one finds that underlying it there is a perfectly logical explanation. It was customary, during that time, for a new Doctor to provide lavish entertainment to celebrate his success. Copernicus, like many other foreign students, knew that by going to Ferrara he could elude his teachers and friends, thus hopefully sparing himself the heavy expenses connected with the attainment of his high honor. Furthermore, Nicolas Copernicus was a reserved young man. He may well have wished to forego the prospect of subjecting himself to considerable public attention.

Upon a careful examination of the diploma Copernicus received from Ferrara, a most interesting fact has been revealed. In addition to being mentioned as a Canon of Frauenburg, Copernicus is described as

"Scholasticus of the Collegiate Church of the Holy Cross in Breslau." So far as is known, the young man had never visited this city, nor would he journey there at any time in his adult life. Again uncle Lucas had apparently used his influence to provide his favorite nephew additional financial security, a fact Copernicus seems to have kept secret from even his few intimate acquaintances. By late 1503, at the age of thirty, Nicolas Copernicus, Doctor of Canon Law, artist, healer, astronomer, mathematician, and master of Greek and Latin, had returned home to Poland. Although still a young man, and a true child of the Renaissance, he would not set foot outside its boundaries for the remainder of his life.

Now that we have a basic grasp of Copernicus' family background and the nature of his scholarly accomplishments, we must inquire as to what manner of man he had become, a question that will merit even greater attention later in this chapter. The truth is, it is very difficult to form a vivid conception of his character, whether early or late in his career. The evidence is quite meager, and, as anyone who has read even casually on the subject knows, the few available facts have been subjected to a number of highly contradictory interpretations. From this writer's point of view, there is no summary by any of his biographers that can be regarded as completely satisfying. Nor have I been able to construct one for myself. It does seem apparent, however, that by the time he began this stage of his career Copernicus displayed certain traits that remained part of his character throughout the rest of his life.

He was obviously a man who kept his own council and appears to have harbored no ambition toward a major position of leadership or power. While Lucas Watzelrode must certainly have entertained the private hope that his gifted nephew would one day succeed to the Bishopric of Ermland, such was not the desire of Copernicus. He obviously enjoyed his duties as Canon and welcomed the financial security derived therefrom; but to become Bishop—never! This lack of interest in high office and the somewhat reclusive nature of his life style have led some scholars to what I believe to be the false conclusion that Copernicus was a "timid canon," so afraid of his private thoughts on the composition of the universe that he nearly carried them to the grave rather than risk their publication. He was indeed a cautious man and in many ways a conservative one, but so too were Ptolemy and Kepler. Of course, when viewed in the same light as the blustering Tycho Brahe and the self-confident, rebellious Galileo, Copernicus pales by comparison. But to be studious and quiescent is hardly a crime, although some of Copernicus' strongest critics have virtually made it appear so. He seems only to have wanted the opportunity to carry on his studies alone and undisturbed. Otherwise, he apparently asked little else of his colleagues and they seemed, from all indications, content to let him go about his business unimpeded. Thus Copernicus belonged not to that category of Renaissance activists referred to earlier as men of action: seamen, explorers, soldiers, and conquistadores. Rather he appears to fit more comfortably into the second category of Renaissance man—the thinkers—those individuals who blazed the trails of inner space carved through the yet to be enlightened recesses of the human mind.

The Uncle's Captive

With his formal studies at an end and his future assured, Nicolas Copernicus was prepared to settle into the comfortable existence of a clerical bureaucrat surrounded by the protective walls of Frauenburg Cathedral. Though already a somewhat withdrawn individual and apparently a man with few close friends, the stars would provide both the company and the necessary intellectual stimulation required by his keen mind. Yet these plans were destined to remain unfulfilled, at least for the next several years. After a few brief months at Frauenburg, Copernicus received word from his uncle that his services were needed at the Episcopal residence in Heilsberg, a small city some forty miles from the cathedral Copernicus had been appointed to serve.

As always, the Bishop had his reasons for intervening. Lucas Watzelrode had invested considerable time, money, and energy in the career of his young nephew, and now that he was aging he wanted Copernicus at his side. Lucas may have also considered the time-consuming duties of a low-ranking cathedral canon too petty for his brilliant nephew; besides, he could use Nicolas' medical training to help maintain his own health. On January 7, 1507, Copernicus was appointed permanent physician to the Bishop by resolution of the Frauenburg Chapter. Copernicus left immediately for Heilsberg where he remained a resident until sometime in 1510.

In those days Heilsberg Castle resembled the stronghold of a medieval prince. The open court was crowded with every possible type from ecclesiastics and noblemen to minstrels, jugglers, clowns, and other retainers. Each day at noon a huge bell tolled the signal for the midday meal, whereupon the Bishop, in full choir costume, led a procession attended by a vicar, chief justice, chaplain, chamberlain, and marshal, followed by a host of others in order of their rank and social position. In the great banquet hall, eight tables were spread, each one signifying a different degree of dignity. A ninth table, used to accommodate jesters and buffoons, occupied the conspicuous central position so that all present might enjoy their mindless antics. While no record exists regarding which of the tables Canon Nicolas was assigned to, it seems probable that he dined at the second, for it was reserved for the higher officials.

Yet beneath this pompous and festive facade there coursed a treacher-out undertow, so forceful that it threatened to eradicate tiny Ermland's continued existence as a semi-independent state. The clouds of war lay constantly on the horizon, forcing the wily Bishop to muster all of his considerable resources to keep the enemy at bay. On one side were the Teutonic Knights, anxious to extend their territory at the expense of their hated archrival. On the other was the Kingdom of Poland, equally ready to grab off neighboring lands if given the slightest encouragement. This perpetually grave political situation was compounded by the weighty spiritual matters brought daily to the Bishop's Palace by troubled churchmen. Young nobles with ambitions also sought the aid of the Bishop, as did the countless humbler citizens of the surrounding countryside. Under the circumstances, it is little wonder Lucas Watzel-rode wanted his trusted nephew in attendance, or that Copernicus worked harder on diplomatic and clerical matters than he did as a physician. Still, he seems to have treated both the rich and the poor, and among both classes built a respected reputation as a kindly and compassionate healer. For several years Nicolas remained his uncle's steadfast companion. As the weeks and months passed, he must have shared more and more, first as confidant, and later as principal advisor, the burdens and anxieties of governing little Ermland.

In January of 1512, the Bishop was invited to the wedding feast of the King of Poland at Cracow, an invitation that a man whose state was in the most delicate of political positions could hardly refuse to accept. Copernicus supposedly accompanied him, and both men are said to have taken an active part in the festivities of their important host. At the completion of the ceremonies Copernicus, for reasons unknown, did not accompany the Bishop on his return journey to Ermland. Lucas developed a severe case of food poisoning and died in his native Torun three days after his arrival, at the age of sixty-four. Whether his nephew could have saved the stricken Bishop is something we will never know; nor is there any record of Copernicus' thoughts on the matter. It would seem, however, that his absence at Lucas' death, no matter how compelling the reason, must have caused him considerable anguish, perhaps even a degree of guilt. On the other hand, it has been argued that Copernicus was in no position to have aided the dying Bishop, for he had not accompanied his uncle to the royal festivities. Whatever the case, the death of the man whom legend says was never known to laugh at last gave Nicolas Copernicus his long-awaited opportunity to renew in earnest an intellectual odyssey whose origins dated back to his student days. It proved a fateful moment; one which was destined to change forever the history of the civilized world.

The Making of an Astronomer

The Cathedral town of Frauenburg (Frombork) stands on the Baltic coast of what was once East Prussia, in Copernicus' time the very outskirts of civilized Christendom. The homes of its citizens clustered round a low hill upon which the fourteenth-century Cathedral still stands. The crenelated ramparts surrounding the commanding structure are strengthened by a series of turrets which, in times of peace, provided lodging for the Chapter members. Copernicus took up residence in one of these square, brick towers at the northwest corner of the Cathedral's defensive fortifications. Contrary to the practice of changing to more suitable quarters as the older members of the Chapter died off, he seems to have spent almost all of this thirty years at Frauenburg living in the same rooms. It was in one of them that he died. In his book titled *Three Copernican Treatises* historian of science Edward Rosen paints a vivid picture of the Canon's lodgings:

> Copernicus' tower was about 50 feet high on a rectangular plot measuring 27 feet by 30 feet. The cellar and ground-floor kitchen, dining space, and maid's room were surmounted by three stories. Of these, the first level, containing a living room and bedroom, was connected by a short flight of steps to a storage area and toilet, and then to the top story, where a workroom, lit by nine windows, gave access to an outside gallery. It was probably in this tower lodging that modern astronomy was born.[2]

From the vantage point of his third-floor gallery, Copernicus' eyes could wander across the horizon above the blue expanse of the Baltic, barely visible in the hazy distance. Between the town and the open sea stretches the *Frisches Haff*, a fresh-water lagoon varying in width from three to four miles and some 50 miles long. On a clear winter night the view of the heavens must have been breathtaking. One wonders how many villagers, long after the death of the now famous Canon, harbored memories of a tall man standing atop a square tower, his robed figure silhouetted against the starry sky as he gazed into the mute heavens beyond.

While there were certain advantages to the location of Copernicus' observatory, there were also some disturbing drawbacks. In the first place, the gallery did not provide a totally unobstructed view on all sides. This Copernicus eventually corrected by constructing a small roofless tower (turricula) from which most of his observations were made. A problem less amenable to solution was that presented by Frauenburg's geographical location. The town lies too far north of the equator for a satisfactory view of the planet Mercury, an object of major interest to the astronomer. Moreover, the celestial bodies sometimes had to be viewed through a dense layer of air distorted by vapors rising from the sea beyond. This proved a nuisance, though a tolerable one, to Copernicus who envied the astronomers of Alexandria because, as he observed, the Nile, upon flowing into the Mediterranean Sea, does not exhale such vapors.[3]

Copernicus' observational instruments were relatively few, and, as was his custom, he made some of them himself. Among his viewing devices were a quadrant for observing the sun, an astrolabe for the stars, and a parallactic instrument, used primarily for observing the moon. There was also the triquetum, an instrument used centuries earlier by Ptolemy, which derived its named from the three strips of wood employed in its construction. This device measured the zenith distances of the planets and stars. Copernicus' triquetum, together with a portrait of the famed astronomer, was sent as a gift to Tycho Brahe in 1584. They were preserved for a time in Tycho's Uraniborg observatory and later removed to Bohemia where they vanished during the Thirty Years' War. Copernicus also used the "Jacob's Staff," a long shaft to which a movable crossbar was affixed. The telescope, of course, was nowhere to be seen, nor would it be readily available to astronomers for another century.

It must be said that in comparison to the instruments employed by some other famous astronomers of the time, those of Copernicus were rather simple. For example, the German mathematician–astronomer Johann Müller (1436–1476), better known by his Latin name of Regiomontanus, had far more advanced instruments installed in his Nuremberg observatory. Copernicus was, of course, aware of the existence of better instruments and had doubtless used a number of them during his student days at Cracow and Bologna. Thus, he clearly realized that the observations he made lacked a certain degree of precision. This seems to have troubled his disciple Rheticus far more than Copernicus himself. He is said to have told the young man that he would be as happy as Pythagoras was when he discovered his famous mathematical theorem if he were able to reduce observational errors of prediction to ten minutes of arc.

The truth is, Copernicus was not a great observational astronomer in the tradition of Tycho Brahe. Though he doubtless loved to gaze into the night sky and contemplate its wonders, he frequently preferred to rely upon measurements taken by other astronomers, Ptolemy included, even though they were far less exact than they need have been for the time. We know, for example, and Copernicus did too, that it is possible to measure an angle in the sky with an error of only three or four minutes arc, even without the aid of a telescope. Why, then did he not seek greater precision, as would be expected of any reputable scientist of the present day?

For one thing, Copernicus was not a modern scientist; rather he was a man caught between the conflicting forces of two great historical ages—the medieval and the modern. The scientific method, as we know it, had not as yet come into widespread use, and neither had mathematics become the universal language of scientific inquiry. Even more significant than this, however, is the fact that Copernicus did not consider himself an astronomer, our current classification of him as such notwithstanding. Even though a part of each one of the last forty-eight years of his life was given over, in one form or another, to astronomical studies, Copernicus himself made a relatively limited number of independent observations. In his single major work, *On the Revolutions of Heavenly Spheres*, published the year of his death, only twenty-seven of those observations are included, although the book is over 400 pages in length. About fifty others were jotted down in the margins of a number of his still extant books. There were doubtless more, but how many we cannot be certain, probably not more than a few hundred. Tycho, on the other hand, made thousands of independent observations during his much shorter lifetime. In terms of extended and meticulous stargazing, Copernicus was obviously quite willing to leave much of the tedious work to others.

If Copernicus did not perceive of himself as an astronomer, at least in the traditional sense of one who makes independent celestial observations, then how did he look upon his work in this field? He preferred, I believe, to think of himself both as a mathematician and philosopher of the skies. It was to the professional mathematician, not the astronomer, that he dedicated his one great treatise, the major part of which makes absolutely no sense whatsoever to the educated layman. Copernicus also saw himself as a philosopher in the sense that once the individual has learned as much about the structure of the universe as is possible through the use of mathematics, it is his duty to postulate a rational operational model of the heavens. This is precisely what he did. Copernicus employed Ptolemy's mathematical method, but he came to a fundamentally different set of conclusions. Exactly what these conclusions were and precisely what their ultimate significance to the rise of modern science has been, still remains something of an unsettled question. Today, over 400 years after the astronomer's death, scholarly debate on the subject is more spirited than ever before.

Since Copernicus chose not to devote the majority of his evenings to the observatory but rather to his study, it is only natural that some inquiry be made about the origins of his belief that the sun is the center of planetary and stellar motion. As has been observed, Copernican experts believe that the heliocentric theory was not expounded by Brudzewski, Novara, or any of the other professors with whom he came into contact. All, with the possible exception of Novara, were solidly within the Aristotelian camp, and even Novara never seriously challenged the idea of an immovable, centrally located earth. Yet where else could Copernicus have learned of the heliocentric hypothesis, if not during his student days at Cracow and Bologna? Did lightning strike in a sudden flash of brilliant intellectual insight, or did he come to embrace the heliocentric doctrine gradually, after a careful examination and eventual rejection of Aristotelian–Ptolemaic cosmology? As with so many other facets of his deeply shadowed life, Copernicus tells us relatively little about the path he trod.

One of the few major clues to the origins of the Copernican system is contained in the dedication of *De revolutionibus*.* Here Copernicus openly acknowledges his debt to the ancients by citing the works of Pythagoras, Herakleides, Philolaus, and others as at least having helped convince him that the geocentric thesis is not without certain faults:

> Therefore I also began to meditate upon the mobility of the Earth. And although the opinion seemed absurd, nevertheless because I knew that others before me had been granted the liberty of constructing whatever circles they pleased in order to demonstrate astral phenomenon, I thought that I too would be readily permitted to test whether or not, by the laying down that the Earth had some movement, demonstrations less shaky than those of my predecessors could be found for the revolutions of the celestial spheres.[4]

He further maintains that his opinion was reinforced by a lifetime of observation and further reading. From what we know of Copernicus' scholarly habits and temperament, there seems little reason to doubt this brief, personal account of historical events. As a student, he may, of course, have experienced a supreme moment of intellectual insight, in which the pieces of the heliocentric universe fell into place, and then spent the remainder of his life in attempting to provide the required scientific proof. Given the admittedly tenuous nature of his appealing but by no means conclusive hypothesis, he would doubtless have been reticent to reveal to his readers such a numinous experience, let alone offer it as the basis for his life's work. Still, it would seem much more in character if Copernicus, as he maintained, "took the trouble to reread all the books by the philosophers which I could get hold of, to see if any of them even supposed that the movements of the spheres of the world were different from those laid down by those who taught mathematics in the schools." [5] His mind was quick, but his methods, from beginning to end, were highly methodical. The quest begun during his last student days after he had mastered classical Greek did not end until the day he died, when the first copy of the *Revolutions* arrived from the printers.

* The book will henceforth be referred to as the *Revolutions*.

Moreover, Copernicus never claimed to have "discovered" the earth's motion in the sense of being the first to formulate it as a hypothesis. He borrowed the concept from the ancients, and a borrower rarely experiences a heightened sense of exhilaration. What he did do that proved unique was to create a *true planetary system* based on the uniform, circular motion of individual bodies, all of whose movements are *directly interrelated* to the motion of every one of the other bodies within that system. None of the ancient exponents of the earth's motion, including Aristarchus, had gone so far; nor had Ptolemy adhered to this rule in his revision of Aristotelian cosmology. This fact alone sets Copernicus apart from all his predecessors.

Sometime between 1511 and 1513, Copernicus first committed his revolutionary hypothesis to writing. His objective, at this time, was not to complete a full scale work but rather to circulate his major ideas privately among a few friends. The resulting manuscript bore the title *Commentariolus* (short commentary), three copies of which have survived. Unlike the *Revolutions*, published some thirty years later, the author intentionally refrained from introducing mathematical proofs, preferring instead to reserve them "for my projected larger work." Nonetheless, the *Commentariolus* incorporates the seven basic propositions which are at the heart of Copernican cosmology:

1. There is no single center for all the celestial bodies or spheres.
2. The center of the earth is not the center of the universe, but only of the moon's orbit.
3. All the spheres rotate about the sun as their midpoint; therefore the sun is the center of the universe.
4. In comparison to the distance of the fixed stars from the sun, the earth's distance from the sun is imperceptible.
5. The apparent motion of the firmament is not the result of the firmament itself moving, but of the earth's rotation on its own axis once every twenty-four hours.
6. What appears to us as the annual motion of the sun is in reality the earth revolving around the sun like any other planet.
7. What presents itself in the heavens as retrogressions and stations are the result not of their [the planets'] own motions but of the earth's. Thus the earth's motion alone is enough to explain the many apparent anomalies in the heavens.[6]

The assurance with which Copernicus enunciated the basis of his new world system at this still relatively early stage in his career also lends strong support to the thesis that the heliocentric theory had been part of his intellectual baggage since Bologna. Yet many of the details still remained to be worked out during long years of tortuous mathematical calculations without the benefit of slide rule or logarithmic tables. The initial draft of the *Revolutions* was not completed until some time between 1529 and 1532, perhaps even later, only to be locked away for several more years, though, as Rosen points out, it seems probable that the author continued to modify and revise its contents up to the very time of publication. There is every reason to believe the Canon's statement written a year before his death, that he kept his hypothesis a secret "almost four times nine years." [7] In fact, it is the long-delayed publication in 1543 of a work whose major tenets circulated privately in manuscript form as early as 1513 that has been more responsible than anything else for the labeling of Copernicus as a timid conservative.

Enter Rheticus

In 1514, very near the time that Copernicus began to privately circulate manuscript copies of the *Commentariolus*, Georg Joachim von Lauchen, who later Latinized his name to Rheticus, was born in the Austrian Tyrol. In his own way he performed the same function for Copernicus that Mencius had for Confucius and Saul of Tarsas did for Christ, albeit on a far less ambitious scale. For just as Confucius and Jesus had their teaching popularized by their respective disciples, so too did the responsibility for the public dissemination of Nicolas Copernicus' cosmology rest largely upon the shoulders of this young man. In this instance, however, the disciple knew and lived with the master, whereas neither Mencius nor Paul ever met the respective individuals they served so well.

As a child Rheticus traveled extensively with his eccentric father and later he gained an excellent education at the Universities of Zurich and Wittenberg. A brilliant student, a German, and a Lutheran, Rheticus was appointed Professor of Mathematics at Wittenberg in 1536, at the age of twenty-two. By this time Wittenberg was a storm-center of the rising Protestantism, whose two major figures, Martin Luther and Philipp Melanchthon, were something less than enthusiastic supporters of the sun-centered cosmology which had become known to them, as to Rheticus, via the academic grapevine. Nevertheless, the deep bitterness and animosity that eventually solidified into the hatred which later characterized Protestant–Catholic relations during the Thirty Years' War and Counter-Reformation had not yet manifested itself. Many Catholics still remained hopeful that the "wanderers" might be won back through the administration of ample doses of kindness, charity, and patience. In Ermland, the doctrines of Luther had been received with only mild dissent on the part of the people, and Copernicus himself took a moderate view of these particular developments. Rheticus' Lutheran superiors also exhibited a degree of restraint found lacking only a few short years

later. Although they were opposed to his strong interest in the heliocentric doctrine, they remained sufficiently tolerant to grant the young
professor a few weeks' leave of absence for the specific purpose of
visiting a Catholic Canon in a Catholic land to learn first hand about
certain cosmological principles that would one day be pronounced
heretical. What is more, Rheticus overstayed the length of his leave
by some two years and still remained in the good graces of the Protestant leadership. In fact, he was promoted to a higher position upon
his return.*

The first meeting at Frauenburg occurred in the summer of 1539.
The great astronomer was sixty-six years of age and Rheticus but
twenty-five. Neither man recorded his impressions of the fateful encounter, but they apparently developed a deep respect and fondness
for one another from the very start. In the contemporary era, when the
young and the elderly are unfortunately separated by great gulfs of
mutual suspicion and outright indifference, it is easy to forget that
some of the most intimate and satisfying relationships are those between
individuals of widely diverse ages. But exactly why Copernicus, who
was known for his retiring manner, took so quickly and genuinely to
the much younger Rheticus is a matter of conjecture. He must, of
course, have been flattered that a young professor of a rival religion
would travel a considerable distance to visit an old man whose work
was known to him only by hearsay. Perhaps, too, he saw in Rheticus
certain characteristics that reminded him of his student relationships
with Brudzewski and Novara. Nor can the possibility be dismissed that
the young scholar represented the son he never had, just as Copernicus
served as the surrogate offspring of his deceased uncle. For Rheticus'
part, the very privilege of living with and learning from the master a
new and exciting cosmological system provided more than sufficient
reason for his protracted absence from the lecture halls of Wittenberg.

It is not only quite possible but highly probable that had Rheticus
not made his unexpected visit to Frauenburg, Copernicus' *magnum opus*
would never have been published. After all, the manuscript of the *Revolutions* had been completed in the main a decade earlier and, except for
periodic revisions, lay dormant as Canon Kopernik approached his bib-
lically allotted three score years and ten. Fortunately, the new association
with Rheticus changed the entire complexion of the situation. A brash
and ambitious young man, he continuously urged Copernicus to make
his system known to the world. As luck would have it, Rheticus possessed

* For a detailed analysis of the reaction of Melanchthon and other Protestant intellectuals to heliocentrism see Robert S. Westman, "The Wittenberg Interpretation of
the Copernican Theory," in *The Nature of Scientific Discovery* ed. by Owen
Gingerich (Washington, D.C., 1975), pp. 393–429. (See also the discussion of the
paper, pp. 430–457.)

an ally, who, for many years previously, had exhorted the reticent Canon —obviously without success—to publish his completed work. This was the astronomer's one intimate friend and a fellow Canon of Frauenburg, Tiedemann Giese. Born in 1480, Giese became a canon about 1504 during the period when Copernicus was still a student in Italy. Along with Andreas Kopernik, Giese was one of the few highly educated members of the Frauenburg Chapter, and over the years the bond between himself and Copernicus grew ever stronger. A broad-minded and tolerant man, Giese proved most receptive to his friend's revolutionary ideas on the cosmos, and he apparently saw no inherent conflict between them and the teachings of the Church. Unlike his friend and colleague, Tiedemann Giese was also a man of considerable ambition, and he rose high in the Church's service, first as Bishop of Kulm in Prussia, and later as Bishop of Ermland itself. Shortly after the arrival of Rheticus, Canon Nicolas, who late in his life decided nothing of consequence without the benefit of Giese's council, left with his young friend for an extended stay with the Bishop at his residence in Loebau Castle.

During the following weeks and months the Catholic Bishop and the Protestant professor put up a united front against Copernicus' persistent objections to their pleas that he publish the *Revolutions*. The aging Canon proved a worthy adversary; victory did not come easily or all at once, for Copernicus didn't relish the prospect of becoming the self-acknowledged father of an unorthodox cosmological system at nearly seventy years of age any more than he had thirty years earlier. The parties first agreed to a compromise, according to the terms of which Rheticus was to publish his own brief account of the secreted manuscript as a way of testing the uncertain waters. Though Copernicus' name appeared on the cover, he was simply referred to as *domine praeceptor* or master teacher in the text. Rheticus' account of the Copernican system was written in ten short weeks under the ever watchful eye of the master, a prodigious feat considering the fact that the young mathematician had not previously examined the lengthy and complex manuscript in detail. Completed by the end of September in 1539, it was titled *Narratio prima* or the *First Report*. At long last the ice had finally been broken, and the Copernican theory appeared in print for the first time in February of 1540.

Whatever the storm of negative reaction Copernicus had anticipated from the publication of his scientific views, none materialized. In fact, just the opposite occurred. The *Narratio* whetted the appetite of the scholarly community sufficiently enough to bring additional requests that Copernicus publish his hypothesis in its entirety. No one, as some uncritical writers have argued, had been fooled into thinking that Rheticus' book was devoted to a presentation of anyone else's work but that of the Frauenburg Canon; and Copernicus now found himself in the even more uncomfortable position of a minor celebrity. Under the circumstances, he risked even greater ridicule if he refused to make his research public at this point than he would if he yielded it up to scholarly examination. Some time in 1540 Copernicus finally gave in to the entreaties of his friends and the nagging infirmities of his advanced age. Too old to supervise the publication of the book himself, he entrusted it to his dearest friend Tiedemann Giese. For his part, Giese, whose knowledge of astronomy and mathematics was that of an amateur, turned the manuscript over to the more expert Rheticus, who agreed to supervise its publication. The opening stage of the Copernican Revolution was clearly at hand. However, before moving on to a more detailed discussion and analysis of Copernican cosmology, it becomes necessary at this point in the narrative to pause for the purpose of inquiring more deeply into the motives behind the Canon's steadfast reluctance to make the details of his work publicly known.

Given the state of the major commitment to science and technology in the modern era, we have often drawn the wrong historical picture regarding the nature of the attitude of institutionalized religion toward scientific activity. Earlier in this work mention was made of the fact that from the beginning of the twelfth to the end of the sixteenth century the Roman Catholic Church took a generally favorable attitude toward the renewed interest in science, and that this change of position from its previously negative stance on the issue derived mainly from the successful melding of Aristotelian cosmology with the principles of medieval Christian theology by St. Thomas Aquinas. Copernicus lived toward the end of this period and was thus shielded by an environment in which religious fanaticism had not yet smothered Renaissance Humanism. The antiscientific position so often associated in the contemporary mind with both the Catholic and Lutheran Churches was almost totally a product of the seventeenth century. Thus, it would seem that Copernicus feared no prospect of religious persecution for his beliefs; neither did he see in his system the seeds of heresy. In fact, one of those who most strongly urged the Canon to publish his manuscript was none other than Nicholas Schönberg (1472–1537), a papal emissary and Cardinal of the Roman Catholic Church. In a letter written to the Canon on November 1, 1536, Schönberg requested Copernicus to communicate his findings to scholars, so that all might have the benefit of his knowledge. Furthermore, Pope Paul III most certainly possessed some information regarding the general contents of the soon-to-be-published manuscript, yet he took no steps to silence its author. Copernicus' loyalty to the Church was unquestioned, and to suppress the printing of a book that did nothing more radical than put forward an innovative working hypothesis of the solar system hardly entered the head of anyone holding an influential position in the Church hierarchy. With the circulation of the

Commentariolus, the broad outlines of the Copernican system had been known in religious as well as in secular circles since 1515, and, if anything, theologians and scholars wanted to learn more about it rather than less. It was not until 1616, when the *De revolutionibus* and all other writings that affirmed the earth's motion were put on the Index, that the Roman Catholic Church officially joined the battle against Copernicanism, and only then because of the intemperate nature of Galileo's exposition of the heliocentric system.

It has long been recognized that Protestant religious leaders adopted a hostile attitude toward the work of Copernicus considerably in advance of the negative Roman Catholic reaction. This situation has given rise to some speculation that Copernicus did indeed fear religious persecution, but more from the direction of Wittenberg than from Rome. An endless spate of anti-Copernican remarks, some of them doubtless true, have been attributed to both Luther and Melanchthon. And there has been considerable scholarly debate on the question of whether or not John Calvin also knew about and disapproved of Copernican cosmology. Nevertheless, the criticism of the heliocentric system voiced by Protestant leaders was quite low-keyed during the years immediately following the Canon's death, reaching the level of invective only as the split between the rival religions erupted into open warfare several decades later.

Yet before dismissing completely the possibility that Copernicus hesitated to publish his views on the composition of the universe because of the fear of arousing religious opposition—a position taken by many twentieth-century historians of science—one additional observation seems to me to be in order. As we are about to see, the vast majority of the material contained in the *Revolutions* fits perfectly within the traditional framework of Aristotelian–Ptolemaic cosmology. But of the few principles that did not agree with those put forth by the ancient astronomers, there is one that provoked greater animus on the part of religious thinkers than all the others put together—the possibility of an *infinite* universe. I am well aware, of course, that many scholars, including the great historian of science Alexandre Koyré, reject out of hand the contention that Copernicus postulates the concept of infinity anywhere in the *Revolutions.* (More will be said about this in the next few pages.) It is my belief, however, that the so-called "timid canon" did indeed realize at least the possibility of infinity, and that even the suggestion of such a profound—indeed revolutionary—principle caused him considerable intellectual discomfort. Not even the specter of a Luther or a dozen like Luther would in the future haunt the Church hierarchy as deeply as the prospect that man might not be the center of Divine Creation, but the resident of a small sphere suspended in the reaches of illimitable space.

While the question of whether or not Copernicus withheld publication of his manuscript at least partially on religious grounds will probably never be resolved to the satisfaction of all historians, there is one point on which virtually every Copernican scholar agrees. In spite of his attainments in the realm of astronomy, the Canon never overcame an ingrained shyness and natural humility which led him to shun publicity both for himself and his work. Moreover, Copernicus truly feared the public derision he thought would result when the details of his system became known. In dedicating his book to Pope Paul III he wrote:

> I can reckon easily enough, Most Holy Father, that as soon as certain people learn that in these books of mine which I have written about the revolutions of the spheres of the world I attribute certain motions to the terrestrial globe, they will immediately shout to have me and my opinion hooted off the stage. For my own works do not please me so much that I do not weigh what judgments others will pronounce concerning them.[8]

Copernicus' choice of the words, "to have me and my opinion hooted off the stage," is of particular interest given an incident that transpired in the 1530s, and about which the Canon probably had knowledge. In the German town of Elblag a troupe of wandering players had given a performance that supposedly ridiculed the notion that the earth moves round the sun. To heighten the effect, according to one account, the terrestrial globe was represented by a pig's bladder whirled around the head of an actor by means of a string, much to the delight of the amused audience. That the effect of such satire was not lost on the common people is a further indication that at least the major principle (the moving earth) of the Copernican system had already become quite well known. No wonder, then, that the retiring Canon reacted with the admonition: "Mathematics are for Mathematicians." It was to them and to them alone that he directed his remarks in the *Revolutions*.

Yet the revolution begun by Copernicus was less a mathematical than a philosophical one. However, by couching his cosmology in the language of mathematics he hoped to avoid, as much as possible, further incidents like the one that took place at Elblag. But even having couched his ideas in a highly esoteric work, Copernicus still dreaded the notoriety he feared their publication would bring. Only after the persistent urging of his friends did the Canon yield: "Therefore when I weighed these things in my mind, the scorn which I had to fear on account of the newness and absurdity of my opinion almost drove me to abandon a work already undertaken. But my friends made me change my course in spite of my long-continued hesitation and even resistance."[9] This is definitely not an expression of false modesty on the aging Canon's part, but rather a reflection of his true feelings based upon a lifetime of quiet contemplation and retiring sociability. As with so many of history's most gifted thinkers, whose modesty matched their intellectual capacity, Copernicus could see greatness in the achievements of others but not in his own accomplishments.

Furthermore, Copernicus recognized a basic contradiction between his belief in the heliocentric system and his attempted method of proving its validity. While he had developed a most interesting hypothesis, and went to great lengths to provide a solid mathematical foundation to support it, he did not succeed in proving it from the scientific point of view. Copernicus admitted as much when he observed: "Although I realize that the conceptions of a philosopher are placed beyond the judgment of the crowd, because it is his loving duty to seek the truth in all things, in so far as God has granted that to human reason; nevertheless I think we should avoid opinions utterly foreign to rightness." [10] His use of the phrase "conceptions of a philosopher"—not of a *mathematician*—is a direct indication of the self-recognized limitations of his mathematical method. Mathematics *tended* to support his conceptual model of the universe no less than it *tended* to support that of Ptolemy. However, it did not *prove* or *disprove* either system in the absolute sense. Copernicus still had to rely upon the traditional argument that philosophically his system was a viable one and that his conclusion is supported by mathematics. It was not until after Kepler, Galileo, and Newton had completed their work on the formulation of exact laws that the mathematical universe could stand alone.

Once Copernicus gave his consent to the publication of the *Revolutions*, immediate preparations were undertaken for the printing. From the summer of 1540 to September 1541 Rheticus stayed with his mentor while the lengthy manuscript, so frequently revised by its author, was recopied for the printer. An additional advantage gained from undertaking this tedious and time-consuming task was that the mathematical calculations could be checked and other significant alterations made. It also helped to better acquaint Rheticus with those parts of the treatise which he had omitted from the analysis of it presented in his *Narratio prima*. Meanwhile, the enterprising Professor had decided upon a publisher, one Johannes Petreius of Nuremberg, a personal friend and a noted printer of books on astronomy and mathematics. Rheticus also solicited support for the printing from the Protestant duke, Albert of Prussia, who compiled by drafting several letters recommending that the *Revolutions* be published. Since Rheticus knew that both Luther and Melanchthon took a generally unenthusiastic view of the heliocentric theory, he apparently felt that the Duke's support would help override any possible objections the theologians might raise. Considering the fact that Rheticus had been granted extensive leaves of absence by these same Protestant superiors to work with Copernicus, this action might be viewed as over cautious. On the other hand, he doubtless wished to allow for every possible contingency given the extreme difficulty already encountered in obtaining Copernicus' permission to publish his book, and the time spent in preparing the manuscript for the printers.

Rheticus returned to Wittenberg for the winter term of 1541, the manuscript copy of the *Revolutions* in hand. Shortly after his arrival, the mathematician was honored by his colleagues by being selected Dean of his faculty—an action that offers additional proof that the era of open-mindedness inaugurated during the early days of the Renaissance had not yet completely run its course. For had the Copernican system been a highly controversial issue among members of Wittenberg's scholarly community at this time, it seems inconceivable that Rheticus would have been elected to this esteemed post. With the arrival of spring and the end of his teaching duties, Rheticus left immediately for Nuremberg where Petreius began printing *De revolutionibus* in early May of 1542. At this point, there occurred one of those unfortunate incidents which do not necessarily change the course of history but which certainly complicate its study for future generations. Rheticus had every intention of personally seeing the manuscript through the press. However, he had earlier applied for the vacancy in the important Chair of Mathematics at Leipzig University, and he received word while in Nuremberg that his application had been favorably acted upon. While the timing of this new appointment must have caused him considerable inconvenience, Rheticus had no reason to view it with displeasure. The printing was proceeding quite well, and he had the good fortune to solicit the aid of Nuremberg's leading Lutheran theologian and a pro-Copernican, Andreas Osiander, who consented to supervise the final stages of publication.

In the absence of Rheticus, and for reasons that are still the subject of much debate, Osiander felt it necessary to include at the beginning of the *Revolutions*—without the permission of Copernicus—certain introductory remarks which are best known today as Osiander's Preface. In it Osiander sets forth the proposition that the following work is merely the exposition of a theory and not to be interpreted as being final:

> For it is the job of the astronomer to use painstaking and skilled observation in gathering together the history of celestial movements, and then—since he cannot by any line of reasoning reach the true causes of these movements—to think up or construct whatever causes or hypotheses he pleases such that, by the assumption of these causes, those same movements can be calculated from the principles of geometry for the past and for the future too. This artist is markedly outstanding in both of these prospects: for it is not necessary that these hypotheses should be true, or even probably true; but it is enough if they provide a calculus which fits the observations.[11]

Osiander's prefatory note was unsigned, and for generations Copernican scholars puzzled over the apparent contradiction between what the author referred to as a "hypothesis" in the preface of his work and his statement in the dedication that "it is his [the philosopher's] loving duty to seek truth in all things" and to "avoid opinions utterly foreign to rightness." The concession Copernicus had supposedly requested at the outset was apparently forfeited only a few pages later. Finally, in 1854, with the publication of the Warsaw edition of *De revolutionibus*, Osiander's authorship of the preface was authenticated and the confusion resolved.*

At first it was widely speculated that Osiander included his remarks as a hedge against potential clerical opposition to the Copernican theory, and perhaps he did. But more recently the belief has arisen that the theologian only reflected in his preface the same conclusion anyone with an open mind would have reached after a careful reading of the book. As has been observed, Copernicus possessed no absolute proof—mathematical or otherwise—of the earth's motion, and thus despite all of his exacting work the heliocentric system remained within the realm of hypothesis for over a century. Of course, had Rheticus not been called to Leipzig at this crucial stage in the printing of the *Revolutions*, the unfortunate historical misunderstanding surrounding the preface would never have occurred. In fairness to Osiander, however, there is little reason to believe that he sought to distract from the value of Copernicus' work or to impute to the Canon views he did not hold. Rather, Osiander saw the *Revolutions* as a part of the great astronomical tradition dating back to the ancient Greeks, and he clearly wanted Copernicus' work to be accepted as an integral part of it.

The confusion over the authorship of the anonymous preface was not the only incident that clouded the publication of the *Revolutions*. A second and more personal injury was suffered by Rheticus, without whose unflagging devotion the book might never have seen the light of day. In advance of his departure from Nuremberg, the mathematician received the text of the dedication which Copernicus drafted shortly before the book went to press: What he read must have both shocked and grieved the young disciple. After attributing to his friends the responsibility for making "me change my course in spite of my long-continued resistence" to publication, Copernicus went on to enumerate

---

* Kepler had suspected as much long before. He once referred to the spurious preface as having been "written by a jackass for the use of other jackasses." Yet the difference could not be detected by the casual observer. Cardinal Bellarmine, after reading the unsigned Introduction, argued to Galileo that Copernicus was philosophically on the side of the Church (see Chapter 7). See also *Three Copernican Treatises*, 3rd ed., pp. 403–406.

those most responsible for his change of attitude. "First among them was Nicholas Schönberg, Cardinal of Capua, next to him was my devoted friend Tiedemann Giese, Bishop of Kulm, a man filled with the greatest zeal for the divine and liberal arts." A little farther on the Canon states that "Not a few other learned and distinguished men demanded the same thing of me, urging me to refuse no longer." [12] After this—silence! Nowhere is the name of Georg Joachim Rheticus mentioned.

One can only guess at the torment this omission must have caused his most active and devoted supporter. Bishop Giese was so embarrassed by his old friend's blunder that he sought to smooth the troubled waters in a written apology to Rheticus, in which he referred to the incident as an "unpleasant oversight" rather than an act of "indifference towards thee." What both men must have known, and what went unsaid, was that the mention of Rheticus' name, a Protestant Professor of Mathematics at Wittenberg, in a dedication to the Pope of the Roman Catholic Church was more than Copernicus, despite his warm attachment to Rheticus, could bring himself to undertake. Still, Rheticus must have suffered from the slight, and perhaps it was at least partially for this reason that the gifted scholar never published his long-lost biography of Copernicus, a work that would almost certainly have cast considerable light on the character of an intensely private man, who preferred the shadows and the twilight to the brightness of the noonday sun.

The Copernican System

The first edition of *De revolutionibus*, published in the spring of 1543, numbered perhaps 400 copies, and took a considerable time to sell out. The primary reason for this seeming neglect is the extreme difficulty of the text, which its author intentionally couched in the complicated language of mathematics. Copernicus had only grudgingly consented to the publication of his work, and now that the long-awaited treatise was finally in the hands of the expectant reader the details of the heliocentric system proved far too complex for the educated layman to follow. Still, a respectable number of Copernicus' scholarly peers did complete the work, though there can be little doubt that many never read it from cover to cover, while countless others abandoned it in a state of confusion, if not abject frustration.*

The book consists of two parts which differ considerably in purpose,

---

* The publication, in 1566 and 1617, of second and third editions proves that the book was in considerably greater demand by scholars than certain contemporary writers have suggested.

character, and importance. The first, and shorter part, makes up the first of six books into which the *Revolutions* is divided. It is addressed to the general reader and provides a fairly clear but highly simplified view of the new world system. The second, and more imposing part, comprising Books Two through Six, is written for the professional mathematician and astronomer. Its complexity, detail, and rigorously scientific form place it on the same level as Ptolemy's *Almagest,* a work much admired by Copernicus and one that deeply influenced the format of his own treatise. A fact that many find surprising is that only about twenty pages, less than 4 percent of the entire volume, are devoted to the new heliocentric cosmology. The remainder of the work consists primarily of a detailed mathematical analysis of the movements of the sun, moon, and planets in an attempt to develop more reliable periodic tables than those constructed by Ptolemy. The significance of Copernicus' contribution in this area is that even those who rejected his cosmological system readily embraced his mathematical calculations and planetary predictions. It is also important to recognize from the outset that Copernicus agreed with Ptolemy on a great number of fundamental astronomical principles, many of which directly supported the basic teaching of Aristotle. But for reasons of time and space, the differences between the two systems will be stressed to a greater degree than will their similarities.

At the beginning of the first book Copernicus states that the world (universe) is globe-shaped as are the celestial bodies contained within. The earth is also a sphere "since on every side it rests upon its center." It is further argued that the movements of the celestial bodies are regular, circular, and eternal "for the motion of a sphere is to turn in a circle . . . where beginning and end cannot be discovered or distinguished from one another." [13] Aristotle himself could hardly have been more pleased with the orthodoxy displayed by Copernicus up to this point; but it is also at this juncture that the reader's faith in the time-honored opinions of the Stagirite is put to its first important test. Part Five of Book One bears the title: "Does the Earth Have a Circular Movement? And of Its Place." Virtually all writers, says Copernicus, concur that the earth is stationary in the center of the universe. "If however we consider the thing attentively, we will see that the question has not yet been decided and accordingly is by no means to be scorned." Every apparent change in the location of an object occurs as a result of its movement, the movement of the viewer, or a combination of the two. The ancients assumed that because the earth seems not to move under us, and the planets and stars appear to move around us, the earth must be at rest. But if one assumes that the earth also has motion, turning daily on its axis from west to east, it would produce an apparent motion of all other objects outside it in the opposite direction, thus accounting for the rising and setting of the sun, moon, and stars, something a number of the ancient Greek astronomers, including Pythagoras, had previously taught.

While the supposition raised by Copernicus is perfectly proper, it immediately raises the traditional argument employed by both Aristotle and Ptolemy in support of the geocentric system; namely, if the earth is in rotation from west to east, then why does an object tossed into the air return to the same spot from which it was thrown instead of landing some distance west of its starting place? Furthermore, why doesn't the earth tear apart and dissipate throughout space if it revolves as rapidly as the Pythagoreans, Aristarchus, and others maintained? (Although the term was not then used or fully understood, the Aristotelians were obviously referring to centrifugal force). Since Copernicus lacked the mechanical laws to refute these arguments on a mathematical basis, he employed the philosophical argument that axial rotation is natural and for that reason would not be self-destructive. Moreover, it is reasonable to suppose that the layer of air surrounding earth also moves with the planet, so that an object thrown upward is carried along in the same manner as are the clouds or a bird in flight. In effect, Copernicus borrowed from Aristotle the concept of natural tendencies to support his anti-Aristotelian hypothesis. But Copernicus raised an even more telling counter argument against the geocentrists, by calling attention to the fact that the alternative to a rotating earth is, as they themselves postulate, a rotating universe. The speed at which the sun, planets, and stars would have to circle the stationary earth in a single day would be incalculably greater than the speed required for the earth's diurnal rotation on its own axis. Why don't the planets and stars then fly apart in the same manner attributed to the earth were it in rotation? The astronomer has cleverly turned the Aristotelian argument on natural celestial motion against its proponents.

To further support this argument, Copernicus challenges the size of the universe as determined by the ancient astronomers and handed down to their successors during the Middle Ages. In chapter six the author observes, "that the heavens are immense in comparison with the Earth and present the aspect of an infinite magnitude, and that in the judgment of sense-perception the Earth is to the heavens as a point to a body and as a finite to an infinite magnitude." [14] Nor, writes Copernicus, is it clear exactly how far this immensity stretches out. In other words, he believes that the earth must be regarded as nothing more than a *point in space* in comparison to the size and distance of the stellar sphere. And the farther away the stars from earth, the greater will be the velocity required to complete a circuit of the heavens in a single day. If accepted, this argument makes Ptolemy's belief in the daily rotation of the stellar sphere that much more difficult to support because the speed required would literally be beyond human comprehension.

Having made these points, it is in the tenth chapter of the *Revolutions* that Copernicus proceeds to fix the order of the planetary orbits, and here that he reintroduced the long-abandoned, sun-centered doctrine of the ancient Greeks to which his name has become irrevocably attached:

> Therefore we are not ashamed to maintain that this totality—which the moon embraces—and the center of the earth too traverse that great orbital circle among the other wandering stars in an annual revolution around the sun; and that the center of the world is around the sun. I also say that the sun remains forever immobile and that whatever apparent movement belongs to it can be verified by the mobility of the Earth; that the magnitude of the world is such that, although the distance from the sun to the Earth in relation to whatsoever planetary sphere you please possesses magnitude which is sufficiently manifest in proportion to these dimensions, this distance, as compared with the sphere of the fixed stars, is imperceptible.[15]

Copernicus next proceeds to construct the basic framework of our modern solar system as shown in Figure 7. At the center rests the immobile sun followed by Mercury, which circles the great star once every eighty-eight days. Venus is the next closest planet with an orbit of seven and one-half months, then earth (twelve months), Mars (two years), Jupiter (twelve years), Saturn (thirty years) and finally, at an undetermined distance, the realm of the stars.

It was this very process of postulating the idea of an orbiting earth and an immobile sun that forced Copernicus to rely increasingly on the conclusion that the universe is much larger than previously believed. Otherwise, he would not have been able to meet what was doubtless the strongest scientific objection raised against his system. To contemporary critics it was logical to suppose that if the earth moves round the sun in a huge orbit, the arrangement and relative brightness of the stars ought to undergo change constantly according to the various positions occupied by the earth on its journey back and forth across the heavens. For example, as a constellation is approached its stars should appear larger, brighter, and farther apart. Conversely, as the earth moves away from the constellation, toward the opposite side of the sun, the stars should grow dimmer and appear to move closer together. But this displacement, or stellar parallax, does not occur, at least not to the naked eye. Thus, either the Copernican theory was wrong or else the stars, as the astronomer maintained, are so far distant from the earth that the orbit described by the planet is as nothing, thereby producing no visible effect to the observer. It is similar to the example cited earlier (see Chapter 2, p. 57) of the individual who picks out an object on the distant horizon and then takes a few steps toward it but can see no noticeable difference in its detail, size and shape. To Copernicus, the earth's orbit

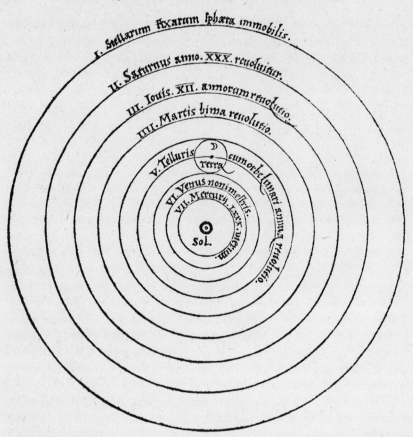

FIGURE 7.   The universe according to Copernicus as shown in Book One
of *De revolutionibus*. (*Courtesy of the Lilly Library, Bloomington, Indiana.*)

round the sun would be comparable to taking a few steps in the direction of a far distant object. The change in position is simply swallowed
up by the immensity of intervening space.

Although the positions of the stars relative to the earth seemingly
remain unchanged, those of the planets do not. From earliest times the
careful observers of celestial phenomena were confounded by the
changes in the directions of planets, later called stations and retrogressions. Aristotle himself had been unable to formulate a satisfactory
explanation that did not violate the time-honored concept of circular
motion; thus it was left to Ptolemy to account for this anomaly with
his complex geometric combinations of epicycles, deferents, and equants.
To Copernicus, however, this phenomenon begged further explanation
and, if possible, a less complex explanation than that supplied by the
Alexandrian. By postulating the earth's motion he ultimately arrived

at one. Unlike the far distant stars, the planets are near enough to one another so that as they move at different speeds in an arc round the sun, the changes in the relative distances between them are clearly observable with the naked eye. Since Mercury and Venus complete their orbits sooner than the earth, they approach and pass it during its annual orbit. In the case of Mars, Jupiter, and Saturn, whose revolutions take much longer to complete than those of the other planets, the earth approaches and passes them. According to Copernicus, this combination of the earth's and the planets' motions is therefore accountable for the ancient anomaly of stations and retrogressions. For example, as the earth moves by Mars, the neighboring planet seems to stand still; then it retrogresses as the earth passes between it and the sun; then it stops again, and finally resumes its original direction from west to east in relation to the fixed stars. This process will then repeat itself a year later.

Although essentially correct, Copernicus' explanation of this anomaly is not as simple as it might at first appear. A problem arose because Copernicus, like Ptolemy and Aristotle, believed that the movement of celestial bodies is perfectly circular. He realized that by simply describing a circular orbit round the sun for each one of the planets all of their movements cannot be exactly accounted for, any more than they can by Aristotle's planetary orbits round the earth. Even the rotation of the earth on its axis once every twenty-four hours, combined with its annual rotation round the sun, leaves certain planetary movements unexplained. As a result, when it came to reconciling the heliocentric doctrine with actual observations, Copernicus had to considerably modify his system from a technical point of view. First, neither the earth nor the planets revolve exactly round the sun as is so commonly believed of Copernican cosmology. Rather they rotate round a point in space removed from the sun by about three times its diameter. Thus the sun actually stands to one side and merely illuminates the whole. It is therefore technically more proper to refer to the Copernican system as heliostatic (a motionless sun) rather than heliocentric or sun-centered.

Secondly, Copernicus writes of the planetary movements in Book Four: "We must confess that these movements are circular or are composed of many circular movements, in that they maintain these irregularities in accordance with a constant law." [16] In other words, the planets in his system continue to move on epicycles just as they do in the Ptolemaic system. Moreover, their motion, like that of earth, is not truly centered on the sun, but on the same center as the earth's orbit round an abstract point in space. Copernicus employed as many as 48 epicycles to account for all the planetary movements in his system, a number slightly in *excess* of those used by Ptolemy. The Copernican system is therefore considerably more complex than most of us have been taught, even though it possesses certain basic advantages over Ptolemy's geocentric model. It was because he lacked Kepler's knowledge of the elliptical orbit that Copernicus borrowed the epicycle from the Aristotelians, an act which many have mistakenly interpreted as proof that there is little if anything revolutionary in his system.

Third, the two motions attributed to the earth—axial rotation and an annual orbit—when combined with the other planetary movements, were still insufficient to account for all the variations in celestial phenomena. Copernicus consequently formulated a third motion for the earth whereby the axis of the planet describes the surface of a cone in a year. This regular but wobbly movement was made necessary, from his point of view, for the purpose of maintaining nearly constant the direction of the earth's axis in space.

A number of other details relating to the structure of the Copernican system could be brought to bear in this discussion, but the basic outlines have been presented, and the remaining details are best left to those wishing to pursue the subject in greater depth on their own.[17] At this point it is important that some attempt be made to determine the nature of Copernicus' contribution to science. The case has been made in the past, and will doubtless be made again in the future, that there was no real reason during Copernicus' time to prefer the heliocentric to the geocentric universe because in terms of explaining celestial phenomena the two are interchangeable. This is particularly true when we put ourselves in the place of the students of the sixteenth century, who frequently heard the two systems expounded by the same professor as alternative hypotheses. It must be remembered that for them it was not only a matter of the intellectual acceptance of Copernican cosmology, which in truth posed few major problems, but of moral truth as well. While Ptolemy's system was better understood because it conformed to common sense experience, that of Copernicus was abstract and discomforting, for it removed man from the center of Divine Creation.

If, however, the Copernican system is evaluated from the contemporary perspective, it indeed becomes difficult to deny its advantages over its geocentric counterpart. It is still widely held that the essential advantage of the heliocentric system rests in its simplicity of operation. Yet Copernicus, as we have observed, saved very little if anything significant in the planetary motions, despite long decades during which he struggled through thousands of hours of mathematical computations. In fact, his system was made more complicated by the requirements that he add to the earth's revolution and rotation a third motion (the precession of the equinoxes) much like the wobble of a top. On the other hand, he contributed both esthetic and systematic qualities found lacking in the Ptolemaic universe.

Copernicus rejected the Alexandrian's system, as much as anything else, for its lack of cohesion and symmetry. He particularly recoiled from the cumbersome celestial machinery bequeathed by Ptolemy to his successors. And even though Copernicus employed as many epicycles as had his predecessor, by postulating the earth's motion Copernicus abolished the use of the equant, that off-centering "trick" for imparting to the planets the appearance of nonuniform motion, and thus clearly established greater overall geometrical simplicity. Additionally, by transferring the center of the universe to the region of the sun, and by supposing that the earth and planets revolve in circles round the same center, Copernicus truly invented systematic astronomy. Although Ptolemy had developed a complex pattern of circular motions for each of the planets, by changing one part of his machinery the rest was not necessarily affected. The chief merit of Copernicus' work is that *all* of his planetary movements and calculations are *interdependent*. A change in the relationship between the earth and one planet's motion cannot be computed without also computing changes between the earth and all of the others. *De revolutionibus* transcends Ptolemy's collection of facts and techniques by setting out in advance a set of rules for dealing with the aberrations in planetary behavior in an organized, systematic manner. Copernicus saw this as perhaps his main accomplishment and spoke of it in the dedication of the *Revolutions:*

> I finally discovered by the help of long and numerous observations that if the movements of the other wandering stars [planets] are correlated with the circular movement of the Earth, and if the movements are computed in accordance with the revolution of each planet, not only do all their phenomena follow from that but also this correlation binds together so closely the order and magnitudes of all the planets and of their spheres or orbital circles and the heavens themselves that nothing can be shifted around in any part of them without disrupting the remaining parts and the universe as a whole.[18]

This represents an enormous scientific stride, although its vindication lay in the future, when the mathematical data could be perfected at the hands of Kepler and Newton. Still, the main conclusion to be drawn from this pioneering step is that in the eyes of Copernicus the universe is fundamentally geometrical. His work therefore represents a clear return to the direction taken by the Pythagoreans where "number is everything." On the one hand, Copernicus had broken with the ancients who equated sense perceptions with scientific truth; on the other, with the scholastics who held that only Revelation, which comes directly from God, gives certainty and that human reason leads only to probable or contingent truth. To Copernicus, the motion of the earth and the other planets is a physical reality and can be understood on a strictly mathematical basis. This belief, more than any other, lay at the heart of the scientific revolution not only in astronomy but in all related fields.

The concept of a systematic, mathematical universe, in which the earth is a planet operating according to principles no different than those which govern its neighbors, struck a fateful and ultimately fatal blow at the two-sphere universe of Aristotle. It will be recalled that the Stagirite had postulated one set of laws for the terrestrial realm and another for the celestial region. Since, according to Aristotle, the two realms are composed of different elements it is only natural that they should behave quite differently. But in changing the arrangement of the celestial bodies—by literally placing the earth in the heavens along with the other planets and the stars—Copernicus destroyed the very basis of Aristotelian cosmology. If the earth is corrupt and mutable, then the planets must also be subject to the same cycle of decay and regeneration. And since Christian cosmology with its Thomistic concept of the Empyrean Heavens and angelic hosts was inextricably tied to the hierarchial model of Aristotle, the revolutionary implications of the Copernican system soon became shockingly clear. The universe is no longer the affair between God and man that St. Thomas had believed it to be. All that was needed to complete the demise of the old order was the formulation of a series of mathematical laws to tie the system together.

Some have even gone so far as to argue that Copernicus, by postulating the heliocentric doctrine, destroyed the principle of a hierarchical universe altogether. The English poet John Donne, in his *Anatomy of the World* published in 1611, saw the new philosophy as putting "all in doubt" by destroying the natural order and harmony established by the ancients:

> And new Philosophy calls all in doubt,
> The Element of fire is quite put out;
> The Sun is lost, and th'earth, and no man's wit
> Can well direct him where to look for it.
> And freely men confesse that this world's spent,
> When in the Planets, and the Firmament
> They seeke so many new; then see that this
> Is crumbled out again to his Atomies.
> 'Tis all in peeces, all cohaerence gone;
> All just supply and all Relation.

It is certainly true that by removing the earth from the center of the world and placing it in the heavens along with the other celestial bodies, Copernicus seriously undermined fundamental Christian cosmology with its belief in the central location of man's home. However, he created a new hierarchy in place of the old, inverting the Aristotelian system by enthroning the sun and surrounding it with subservient planets and stars.

> In the center of all rests the sun. For who would place this lamp of a very beautiful temple in another or better place than this wherefrom it can

illuminate everything at the same time? As a matter of fact, not unhappily do some call it the lantern; others, the mind and still others, the pilot of the world. Trismegistus calls it a 'visible god'; Sophocles' Electra, 'that which gazes upon all things.' And so the sun, as if resting on a kingly throne, governs the family of stars which wheel around.[19]

It is almost as if man had once again become a sun-worshipper, as in the days of ancient Egypt when this brilliant object was venerated as nature's ultimate source of power and life. This observation, as in the case of the pharaohs of old, who chose the sun as their symbol of power and supreme authority, was not lost on the new absolute monarchs of Europe. Little more than a century after the death of Copernicus, King Louis xiv of France, mindful of the new advances in science, also chose for his symbol the sun, and presented himself as the sole source of light and power throughout his earthly kingdom: *le Roi Soleil*. At Versailles Louis was constantly surrounded by nobles, both greater and lesser, who, like the planets themselves, served as satellites to their divinely ordained master. Thus, unlike Newton, Copernicus did not completely break with the Aristotelian concept of a hierarchical universe; he remained too strongly tied to the tradition of ancient astronomy for that.*

Despite the revolutionary implications of the heliocentric system, many of which are much more obvious today than in Copernicus' time, his cosmology suffered certain inherent weaknesses. For one thing, we must keep in mind that the mathematician–astronomers of his day lacked the mechanical knowledge to prove the actual rotation of the earth on its axis; and they also lacked specific knowledge of the force under which the planets move and by which they remain in their orbits. The idea that this is one and the same force which causes a stone (or an apple?) to fall to the ground was in the days of Copernicus completely foreign to the realm of ideas. If one had asked Copernicus precisely why the earth and the planets move, he could have done no better than fall back upon Aristotle's dictum: "Because they are spheres and a sphere placed anywhere in space rotates by nature without requiring anyone to turn it." The very fact that he remained tied to the Aristotelian concept of the perfect circle prohibited him from fully developing the revolutionary implications of his system. Johannes Kepler, the discoverer of the elliptical orbit, hit the nail on the head when he re-

---

* Perhaps in recognition of the fact that the importance of the sun in his system had increased by a ratio directly proportionate to the decline in the earth's significance, Copernicus hastened to compensate to at least some extent for this loss: "The earth is by no means cheated of the services of the moon; but as Aristotle says in the *De Animalibus*, the Earth has the closest kinship with the moon. The Earth moreover is fertilized by the sun and conceives offspring every year."

marked: "Copernicus failed to see the riches that were within his grasp and was content to interpret Ptolemy rather than Nature."

Neither was Copernicus' belief that the center of the planetary orbits converge at a point in space very convincing. Even though Newton's law of universal gravitation would not be conceived for a century and a half, most astronomers felt an instinctive affinity toward the proposition that the motions of the planets were somehow directly related to concrete objects, rather than to some abstract point in space. This feeling was later vindicated by Newton's discovery which forced a major revision of Copernicus' celestial framework.

Moreover, the point has been made that Copernicus made relatively few observations of his own, relying instead upon the work of his predecessors. It was beyond the power of any astronomer of Copernicus' time, no matter how brilliant, to decide which of those observations were trustworthy and which were not. Copernicus, of course, tried to be as critical in his thinking and calculations as possible, but the errors proved too numerous and deeply embedded to be removed. For example, the variations in the length of the year and in the rate of change of stellar longitudes, assumptions on which Copernicus based much of his work, were later proved nonexistent, the result of centuries of accumulated error. Numerous other anomalies and errors of fact could also be cited, but there is no point in further pursuing them at this point. Despite his shortcomings, the great service that Copernicus rendered to modern science was to advance certain basic mathematical principles, later proven to be fundamentally correct, for assuming that the sun is at the center of our planetary system and for possessing the courage to follow his reasoning to its logical conclusion based upon the data then available.

Physically, he may have spent the major part of his adult life behind the secure walls of a medieval cathedral, but intellectually his mind probed the very limits of the then known universe, restructured its components in a revolutionary manner, and, still unsatisfied, reached beyond toward infinity. I, for one, cannot agree with those who maintain that the Copernican system represents little more than a last ditch attempt to patch up the out-dated celestial machinery of Ptolemy by reversing the arrangement of its cosmic wheels. Yet neither did Copernicus make a clean sweep of the methods and conceptions which astronomy had employed through the ages. He did, however, inaugurate a new era by breaking with the past and taking full advantage of the opportunity to contemplate the heavens with a largely unbiased mind. By virtue of his innovations Copernicus shifted the focus of scientific inquiry in a radically new direction. His ideas were revolution-making and had a decidedly catalytic effect upon the work of his successors. Perhaps as the best measure of his ultimate contribution to the history of scientific thought one might consider this question: Who, if not Copernicus, is responsible for the genesis of the revolution which ultimately dethroned medieval science and established modern science in its stead?

The Question of Infinity

Earlier in this chapter the question was raised as to whether or not Copernicus contemplated the possibility of an infinite universe in contrast to the closed world postulated by Aristotle and his successors. It has long been the custom among many of the major interpreters of Copernican cosmology to conclude that because the astronomer retains the eighth sphere of the Ptolemaic system—that of the fixed stars—his world remains a finite one, circumscribed by a definite material boundary. In support of this proposition the following passage taken from the *Revolutions* is frequently cited:

> . . . then the order of the spheres will follow in this way—beginning with the highest: the first and highest of all is the sphere of the fixed stars, which comprehends itself and all things, and is accordingly immovable. In fact it is the place of the universe, i.e., it is that to which the movement and position of all the other stars are referred.[20]

Using this statement as the basis for his contention that the Copernican universe is indeed finite, historian of science Alexandre Koyré writes:

> It seems to be psychologically quite normal that the man who took the first step, that of arresting the motion of the sphere of the fixed stars, hesitated before taking the second, that of dissolving it in boundless space; it was enough for one man to move the earth and to enlarge the world as to make it immeasurable—*immensum;* to ask him to make it infinite is obviously asking too much.[21]

While it may indeed have been too much to ask of one man, the question remains: did Copernicus believe in an infinite universe?

In further support of Koyré's thesis the diagram of the universe printed in the first book of *De Revolutionibus* (see Figure 7) has freqeuntly been cited. Here the sphere of the fixed stars is represented by a circle which encloses the revolutions of the six planets round the stationary sun. Above this circle is the caption: *Stellarum fixarum sphaera immobile* or sphere of the fixed stars. Thus, from these and other similar examples it would appear that Copernicus accepted Aristotle's doctrine of finitude, by attaching the stars to a crystal sphere which encloses a limited universe.

It must be remembered, however, that Aristotle had been confronted with an entirely different physical problem than had Copernicus. The Greek natural philosopher had no other choice but to limit the size of the universe because of his belief that the heavens turn completely round the central earth once a day. If one supposes that some parts of the heavens are infinitely distant from their center, then those parts would by necessity have an infinite distance to travel. Moreover, they would be required to move at an infinite speed in order to complete a circuit once in twenty-four hours. Aristotle, as did Ptolemy after him, rejected the possibility of infinite speed: the universe could be nothing but finite. He also realized that in an infinite universe there is no center, a proposition that would have destroyed the very idea of a closed geocentric system.

Copernicus agreed with Aristotle's conclusion that infinite velocity is impossible. In fact, as we have noted, this was one of the primary considerations which led him to the conclusion that the stars are fixed in their places and do not move about the earth. Rather it is the earth which moves, and at a much slower rate of speed than is required of Aristotle's *primum mobile* and stellar sphere. We also observed that Copernicus believed the stars to be much farther away from earth than did Aristotle because no stellar parallax had been detected. Both Copernicus and his followers were forced to draw the conclusion that within this vast space of the starry sphere not only is the earth a point, but the earth's orbit is itself a point relative to this vast extent. To the Aristotelian opponents of Copernicus this conclusion was incredible; to his disciples it was staggering; and to traditional religious doctrine it proved deadly. Still, as Koyré points out, the tremendous expansion of the Aristotelian–Ptolemaic universe does not necessarily bring us any nearer to true infinity. A universe can be vast in size and yet remain finite. Nevertheless, from a psychological point of view, Koyré himself admits that "it is somewhat easier . . . to pass from a very large, immeasurable and ever-growing world to an infinite one than to make this jump starting with a rather big, but still determinably limited sphere: the world-bubble has to swell before bursting." [22] A case can be made, however, that Copernicus not only inflated the bubble of the universe to hitherto undreamed of proportions, but that he stretched it beyond the breaking point, scattering the stars through a universe whose uncharted boundaries—if they exist at all—may forever remain a mystery to the mind of man.

Simply because Copernicus' diagram in the *Revolutions* pictures the fixed stars as a circumscribing sphere, it does not necessarily follow that he embraced the Aristotelian belief that the stars are attached to the inside of a crystal globe. During the centuries following Aristotle's death a number of major astronomers raised serious questions about the stars all being located equidistant from earth. Although Ptolemy did not directly broach the matter, his silence on the subject is most interesting. It is extremely difficult to understand how the planets in the Alexandrian's system could more on epicycles and equants and still remain affixed to their respective spheres of crystal. And if the crystal planetary spheres do not exist, then there was good reason to question the existence of a crystal stellar sphere as well. In fact, Aristotle seems to have been almost the only scientific thinker of consequence to believe in the actual existence of the celestial shells. Others, Ptolemy included, simply looked upon them as abstract geometric constructions and little more. Significantly, those who continued to accept the doctrine of the crystal spheres during the Middle Ages did so primarily on theological as opposed to scientific grounds and, in the case of Thomas Aquinas, were largely confined to the clergy.

A number of astronomers of Renaissance Europe including Sacrobosco (John of Holywood), Christopher Clavius, and Nicolaus of Cusa discussed the eighth sphere as an extended zone rather than as a crystal shell. They termed the spherical outside limit of the zone the "convexity" and the inner limit was called the "concavity." In other words, the stellar sphere became an extended layer or large band in which the stars were scattered about at *different* distances from the earth. This proposition had been well established during the century before Copernicus, so that few if any responsible astronomers of his day regarded the stellar sphere as a crystal shell. The concentric circles in the Copernican diagram, like the epicycles of Ptolemy, became geometrical representations and nothing more. Rheticus, in the *Narratio prima*, the first published work on Copernican cosmology, says nothing of a crystal sphere to which the stars are attached. But he does discuss the *inner concavity* of the stellar sphere and in doing so "he implicitly accepted and used the idea that the eighth sphere is an extended space which begins with or beyond the orbit of Saturn," according to historian of science Grant McColley.[23] Although still not an unqualified endorsement of the concept of infinity, Rheticus worked directly from his master's manuscript and presumably under his scrutiny. Clearly Copernicus was moving in the direction of a vast universe whose limits were unknown, and, for that matter, remain so.*

Earlier mention was made of the fact that in chapter six of the *Revolutions* Copernicus states that in comparison to the immense heavens the earth is miniscule—"as a point to a body, and as a finite to an infinite magnitude." This statement is followed in chapter eight by this pronouncement:

> But they say that beyond the heavens there isn't any body or place or void or anything at all; and accordingly it is not possible for the heavens to move outward: in that case it is rather surprising that something can be held together by nothing. But if the heavens were infinite and were finite only with respect to a hollow space inside, then it will be said with more truth that there is nothing outside the heavens, since anything which

---

* See also the discussion of this point by Salomon Bochner in *The Copernican Achievement* ed. by Robert S. Westman (Berkeley: 1975), pp. 45–48 and by Owsei Temkin in *The Nature of Scientific Discovery*, ed. by Owen Gingerich, p. 122.

occupied any space would be in them, but the heavens will remain immobile. For movement is the most powerful reason wherewith they try to conclude that the universe is finite.[24]

Since Copernicus had earlier done away with the moving heavens, the most powerful physical objection to infinity has been removed. At the very least, the Canon has put forth an infinite universe as a promising hypothesis. He then suggests that the dispute "as to whether the world is finite or infinite" be left "to the philosophers of nature." But he quickly adds: "Why therefore should we hesitate any longer to grant to it [the earth] the movement which accords naturally with its form, rather than put the whole in commotion—the world whose limits we do not know and cannot know?" [25] If not an unreserved advocate of infinity, the astronomer is most certainly in favor of keeping an open mind on the subject. Since absolute mathematicial proof was lacking, the question of infinity must still remain within the realm of philosophical speculation. In this instnce, however, he was doubtless somewhat relieved, because he certainly did not want to be the one to shatter the inner dome of heaven beyond repair. By stating his views on infinity in hypothetical terms, the Canon was spared this awesome responsibility.

Yet Copernicus' removal of the major physical obstacles to an infinite universe still left him with the burden of theological opposition, for he knew full well that one could not even pose the question of infinity without attacking, at least indirectly, the time-honored Christian belief that the universe is an affair between God and man. Thus he published his book only after a prolonged hesitation, and even then went only as far as his conscience and limited scientific knowledge would allow. The so-called timid Canon had gone far enough, however, to strike the first of a major series of fatal blows not only to the science of his day, but against the human egotism that lay at the heart of Thomistic philosophy. Both rested upon the assumption that the universe had been created for man—that the sun and earth were his, that the flowers bloomed for his pleasure, that even the stars were placed in the heavens for his instruction alone. The inescapable conclusion to be drawn from Copernican cosmology is that the human species and its place of habitation occupy a modest, perhaps even minor, position in an illimitable universe of planets and stars. In his heart Copernicus must have known this, but he was unable to put such thoughts into words, at least on the printed page. Yet as the implications of Copernicus' work became increasingly clear during the decades following his death, the majority of philosophers and theologians, both Protestant and Catholic, not only found the proposition of infinity totally distasteful but an outrage to the ancient Biblical covenant between God and his highest creature, man. Perhaps late in the evenings, over a bottle of good wine, in the safety and warmth of his cathedral tower apartment, the Canon discussed the idea of an infinite universe and its theological implications with his only trusted friends, Rheticus and Tiedemann Giese. This, of course, is something we shall never know. What we do know is that at the very least the concept of infinity is implied in the Copernican system, and that from the sixteenth century onward astronomers were irresistibly drawn in its direction.

Just as the voyages of Columbus had broadened man's conception of his terrestrial home, the work of Copernicus allowed a new freedom of cosmological thought. Mathematicians in particular found the heliocentric system liberating to the spirit, and no sixteenth-century thinker took to Copernicus' cosmology with greater enthusiasm than the English gentleman Thomas Digges (1546–1595). The son of surveyor and mathematician Leonard Digges, young Thomas followed in his father's footsteps by becoming active in the movement to teach practical mathematics to the uninitiated. He also became skilled in astronomy and made a series of observations on a new star or nova which first appeared during 1572 in the familiar constellation of Cassiopeia. Already a dedicated Copernican, Digges believed the intensely brilliant star might be near enough so that as the earth changed positions an accurate recording of its parallax could be made, thus absolutely confirming terrestrial motion. But Digges, to his great disappointment, could discover no perceptible parallax and hence concluded that the object must be located in the depths of the "immutable" heavens. In the meantime, however, the young mathematician made an even more important contribution to the propagation of the heliocentric system—that of an infinite Copernican universe.

In 1576 Digges undertook a revision of a twenty-year-old work written by his father titled *A Prognostication Everlasting*, a perpetual almanac based upon the centuries-old mathematical calculations of Claudius Ptolemy. Unable to countenance the unchallenged perpetuation of the Ptolemaic system any longer, Digges appended to his father's book a short work of his own bearing the cumbersome Elizabethan title *A Perfect Description of the Caelestiall Orbes according to the most aunciene doctrine of the Pythagoreans lately revived by Copernicus and by Geometricall Demonstrations approued*. The appended material consists primarily of the translation of the essential parts of Book I of *De revolutionibus*, but with an added dimension. Digges inserted into the text his now widely printed diagram of the infinite Copernican universe. When compared to the diagram of the Copernican universe in *De revolutionibus* (see Figure 7), it is evident that Digges introduced only a single but extremely significant modification: He added a new dimension to the heavens by placing the stars above as well as below the line by which Copernicus represented the stellar sphere. In fact, in the original

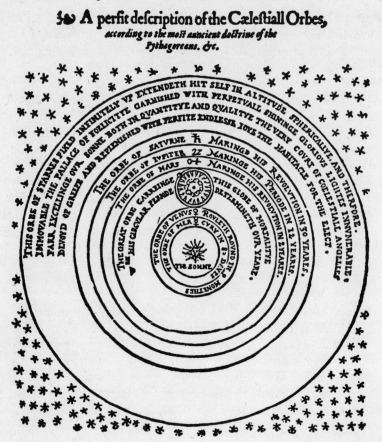

FIGURE 8. The Copernican universe of Thomas Digges, from a *Perfect Description of the Caelestiall Orbes*, 1576.

edition of Digges' work, the stars cover the page to its borders. The universe of Digges is no longer the closed world of the ancients and of the Middle Ages; within the outer sphere he has written:

> This orbe of starres fixed infinitely up extendeth hitself in altitude spherically, and therefore immouable the pallace of felicitye garnished with perpetually shininge glorus lights innumerable, farr excelling our sonne bothe in quantity and qualitye the very court of coelestiall angelles devoid of greefe and replenished with perfite endlesse love the habitacle for the elect.

The stars have burst through the vault of heaven and are scattered throughout the immensity of space postulated a generation earlier by Copernicus. They are huge to the point of incredulity and know no bounds.

And yet, as Koyré points out, the fact cannot be overlooked that the concept of infinity as presented by Digges is still dominated by theological considerations. His stars extend upward into the firmament where angels dwell and God makes His permanent abode. So while Digges has conceived of the abolition of an absolute crystal boundary between the celestial and the terrestrial realms, the physical world, while greatly extended, is somehow separated from Paradise. Neither does he speak of planets beyond Saturn, so that the infinite stars in his system are still composed of a substance or substances not found on earth, as in the Aristotelian system. It is also obvious that the contradiction between an infinite Copernican universe and the continued belief in the existence of a physically real heaven was not easily resolved in the sixteenth-century mind. Neither Copernicus nor Digges had been able to think of the sun and its planets as a solar system no different than thousands or millions of others scattered throughout the universe. In an infinite universe, no part of space can be distinguished from any other part. There is no center, no top, and no bottom. Our sun can be the center of a solar system but not the center of an infinitely extended system of physical bodies—infinity has no dimensions. The inability— or lack of willingness?—on the part of Digges and Copernicus to recognize this fact shows their still strong attachment to the religious philosophy of the times. It remained for others to deal with the paradox of Digges' infinite universe in which the Throne of God somehow managed to remain fixed far above the sun and its circling planets.

Ironically, the man who first took the stars out of a theological heaven and placed them in an infinite astronomical sky was, like Copernicus, a member of the Roman Catholic clergy. His name was Giordano Bruno, and he was born about 1548 in the small Italian town of Nola. Bruno joined the Dominican Order while still a boy and received his early education at the University of Naples. It is rather difficult to believe that anyone could have possessed intellectual and emotional qualities any less suited to the monastic life than Giordano. A highly contentious and restless freethinker, Bruno spent a turbulent eleven years as a monk, and finally, after revolting against the teachings of the Church on a broad range of subjects, he was forced to flee Italy. He took up the life of a wandering scholar and teacher, gradually making the rounds of the important European capitals. During the 1580s he visited London, where greater freedom of opinion prevailed than on the Continent. He may have become personally acquainted with Digges; if not, he almost certainly knew the mathematician's work on the Copernican universe. Bruno's writings flourished in this receptive environment, and it was during his two-year stay in England that his first two essays on cosmological problems were completed and published in 1584.

Bruno based his belief in an infinite universe on two major sources—

the Epicurean theory derived from the ancient Greek atomist Lucretius, who taught that space extends without limit in all directions, and the Copernican doctrines that the universe is immense and that the earth is in motion. But Bruno went much further in his mystical speculations than either Copernicus or Digges. Bruno's vision of the universe correctly holds that an infinite space has no central point. He taught that our sun is only one of an infinite number of stars strewn through infinite space. These stars also have their planets so that an infinite number of solar systems exist. Bruno also distinguished between the "world" and the "universe," something Aristotle and Copernicus had not done. To him the world signifies our solar system, while the universe consists of the totality of these countless worlds. According to Bruno, there is every reason to believe that life of the type that exists on earth also exists on other planets; to deny such a possibility is to limit the powers of God which is impossible, for they, like the universe of His creation, are also infinite. In fact, it is beyond finite man's ability to fully grasp what infinity truly means. In his vernacular dialogue titled *De l'infinito universo e mundi (Of the Infinite Universe and World)*, Bruno succinctly summarized his views:

> There is a single general space, a single vast immensity which we may freely call void: in it are innumerable globes like this on which we live and grow; this space we declare to be infinite, since neither reason, convenience, sense-perception nor nature assign to it a limit. For there is no reason, nor defect of nature's gifts, either of active or passive power, to hinder the existence of other worlds throughout space, which is identical in natural character with our own space, that is everywhere filled with matter or at least ether.[26]

The implications of Bruno's remarks could not be more clear. He transforms our earth and the solar system of which it is a part into insignificant specks in the eye of a boundless universe. The compact and orderly cosmos that prevailed from ancient through medieval times has become a vast unknowable chaos. The anthropomorphic God of Judaeo-Christian tradition has lost His celestial throne to the pantheistic concept in which God is identical with the various forces and workings of nature. Bruno clearly presages the modern outlook on the universe; through his writings the Copernican departure from traditional cosmology reaches its maximum. It must be noted, however, that Bruno's approach was hardly scientific. Unlike Newton, who employed mathematical laws to arrive at his concept of a mechanistic universe, Bruno forever remained a mystic, scorning those who would not soar with him into the rarefied atmosphere of speculative philosophy and intellectual fancy. Yet whatever Bruno's methods and motives, his beliefs were essentially correct, even though the three greatest astronomers of his own and the succeeding generation—Brahe, Kepler, and Galileo—rejected his concept of infinity and the plurality of worlds outside our solar system.

Bruno foolishly let himself be lured back to his native Italy in 1591, where he was imprisoned first in Venice and later by the Roman Inquisition. Numerous charges were leveled against him, the most important being his denial of the concept of the Trinity. When offered the opportunity to abjure, he steadfastly maintained that he had nothing to recant, for, as he told the authorities, "I have simply told the truth." It took the Office of the Inquisition an unusually long time to reach a final determination in the case of the apostate monk, but finally, after eight years, Bruno was judged "an impenitent and pertinacious heretic." He was burned at the stake on the Campo dei Fiori in Rome in 1600. Nowhere in the charges brought against Bruno was there any mention of his Copernican views. But once martyred, as has so often been true of unorthodox thinkers, Giordano's beliefs, including those on cosmology, became even more widely known than when he was alive. It is not without significance that the year of his death virtually marks the beginning of the Roman Catholic Church's repressive attempts to control what it considered to be unorthodox scientific teachings, namely, the concept of the infinite heliocentric universe. But this action, combined with the Copernican emphasis on mathematical proof, only stimulated astronomers to insist upon the physical validity of propositions even more strongly than before. After several generations of intensive scientific activity the cosmological views of Copernicus, Digges, and Bruno would ultimately be vindicated. The genie of modern science had finally been released from its ancient container, and the Church, despite a major and prolonged counter-effort, could never again control it.

5. Tycho Brahe: The Imperious Observer

The Rebellious Scion

JUST AS THE GREAT OCEAN-GOING NAVIGATORS of the fifteenth, sixteenth, and seventeenth centuries flew the flags of various Western European nations, the most famous astronomers of the age also represented the intellectual genius of many different countries: Newton was an Englishman, Galileo an Italian, Kepler a German, Copernicus a Pole, while Tycho Brahe, the greatest naked-eye astronomer in all history, was a native of tiny Denmark, then twice its present size.

Copernicus had been dead for some two and a half years when Tycho was born on December 14, 1546. One of his two greatest contemporaries in science, Galileo, was born eighteen years later; the other, Johannes Kepler, Tycho's brilliant associate, was about twenty-five years his junior. However, unlike Kepler, who descended from a family of both declining economic means and genetic vitality, Tycho Brahe was the son of an old, distinguished, and wealthy branch of the Danish nobility, whose members had served their king for countless generations both in government and on the field of battle.

Tycho's place of birth was the family seat located at the town of Knudstrup in Scania (Skane), the southernmost province of the Scandinavian Peninsula, which at the time still belonged to Denmark. His nobleman father, Otto Brahe, was Governor of Helsingborg Castle, located directly across the Sound from Elsinore of Hamlet fame. The post of governor was an important one because the man who held it had the responsibility of daily monitoring all shipping entering and leaving the Baltic Sea. Of Tycho's mother, Beate Bille, also of noble descent, little is known, except that she bore about a dozen children, ten of whom survived infancy; after her husband's death Beate was appointed Mistress of the Robes to Queen Sophia of Denmark, wife of King Frederick II.

Tyge (Latinized Tycho) was the second child of the Brahe marriage but the first son. As was the Scandinavian custom, the boy was named for his grandfather, a quarrelous patriarch whose short temper and love of wealth both became an important part of the legacy bequeathed to his grandson. Tycho had a still-born twin brother, the thought of whom must have preoccupied him rather frequently because he commented on the subject on several occasions in his writings. At the age of twenty-four Tycho composed a Latin poem which was published in 1572 and subsequently inscribed on his brother's tombstone. That Tycho already possessed a strong sense of his own destiny when he wrote the poem is obvious. Speaking as though it was his deceased brother who had written the Latin verse, Tycho states: "I was unborn in my mother's womb, when death became my door to life. There was another [Tycho] enclosed with me, a brother who still lives, for I was a twin. God granted him a longer life than me, so that he might see strange things on earth and in the heavens." [1] At this early stage in his career even Tycho, who already placed a high value on the accurate scientific data compiled from his celestial observations (and rightly so), could not have known just how prophetic this inscription was to become. For his detailed observations of the heavens, like the cosmology of Copernicus, comprised an important part of one of those great turning points that occur in every age, new ways of seeing and asserting the coherence of the world.

Of all the major figures in the scientific revolution in astronomy, none was more spontaneous or unpredictable in his actions than Tycho Brahe. Life for him was an adventure, and his story is made even more fascinating as the result of an unusual set of historical circumstances which helped to shape his unique personality and mercurial temperament. The first major incident occurred not long after Tycho's birth, although certain of its aspects are still debated among his biographers. Tycho's father, Otto, had a childless brother named Jörgen who, although married, had no hope of producing a family of his own. According to one account, Otto and Beate, out of compassion for the lonely brother, had promised Jörgen Brahe that if a son were born to them, the child would be placed in his care to be raised as his own. However, after Tycho's birth the proud parents repented their ill-considered decision, much to uncle Jörgen's dismay who, not to be thwarted, kidnapped Tycho when he was about a year old. [2] Another account maintains that Jörgen Brahe coveted Tycho from birth and, unable to endure his childless condition any longer, simply stole the infant from his crib when both parents were absent from the family castle. [3] In any case, there is little doubt that the child was taken illegally, an act which caused the Governor of Helsingborg to issue a number of extremely serious threats, including murder, against his brother, who remained intransigent behind the locked gates of his own castle in nearby Tostrup.

Fortunately, after the parents' initial shock was overcome, tempers cooled sufficiently so that the family crisis was eventually resolved to the satisfaction of all concerned. Nature luckily intervened at just the right moment, for shortly after Tycho's abduction Beate Brahe gave birth to another son, Steen, the second of five Brahe brothers. Steen's arrival cast an entirely different light on the situation, and the parents chose not to dispute Tycho's kidnapping any longer. After all, the child was in loving hands, and he would doubtless inherit his uncle's considerable fortune upon the latter's death. Moreover, Tycho could visit his parents frequently, while giving comfort to Otto's lonely brother. The antagonism that had resulted from the abduction was so successfully resolved that some years later Otto and Beate chose to name their eighth child, a boy, Jörgen, after the offending uncle.

What young Tycho may have thought of his kidnapping and how he reacted to it, we do not know. Because of his tender age, it must have been several years before he completely understood the nature of his uncle's action, and by then he was almost certainly accustomed to looking upon Jörgen no differently than had he been his natural father. Perhaps in retrospect Tycho simply viewed the incident as another manifestation of the many prerogatives of noble life; his uncle Jörgen, by virtue of his high social position, simply took that which had been denied him by more conventional means. If so, it proved to be a lesson in the use of privilege that Tycho never forgot. Besides, as Jörgen's only child, Tycho received far more attention than he could have expected in a home crowded with nine brothers and sisters. This special treatment probably spoiled the child from the beginning, which in turn nurtured the development of an obstinacy and stubbornness of character which we shall encounter time and again. And because of an unusual accident which indirectly resulted in the premature death of his ill-starred uncle, the purloined son became sole heir to a considerable estate much sooner than he or anyone else had expected.

Little more is known of the great astronomer's life between the ages of one and twelve than of the childhood of Copernicus. Until 1559 Tycho appears to have been educated by private tutors at his uncle's castle, learning, reading, writing, and Latin, with some instruction in the *belles lettres* and poetry. While Tycho considered himself an astronomer first, last, and always, like many other well-educated men of the Renaissance he developed competency in more than one field. As a result of his youthful introduction to classical poetry, he composed Latin verse by the ream, some worse and some considerably better than that written by his educated contemporaries. In later life, he had the advantage of owning his own printing press, a circumstance to be envied by any modern writer. Whatever Tycho wrote, he printed, being thus spared the trauma of first securing the blessing of a publisher and then enduring the pain wrought by the merciless pruning of his manuscripts by zealous editors.

Like his brother Otto, the Governor of Helsingborg Castle, Jörgen Brahe was not only a wealthy country squire, he also held an important position in the service of his king—that of vice-admiral in the Danish navy. It was taken for granted that Tycho, in true Brahe fashion, would also one day hold a responsible position in public affairs, and with this end in mind Jörgen sent his foster-son to the University of Copenhagen for further study in April 1559. The institution had been founded in the 1440s with the Pope's consent by King Christian I, but it languished for lack of proper funding until the Reformation swept over Denmark during the first half of the sixteenth century. The value of the University as a center for Protestant education then became obvious, and a revitalization effort was undertaken, financed by the State's confiscation of properties formerly owned by the Catholic monasteries. At the time of Tycho's matriculation at the age of twelve, the University employed some fifteen professors divided into the four faculties of theology, medicine, philosophy, and law—the latter field being the one in which Tycho was to prepare for a career in government service.

Although a majority of its students entered the University at about the age of eighteen, boys as young as eight years old were admitted. Even though Tycho was obviously on the young side, his prior preparation by private tutors stood him in good stead with his fellow students. In fact, he was probably much better grounded in academics than most of his peers. As had been the case in Europe for several centuries, students at Copenhagen were instructed in the seven liberal arts: the linguistic *trivium* comprising grammar, rhetoric, and dialectic succeeded by the mathematical *quadrivium* of arithmetic, astronomy, geometry, and music. The classical languages were strongly emphasized, and Tycho owed his mastery of Latin and Greek to the excellent instruction he received in these subjects at the University between 1559 and 1562.

It would be pleasing to one like myself, who has chosen the field of education for his lifework, to be able to trace Tycho Brahe's early interest in the science of astronomy back to the lectures given on the subject by one of his professors while he was enrolled at the University of Copenhagen. Alas, such is not the case, although some credit must go to Dr. Johannes Franciscus, who lectured on Euclid's *Geometry* in addition to astronomy and the planetary theory of Ptolemy, for reinforcing Tycho's all-consuming passion for astronomical observation once it had been aroused. Unlike Copernicus, who, true to his character, appears to have moved into the field of astronomy through a series of methodical steps as a young adult, Tycho underwent a virtually instantaneous conversion at the still tender age of 13. The less-than-spectacular incident responsible for triggering this sudden turn of events took place on August 21, 1560, during Tycho's second year at the University. Prior to this date the announcement had been made that a total eclipse of the sun, to be only partially visible from the northern latitudes of Copenhagen, would occur within a few weeks. Tycho's curiosity was aroused, and he anxiously awaited confirmation of what to him was an extremely bold prediction. Upon witnessing the eclipse at the exact time forecast, his seventeenth-century biographer, Pierre Gassendi, tells us that to Tycho it "seemed something divine that men could know the motions of the stars so accurately that they could long before foretell their places and relative positions." Tycho's previous introduction to astronomy through Franciscus' lectures took on an entirely new and profounder meaning. He began to purchase treatises on astronomy, including the *Ephemerides* of Johann Stadius and the only major printed book on the subject, Ptolemy's *Almagest*. For the latter he expended the considerable sum of two Joachims-Thaler, even though it is doubtful whether at this early stage in his educational development Tycho could master very much of its complex contents. Still, he was well aware of the volume's historic importance, and just to be able to thumb through its pages, brimming with a wealth of ancient astronomical knowledge, was inspiration enough in itself. He knew that one day he would fully understand Ptolemy's work and perhaps even publish a book on astronomy of his own. Thus the die was irrevocably cast—in his mind's eye Tycho Brahe was already envisioning himself as an astronomer of the first order.

Jörgen Brahe kept a close watch on his nephew's progress at the University and from all indications took an extremely dim view of Tycho's continued interest in astronomy. In 1562 the decision was made to send Tycho abroad for further education in Germany, where it was hoped his interest in the heavens would wane as he studied the more mundane subject of Law, considered far more worthy of a nobleman's son. After weighing several alternatives, Jörgen's advisers finally recommended the University of Leipzig; and to make certain Tycho's attention did not wander from his legal studies his uncle selected a young man of great brilliance and promise in his own right, Anders Sörensen Vedel, to accompany Tycho as his tutor. Vedel's scholarly interests lay in the field of history, and in later years he held the distinguished post of Royal Historiographer, during which time he gained widespread acclaim in scholarly circles for his translation of Saxo Grammaticus as well as for the collection and preservation of numerous Nordic sagas. Although he did not know it at the time, the choice of Vedel as his companion proved crucial to the future development of Tycho's chosen career. Since Anders was only four years Tycho's senior, their relationship became less that of master and student than one of mutual respect and abiding friendship. They corresponded regularly throughout their lives, and Vedel was an ever welcome guest at Tycho's famous observatory on the island of Hveen.

After they had settled in Leipzig, Vedel made every attempt, at least for a time, to carry out Jörgen Brahe's instructions to the letter by closely following Tycho's progress in the study of the law. No ordinary adolescent, Tycho employed every means within his grasp to circumvent Vedel's careful supervision, not to mention his uncle's strict orders. For a number of months the young nobleman lived a double life: by day he attended the endlessly boring round of lectures on jurisprudence, while by night, as Vedel slept, Tycho's sharp eyes were turned toward the heavens as he studied the positions of the planets and stars. Every bit of his modest allowance went for books on astronomy and small pieces of equipment that could be easily hidden from his suspicious tutor. One of his most prized possessions was a small globe, about the size of an orange, upon which he plotted the major constellations, using as his guide the *Ephemerides* of Stadius supplemented by his own secret observations.

Tycho's game of hide and seek with Anders Vedel lasted for more than a year. As the months passed, however, Vedel became increasingly aware of his charge's passion for astronomy, and he finally succeeded in bringing Tycho's nocturnal activities to light. Tycho must have awaited Anders' judgment with considerable misgivings; but he need not have worried. Young as Vedel was, he had the good sense to recognize Tycho's rare combination of ambition and natural talent and to give in to it without debate, thus cementing their lifelong friendship. From this point on, Vedel went out of his way to introduce Tycho to members of Leipzig's scholarly community who could help increase his knowledge of mathematics and astronomy. Vedel also sanctioned Tycho's purchase of larger and more accurate observational instruments, in addition to numerous books and pamphlets on the stars.

Earlier mention has been made of the eclipse of 1560, which so fascinated Tycho that he became an instant and irrevocable convert to astronomy. While he pursued his studies at Leipzig, a second celestial event took place which determined the exact course his study of the heavens would take. Between August 17 and 24, 1563, Tycho observed the close conjunction of Saturn and Jupiter, as had been predicted in the Alphonsine Tables (based largely on Ptolemy's observations) published in 1252, and the Prutenic or Prussian Tables, compiled by Erasmus Reinhold from Copernicus' calculations, published in 1551. To Tycho's amazement, the time of conjunction as predicted in the Alphonsine Tables was in error by an entire month, while the Prutenic Tables proved to be inaccurate by a matter of several days. How could it be, Tycho must have asked himself, that he, a young, self-trained astronomer, possessed of very simple instruments, could predict certain celestial events even more accurately than the greatest astronomers of the past? The answer, of course, is that with the exception of the ancient Greek Hipparchus, who flourished during the second century B.C., these men did not think of themselves as astronomers in the more modern context in which Tycho had come to understand the word. We must keep in mind that up to this time Tycho had received little formal instruction in the field; as a result he simply had not learned that the accurate observation of planetary and stellar positions was of secondary importance to his predecessors, whose interests were largely speculative and philosophical rather than empirical. For the true observational astronomer the heavens remained completely virgin territory. Having come to this startling realization, Tycho became determined to devote his career to the gathering of precise and continuous observational data about his new and enchanting mistress, the heavens. It is for his unrivaled achievements in this area that the man is primarily remembered and honored today. While it is true that Tycho subsequently developed his own model of the heavens—a rather ingenious compromise between the systems proposed by Ptolemy and Copernicus—it was later rejected by his own brilliant associate Johannes Kepler largely because, as we shall see, Tycho's scientific data proved so accurate that it conflicted with the very system he himself created. To put the matter simply, Tycho was not a great metaphysician like his gifted predecessor Nicolas Copernicus. The compilation of extremely accurate celestial observations proved his forte, and he did this so well that Kepler, both richer in imagination and intellectually more audacious, used his observations to revolutionize astronomy by developing its first three mathematical laws without which Newton's grand synthesis would have been impossible. In a very real sense, the work of Tycho Brahe served as the bridge over which the transition from ancient to modern science proceeded.

In May of 1565 Tycho received a letter from his uncle instructing him to return home at once. As Vice-Admiral, Jörgen Brahe had been chosen by the king to assume command of a squadron of Danish ships which was soon to put out to sea. Denmark was at war with Sweden, and Jörgen probably felt that Tycho's proper place was at home overseeing Tostrup Castle and caring for his aunt Inger during the Vice-Admiral's absence. Tycho must have received the news with deep concern, for he realized how extremely difficult it would be for him to pursue his studies of the sky in the home of his uncle. Jörgen would not be away forever, and even during his absence the servants would undoubtedly gather information about Tycho's strange activities. However, fate seems to have had a way of fortuitously intervening in the lives of the great astronomers at precisely the right time, and this was just such a moment. Copernicus, it will be recalled, had he not lost his father as a boy, would quite probably have been forced into the family occupation of wholesale merchant. Now Jörgen Brahe—the main obstacle between Tycho and his passion for astronomy—also met with a tragic and untimely end, but one which conveniently paved the way for his nephew's chosen career. Shortly after Tycho's return home, King Frederick II, while in the company of Jörgen Brahe and a number of other nobles, was crossing a bridge when his horse startled and plunged with its famous rider into the frigid waters below. Uncle Jörgen, with typical Brahe daring, jumped in, rescued the king, and luckily escaped with his own life. But shortly thereafter he caught a severe cold which developed into pneumonia from which he died on June 21, 1565. Tycho, now nineteen years old, was heartsick, for he had developed a deep love for his uncle despite their strong differences of opinion over Tycho's future. On the other hand, Tycho now became his own man supported by a substantial inheritance that made his independence complete.

At the urging of his maternal uncle Steen Bille, Tycho remained in Denmark for another year. But there was still so much to learn about astronomy, and several of Europe's great centers of scientific scholarship beckoned. He finally decided to leave during the summer of 1566. "Neither my country nor my friends keep me back," he wrote revealingly to a close acquaintance. "There will always be time enough to return to the cold North to follow the general example, and, like the rest, to play in pride and luxury for the rest of one's years with wine, dogs, and horses. May God, as I trust he will, grant me a better lot." [4] As the passage from Tycho's letter so clearly indicates, the young scion had already formed a very negative opinion of the preoccupations of his fellow members of the Danish nobility. We would be in error, however, to conclude that Tycho's attitude sprang from a deep-seated sense of social justice; his feelings were based on grounds other than those of opposition to great privilege and monopolistic wealth. In fact, Tycho so exploited his own privileged position toward the end of his life that he went into self-imposed exile when his new King, Christian IV, refused to indulge his huge appetites in the manner to which he had become accustomed under Christian's father Frederick. Rather, what Tycho disapproved of, as his letter strongly suggests, was the nobility's idleness and empty-headed pursuit of activities having little redeeming social or intellectual value. To the majority of Danish noblemen, and to their counterparts elsewhere in Europe, hunting, gaming, and drinking were ends in themselves, while to Tycho they were mere frivolities to be indulged in only for occasional amusement and relaxation. As if to drive the point home, he framed the motto which was specially printed in his books around the family coat of arms: "Arms, descent, estates perish; a noble mind and learning alone achieve lasting rank." We see in him an attitude similar to that displayed by certain major thinkers of the Enlightenment such as the Baron de Montesquieu, who, favored by high birth and great wealth, wanted to use these advantages for something more worthwhile to society than the conspicuous consumption of luxury goods and endless rounds of fetes and masqued balls. For their part, many of Tycho's peers mocked his pursuit of the stars as a worthless and undignified undertaking for a man of his position. One who indulged in such nonsense lowered himself by assuming the role of a craftsman or, even worse, a common laborer. Philosophical inquiry on the part of a nobleman, although not particularly esteemed, was to be tolerated; but the construction of large instruments for the purpose of recording "meaningless" data while in the constant company of common assistants went far beyond the bounds of accepted practice. Luckily for Tycho and the future of astronomy, he was destined to have an island all to himself; had this not been the case, it is doubtful that he would have remained in Denmark throughout the most fruitful period of his remarkable professional life.

Nova

Tycho took his leave from Denmark for the second time in 1566. For the next six years he continued his astronomical studies at the Universities of Wittenberg, Rostock, Augsburg, and Basle, the latter being his favorite center of learning in Europe. Wittenberg, the first stop on Tycho's leisurely itinerary, brought him into contact with the noted Professor of Medicine Casper Peucer, the son-in-law of Philipp Melanchthon, who, like Tycho, also nurtured an avid interest in both astronomy and mathematics. But after only five months at the great center of Protestant learning, Tycho reluctantly severed his relationship with Peucer because of an outbreak of the plague which forced the young nobleman to seek healthier surroundings further to the north. He next visited Rostock whose university, located near the Baltic, was highly popular among Scandinavian students and those from the Low Countries. It was there that the best-known incident of his private life took place.

In December 1566 Tycho attended a betrothal dance held at the home of a Professor Bachmeister in honor of his daughter and her fiancé. Manderup Parsbjerg, a Danish countryman, who later rose to the high position of chancellor in his king's service, was also in attendance. Although the details are somewhat sketchy, the two young noblemen, emboldened no doubt by liberal quantities of wine and beer, tangled over the question as to which of them was the better mathematician. They were quickly separated by friends, and it is doubtful that anything further would have come of the matter had the two not met under similar circumstances at a Christmas party given a week later. There the quarrel was renewed and this time both claimed satisfaction with the sword. They met near a graveyard under the most absurd of physical conditions; it was so dark neither man could see the hand in front of his face, let alone the weapon in that of his opponent. It would probably have been better, at least from Tycho's point of view, had the two waited until dawn to settle their differences, but apparently the mere suggestion of a more reasoned course of action smacked of cowardice, and neither wanted to be accused of that. The duel, if indeed it can be dignified by giving it the name, lasted only a few moments, but it left a permanent mark. Parsbjerg's sword sliced off the bridge of Tycho's nose. In an attempt to repair the damage Tycho, true to his character, scorned a common substitute made of wax. Instead, he employed a craftsman to fashion an elegant replacement of gold and silver alloy, which he wore for the rest of his life. The prosthesis was painted the color of flesh and seems to have been difficult to tell from the real thing. Gassendi maintains that Tycho always carried a small container in his pocket filled with some ointment or glutinous composition which he frequently rubbed on his nose in order to keep the false bridge in place. It is interesting to contemplate what the effect on the history of astronomy might have been had Parsbjerg's thrust struck Tycho across an eye rather than across his nose. With only one good eye Tycho's effectiveness as an astronomer would have been considerably diminished. Perhaps he too realized this and accepted his minor disfigurement more graciously as a result. The two men later met at court from time to time, and Tycho appears to have harbored no permanent grudge against his once youthful enemy. Ironically, had Tycho not been so impatient to defend his honor time certainly would have proven what his sword could not—that he was, after all, the better mathematician of the two.

Ever since the days of his youth Tycho had shared the widespread belief of his time in astrology, which scholars still considered inseparable from astronomical studies. In fact, of the major figures in the scientific revolution in astronomy only Copernicus maintained silence on the matter. Yet even he must have been influenced to some degree by its teachings. The Canon studied at the University of Bologna where the first professorship of astrology in Europe was established during the late Middle Ages. And, as was pointed out in chapter three, one of the most popular sayings among members of Bologna's medical faculty ran: "A doctor without astrology is like an eye that cannot see." This belief was based on the widely taught doctrine that each part of the human anatomy is controlled by a specific sign of the zodiac so that the physician, if he wishes to be successful in the practice of his art, must have specific knowledge of the planetary movements at all times. Furthermore,

Rheticus made use of astrology to gain support for the Copernican Revolution. The *Narratio prima*, his introduction to the Copernican system, contains a digression on astrology, and since the work was written under the careful scrutiny of the master it is only reasonable to conclude that, at the very least, Copernicus had no major quarrel with its practice. Everywhere he studied, Tycho, like Copernicus, encountered men dedicated to the belief that the principles which govern the movements of the heavenly bodies also dictate the conditions of human life. As a boy, astrology had a natural appeal to his highly developed sense of imagination, and in 1566, during his stay at Wittenberg, he cast his first horoscope in the form of a poem predicting the death of Sultan Suleiman during an upcoming eclipse of the moon. To his embarrassment, Tycho later learned that the Sultan had expired even before the poem was written; but he was able to laugh over the incident years afterward.

While Tycho never completely abandoned the practice of judicial astrology, he grew increasingly skeptical of its more popular aspects. With the development of the printing press, astrological calendars were being sold by the thousands throughout Europe during the sixteenth century, while charlatans, in quest of easy money, moved into the field by the score. One of Tycho's more troublesome responsibilities entailed the casting of horoscopes of the King of Denmark, his children, and certain members of the nobility. Like Kepler and Galileo, who also practiced astrology at the request of wealthy patrons, Tycho increasingly did so with tongue in cheek. On at least one occasion he felt compelled to add the following clause to a recently completed horoscope of one of the king's sons: "The free will given by God to man is stronger than the heavenly influences." Yet like Kepler, he remained a lifelong adherent of astrology in the ideal sense, despite his sharp criticism of its popular misuse. Tycho believed in the existence of a basic affinity between celestial movement and terrestrial events. He explained this view in a public lecture delivered in Latin at the University of Copenhagen in 1574:

> To deny the forces and influence of the stars is to undervalue firstly the divine wisdom of providence and moreover to contradict evident experience. For what could be thought more unjust and foolish about God than that He should have made this large and admirable scenery of the skies and so many brilliant stars to no use or purpose—whereas no man makes even his least work without a certain aim . . . we, on the contrary, hold that the sky operates not only on the atmosphere but also directly upon man himself. Because man consists of the elements and is made out of earth, it is necessary that he be subjected to the same conditions as the matter of which he consists. Since, furthermore, the air which we inhale and by which we are fed no less than by food and drink, is affected in a different way by the influence of the sky, it is unavoidable that we should at the same time be affected by it in different ways.[5]

Tycho's views on astrology, regardless of their ultimate scientific merits, reveal his deeply rooted cosmological belief in the coherent interrelationship of all that exists. Einstein expressed this same insight in a somewhat different manner in his famous statement that "God does not play dice with the world." This belief, more than any other, has served as the basic source of inspiration for the greatest scientists in all fields the world over.

The damage Tycho sustained to his anatomy during his visit to Rostock was at least partially compensated for by his increasing skill as an astronomer. While in the city he viewed a major eclipse of the sun whose movements, along with those of the moon, became lifelong subjects of interest to him. During the summer of 1567 the young astronomer returned to Denmark for a brief visit where he found that in spite of the low esteem in which the nobility held his profession, there were those who did have his personal interests very much at heart after all. His uncle Steen Bille and aunt Inger, Jörgen Brahe's widow, had extracted a written promise from King Frederick that the first canonry to become vacant at the local cathedral of Roskilde would go to their nephew Tycho. Unlike Copernicus and the other canons of Frauenburg, who lived near their church and functioned as a corporate body, all those appointed by the Danish Crown were laymen whose appointments were recognized by everyone as comfortable sinecures and little more. About the only duties associated with the office involved keeping the church in a reasonable state of repair and dispensing limited amounts of charity to the local youth. The vacancy promised to Tycho occurred in 1568, and he was appointed to the office without opposition. Unfortunately, however, even the minor responsibilities associated with Tycho's lucrative appointment were left unattended by him, and this inexcusable neglect eventually became a source of bitter contention between the headstrong astronomer and his sovereign, Christian IV. Yet Tycho was destined to receive a number of other major grants from the crown, including a 2000-acre island, before this sad episode marred both his reputation and noble standing in Denmark.

Tycho's love of German universities drew him southward again in the summer of 1568. He spent some time in Wittenberg and then proceeded on to Basle, where he was matriculated at the University later in the same year. By early 1569 he was on the move once more, this time in the direction of Augsburg, the reputation of whose skilled instrument makers seems to have captured Tycho's attention. The portable devices he used during the early stages of his career had served their owner well, but as an accomplished astronomer Tycho needed larger and more accurate equipment, which, as canon and heir, he was now well able to afford.

Shortly after his arrival in Augsburg, Tycho became friends with two distinguished brothers, Johann Baptiste and Paul Hainzel, the former burgomaster, the latter an alderman of the city. They knew in advance of Tycho's already growing reputation as an astronomer and solicited his help on a number of interesting projects. One of their most ambitious undertakings involved the construction of a gigantic quadrant for the purpose of measuring the altitudes of celestial bodies. The ultimate product of their planning was an instrument approximately 19 feet high fashioned from oak with metal bracings and dial. Far more accurate than any of his other devices, when completed the instrument enabled Tycho and the Hainzels to measure every minute of arc and to estimate the intervening seconds with considerable accuracy; the first time, so far as is known, this had ever been done. It required the uninterrupted labor of 20 craftsmen a month to complete the project, after which the quadrant was assembled atop a hill on the Hainzel's country estate near Augsburg. While it appears that Tycho himself used the quadrant very little, the experience of helping to plan and supervise the building of the great instrument proved invaluable when he later occuped his own observatory, for which he personally designed every one of the dozens of unmatched observational devices.*

A second and even more audacious project undertaken during Tycho's two years in Augsburg involved the design and construction of a huge globe, five feet in diameter, on which the astronomer planned to accurately fix the positions of all the stars as determined by his future observations. The globe's substructure, composed of wooden plating attached to metal rings, was to be covered with thin sheets of guilt brass suitable for engraving. Years passed before the great sphere finally arrived at Hveen, and only then because of Paul Hainzel's unflagging determination to make good his promise to Tycho that he would superintend its completion. Tycho also designed and commissioned the crafting of several other superb instruments during his stay at Augsburg, while at the same time keeping up an active schedule of nightly observations. By the time he left the city, after receiving word of his father's grave illness in 1570, Tycho Brahe had developed the necessary tools, both mechanical and intellectual, to write an entirely new and very crucial chapter in the history of modern science.

Tycho arrived home in December 1570 only to learn that his father's condition was terminal. Otto Brahe died on May 9, 1571 at the age of fifty-three, whereupon Tycho and his oldest brother Steen jointly inherited the family property at Knudstrup. Tycho's uncle Steen Bille, ever mindful of his nephew's talent and sense of alienation from his social peers, invited the lonely young man to live with him at nearby Heridsvad

---

* Tycho later learned, much to his dismay, that the Augsburg quadrant was irreparably damaged during a severe thunderstorm some five years after its completion.

Abbey, a former Benedictine monastery confiscated by the crown during the aftermath of the Reformation. Bille appears to have been the only close relative who appreciated Tycho's interest in the science of his day, largely because Steen himself devoted considerable time to scientific and technological pursuits. He established Denmark's first paper-mill and glass-works and arranged to have a laboratory installed at Heridsvad so that he and his nephew might conduct a broad range of experiments. Perhaps Steen cherished the hope that, through their combined knowledge of alchemy and astrology, he and Tycho might bring to a successful conclusion the alchemist's unending quest for the secret of transmuting lead into gold. Whatever his underlying motives, Steen Bille was most successful at turning Tycho's attention away from his personal problems. His nephew became so deeply absorbed in experimentation that for a time Tycho all but abandoned his pursuit of the stars.

The major turning points in Tycho Brahe's life seem always related to the appearance of some unusual celestial phenomenon: the eclipse of the sun in 1560, for example, which first awakened his interest in astronomy, and the conjunction of Saturn and Jupiter in 1563, which determined the nature of his life's work. The third and most important of these sporadic events occurred on the evening of November 11, 1572 during Tycho's residence at Heridsvad. Tycho had worked late in the laboratory and was on his way to the house for supper when, glancing skyward, he was astounded to see a new star in the constellation Cassiopeia in a place where none had previously appeared. Literally unable to believe his own eyes, he hurriedly called out some domestic servants who confirmed the observation. Still not completely convinced of the existence of what he saw, Tycho approached some peasants passing by the estate, and they too assured him of the star's presence. What Tycho observed, of course, was a nova, or more correctly a supernova, an exploding star that increases in luminosity up to several hundred thousand times in a matter of a few days, burns brightly for several months, and then gradually disappears as it burns itself out. The new star remained visible for some eighteen months, and during its period of greatest brilliance people with good eyesight could even glimpse it during the middle of the day. The attention of scholars from all over the world was quite naturally focused on Cassiopeia's unannounced and, from the Aristotelian point of view, illegitimate offspring. Soon the common people were also agog over the nova, though mainly because of its supposed astrological significance. Tycho, like many other astronomers, also took an astrological interest in the star, but his primary concern centered on the relationship between the unexpected appearance of the yet to be explained celestial phenomenon and the basic Aristotelian doctrine of the immutability of the heavens.

While he deeply respected the genius of Copernicus, Tycho, unlike his contemporary Thomas Digges, rejected both the possibility of an infinite universe and the belief that generation and decay might take place in the celestial region beyond the immediate vicinity of the earth and the moon. And yet the appearance of a new star in the midst of the unchanging heavens—if indeed it was a star—would clearly challenge the Christian idea borrowed from Aristotle that the heavens will remain unchanged to eternity. To his credit, Tycho approached the problem with the detachment of a true scientist. As luck would have it, he had just finished making a new sextant which would enable him to accurately measure the distance of the strange object from the major stars of its constellation. Through a series of ongoing observations one could then determine whether the object in the sky was fixed, as all stars appear to be to the naked-eye observer, or whether it displayed parallax and might therefore be some unusual form of comet or meteor moving slowly but perceptibly through the sublunar regions.

From 1572–73 Tycho carefully measured the distance of the nova from the nine brightest stars of Cassiopeia on dozens of separate occasions, which included the taking of multiple observations at different intervals on the same night. He even took the precaution of leaving his sextant clamped in place between sightings taken on the same evening to insure against human error in realigning the instrument. Other astronomers were doing the same thing but with far less sophisticated instruments. Michael Maestlin, one of the leading astronomers of his time and Kepler's teacher at Tübingen University, simply held a string at arm's length from his eyes so that it intersected a fixed star at either end and passed across the new star in the middle. After a number of sightings, each lasting for several hours, Maestlin concluded that the new star did not move after all. Apparently Digges employed a similar method in England and reached the same conclusion as Maestlin had in Germany. However, there were others who believed just the opposite. The most common explanation given by the die-hard Aristotelians was that the star was in fact a tailless comet moving at such a slow rate of speed that it appeared to be all but motionless to the terrestrial observer. Doubtless, several of those who made this claim were sincere in their beliefs, for if one watches a given star long enough it does indeed seem to move. At any rate, it was the best the Aristotelians could do under the circumstances.

This major conflict of opinion could be resolved only if someone proved capable of providing overwhelming evidence in support of one side or the other. At least fifty different publications resulted in which numerous claims were made, ranging all the way from the ridiculous assertion that all Europe was suffering from a massive case of optical illusion, to the sublime forecast of Christ's second coming. None, however, carried sufficient weight to win over its opponents, particularly those precious few who, touched for the first time by the modern scientific spirit, were willing to rethink their positions in light of new and convincing evidence. Tycho discussed the nature of his celestial observations with a few close friends and in his correspondence; by 1573 he had definitely concluded, along with Maestlin and Digges, that the object in the sky was in fact a new star, and that it did indeed stand still in the midst of the supposedly immutable heavens. But despite considerable urging to publish these findings, Tycho displayed a degree of reticence quite out of character for one who had so openly attacked the social conventions of his times while also craving recognition as a great astronomer. Yet we must keep in mind that this was a young Tycho Brahe, aged twenty-six, still in the early stages of his brilliant career, a man not yet surrounded by a score or more of submissive assistants or secure in the splendid isolation of his beloved island retreat on Hveen. The writing of books, especially a work on astronomy, was still considered an undignified occupation for a nobleman; and perhaps Tycho felt that he had sufficiently fanned the fires of class antagonism at this point. However, in the end, like Copernicus, who published *De revolutionibus* only because of the persistence of his friends, Tycho brought forth his *De Nova Stella* (*On the New Star*) in 1573, a little octavo of fifty-two pages, about half of which were devoted to "hard, obstinate facts" about the new star.

Brief though this volume is, it established Tycho's claim to fame in his own lifetime and would have won him a lasting place in the history of astronomy even had he never made another observation. For the first time the scholarly world was introduced to the work of an astronomer whose careful attention to detail, supported by rigorous scientific documentation, could leave no doubt as to the behavior of the phenomena under examination. Tycho was most careful in that he separated his theoretical belief as to the origins of the nova from the incontrovertible empirical data he had gathered over many months of rigorous observation. For those possessed of an open mind, he clearly established the nova of 1572 as a motionless star located in what was previously believed to be the inviolable celestial realm.* On the other hand, Tycho speculated, but could not prove, that the new star was composed of vapors that had congealed from the Milky Way. He even thought he saw a new hole in the galaxy resulting from the removal of the stellar matter of which he believed the nova was composed. What he could not have known is that the Milky Way is much more than a luminous band of clouds visible

* Tycho's careful analysis of the exact position of the star enabled astronomers of the twentieth century to locate the remnants of the now invisible supernova. In 1952 the radio telescope at Jodrell Bank in England discovered that Tycho's star, as it is now commonly called, is still a strong source of radio emission. See Sir Bernard Lovell, "Tycho Brahe," *History Today*, Vol. XIII (Oct., 1963), p. 678.

in the night sky. The discovery that it is composed of countless thousands of individual stars would have to wait for an ingenious Italian, who turned his telescope away from the familiar terrain of the terrestrial realm to focus it upon the uncharted reaches of the silent heavens.

In 1577, five years after the appearance of the nova, Tycho's supreme confidence in the value of practical astronomy was publicly vindicated a second time, much to the growing discomfort of the Aristotelians. In November, a great comet appeared in the sky and remained visible for more than two months. These little studied phenomena had traditionally been explained away as cloudy exhalations from the earth's atmosphere moving slowly through the upper regions of the air, but well within the orbit of the moon. Tycho subjected the comet to intensive measurements with both a sextant and radius, carefully computing the distance of its head from various fixed stars. His ultimate objective was to find the comet's parallax which could then be compared with that of the moon: if it was inside the lunar orbit, as the Aristotelians maintained, the degree of motion would be somewhat greater than the moon's which Tycho placed at about four degrees. He soon discovered, however, that the comet had no perceptible parallax whatsoever, a fact which convinced that this was no sublunar body. According to his calculations the comet must have been at least six times farther away from the earth than the moon. In other words, Tycho assigned it to the region of the fixed stars. He thus struck another severe blow to the Aristotelian doctrine of the solid, concentric, crystalline spheres to which the planets were still thought by many to be attached. By moving across the plane of the planetary orbits the comet would have had to pass through the crystal spheres and in so doing violate Aristotle's concept of celestial immutability. In and of itself, Tycho's discovery was not absolutely conclusive; certainly it did not administer the *coup de grâce* to Aristotelian cosmology, as has been suggested by one widely read historian of science.[6] One could still remain in the Stagirite's camp without thinking of the planetary orbits as solid spheres of crystal. On the other hand, Tycho's observations and conclusions clearly deepened the wrinkles in the Aristotelian brow by casting additional doubt on another major feature of the old cosmology. By placing the comets in the superlunary region Tycho helped keep open the question raised by Copernicus of the mechanism by which the solar system operates. Although he strongly disagreed with certain major aspects of the heliocentric doctrine, neither could Tycho, in deference to his own scientific findings, support the Aristotelian–Ptolemaic model. Finally, as we shall see, he designed his own celestial machinery built with major components salvaged from the two rival systems, whose leftover parts Tycho consigned to the scrap heap of erroneous conjecture.

Lord of Hveen

Although Steen Bille must have cherished the hope that his gifted nephew would stay on at Heridsvad indefinitely, after the publication of his book Tycho began seriously to consider returning to Germany. However, a number of obstacles postponed his departure, not the least of which was a severe attack of the ague which confined him to bed during the better part of the summer of 1573. At about the same time the young astronomer contracted a second malady—this one of the heart—for he had developed a fondness for a local girl, named Christine. Virtually the only thing known for certain about her is that she was the daughter of a commoner; her father was either a peasant farmer on the Brahe estate at Knudstrup, or else she came from around Heridsvad Abbey where she was employed as a domestic servant by Tycho's uncle, Steen. Perhaps Christine had the responsibility for helping to nurse Tycho back to health during his period of convalescence; if so, she doubtless had ample opportunity to observe the more irascible aspects of his volatile nature. Despite the unwritten but obvious social prohibition against a marriage between a nobleman and commoner, Tycho characteristically brushed this consideration aside and took Christine as his wife. This action enraged friends and relatives alike; at one stroke he had succeeded in destroying much of the good will that resulted from the scholarly acclaim won by the publication of his famous book. Only his sister Sophie and uncle Steen seem to have taken the marriage in stride.

According to ancient Danish law, a formal church ceremony was not considered necessary to legalize a marriage. A woman who publicly lived with a man, kept his keys, and ate at his table for three years was considered his lawful wife. Although the custom was beginning to undergo change during the 1560s, there was still nothing illegal or immoral in foregoing the sanction of a clergyman, which is apparently what Tycho, with his ingrained distaste for ceremony, decided to do. He "sinned" only in the sense that he married down—or what the French nobility of the seventeenth century called derogation—and for this he was never forgiven. For the rest of her life Christine Brahe was commonly referred to as Tycho's concubine (or worse) by her husband's detractors, while their eight children were labeled as bastards. Had Christine been of gentle birth, such epithets would never have been thought of, let alone uttered in public.

Again, Tycho's defiance of the conventions of his day seem hardly directed toward the cause of social justice. The most common explanation for his heterodox behavior is that he felt no woman of noble birth would marry a man without a proper nose of his own, no matter how famous or high born he might be. While this story may have an element of truth to it, it seems a far too simplistic explanation for a number of reasons. The most important is that in an age when the techniques of modern medicine were still very much in their infancy, permanent physical disability brought on by disease or injury was far more commonplace than today. Many a wedded nobleman would have gladly exchanged the bulk of his wealth for a healthy set of teeth, a pair of strong eyes, or a countenance untouched by the terrible ravages of the pox. In appearance, Tycho was the equal of the majority of his peers being of medium height and stout build, broad-shouldered, with a high forehead and red hair. He wore a long, thick moustache which curled over a meticulously groomed auburn beard. And although in his portraits his famous nose appears a bit too square at the bridge, he is far from an ugly man, nor did his contemporaries consider him as such. Rather, I suggest that Tycho's choice of a wife rested more on the grounds of his character and temperament than on personal misgivings about his physical appearance. Tycho wanted a submissive but responsible companion, someone who was capable of taking charge of his household and raising the children while he tended to other, less mundane matters. He knew that so exacting a nature as his, coupled with an eccentric manner of living, would be intolerable to a woman of cultivated tastes and highly refined social graces. Better a wife of peasant stock, used to taking orders and assuming responsibility, than a pretty bauble that might burst under the strain of trying to cope with a selfish, egotistical genius. From the day she first entered Tycho's home until her death in 1604, three years after her famous husband, Christine Brahe never, so far as is known, failed in her duties either as a wife or mother. Under the circumstances, Tycho could hardly have done better in his search for a faithful companion.

Early in 1575 Tycho decided to visit southern Europe once again; this time for the purpose of selecting a place where he might permanently settle down with his family to undertake his lifework. He began his search by paying a visit to the city of Cassel where he met a kindred spirit in the Landgrave Wilhelm IV of Hesse, a distinguished astronomer who came to play a major part in determining Tycho's future. The two spent some pleasant evenings exchanging information about the nature of celestial phenomena while taking readings on planetary and stellar positions with the finely crafted instruments under a moveable roof in Wilhelm's castle. The Landgrave, if we can believe a frequently repeated story, must have been as starstruck as his Danish guest. One evening a servant supposedly rushed into the observatory to inform his master that the castle was on fire. Wilhelm happened to be viewing the nova of 1572 and, not wishing to be disturbed, calmly told the servant to look elsewhere for assistance in putting out the flames. It is hard telling just how long Tycho might have remained at Cassel had not tragedy overtaken his new found friend. Not long after his arrival the Landgrave's little daughter died unexpectedly, and under the circumstances Tycho felt compelled to take his leave. Even though his visit with Wilhelm was brief, Tycho, in addition to his keen intelligence, was capable of displaying great charm in the company of those he genuinely respected. He made a favorable and lasting impression on Wilhelm, and, although the two men never again met face to face, they maintained a frequent correspondence over the years.

After traveling through Germany, Italy, and Switzerland, and meeting many astronomers, Tycho finally returned home with plans to settle in Basle, the principal city of Switzerland, located in the northwest on the Rhine River. The city's central location seemed most convenient, and its university, where he had already spent some months during the winter of 1568–69, was one of early modern Europe's most important centers of learning. However, unbeknownst to Tycho, some of his friends, disturbed by the news that he planned to leave Denmark, made direct appeals to King Frederick ii, urging him to thwart the astronomer's intention to settle abroad. Landgrave Wilhelm, the most influential among them, even dispatched a special embassy to Copenhagen, partly for the purpose of exhorting the king to make certain gestures that would keep Tycho in his native land. To do so, Wilhelm reasoned, would not only advance the cause of science but would also bring much credit to the king and his country. The Landgrave seems to have recognized what we have grown increasingly aware of in the twentieth century: a nation's treasure consists not only of its natural resources and other forms of material wealth, but also includes an even more precious resource, the intellectual capability of its native citizenry.* King Frederick, for his part, was a patron of learning and only too anxious to keep a man of Tycho Brahe's promise at home.

His mind made up, Frederick dispatched a royal messenger with instructions to travel day and night until he located Tycho. As the astronomer lay in bed one early February morning in 1576 thinking, in his own words, "about my German trip and at the same time wondering how I could disappear without attracting the attention of my relatives," the king's messenger arrived with instructions for Tycho to come at once to the royal hunting lodge located a short distance from Copenhagen. Tycho did as commanded and upon his arrival was granted an immediate audience with his king. Frederick, who was genuinely fond of Tycho, after a long conversation, during which he expressed his deep concern over the young nobleman's plans to quit Denmark, made Tycho

---

* Today we have even gone so far as to attach a popular name to the intellectual migration from one country (usually poor) to another country (usually richer); we call it the "Brain Drain."

what is still unquestionably the most generous offer ever tendered by the government of any nation to a native scientist. The chief part of Frederick's gift was the island of Hveen, situated in the Sound, some fourteen miles from Copenhagen. There the king would help underwrite much of the cost of constructing a home and observatory which would enable the astronomer to devote himself to his studies undisturbed by the affairs of court and State. Such generosity exceeded even Tycho's wildest dreams, and within a short time the fairy-tale offer was accepted. On the morning of May 23, 1576 Frederick II, King of Denmark, affixed his signature to one of the most unusual and important documents in the history of modern science. Despite its unmistakably medieval tone, one can hardly read the following lines, excerpted from the agreement between citizen and sovereign, without thinking that here lies the seed of the modern power system, the creation of which would not have been possible without a formal alliance between the scientist and the State:

> We, Frederick the Second, etc., make known to all men, that we of our special favour and grace have conferred and granted in fee, and now by this our open letter confer and grant in fee, to our beloved Tyge Brahe, Otto's son, of Knudstrup, our man and servant, our land of Hveen, with all our and the crown's tenants and servants who thereon live, with all rent and duty which comes from that, and is given to us and to the crown, to have, enjoy, use and hold, quit and free, without any rent, all the days of his life, and as long as he lives and likes to continue and follow his *studia mathematices,* but so that he shall keep the tenants who live there under law and right, and injure none of them against the law or by any new impost or other unusual tax, and in all ways be faithful to us and the kingdom, and attend to our welfare in every way and guard against and prevent danger and injury to the kingdom. Actum Fredericksborg the 23rd day of May, anno 1576.
>
> "FREDERICK" [7]

Later on the same day Frederick instructed his chancellor of the exchequer to pay Tycho some 400 dalers towards erecting a building on Hveen, for which it was understood that the young nobleman was to furnish certain construction materials. A year later, as a further measure of his generosity and confidence in the young astronomer, Frederick granted Tycho the lucrative manor of Kullagaard in Scania, to be held so long as the king deemed fit, on the condition that Tycho keep the lighthouse of Kullen in proper working order. In 1578 Tycho gained the use of estate in Norway; and all this in addition to an annual royal pension of eleven farms near Helsingborg, free of rent, plus the income from an 2000 dalers, not to mention the revenue derived from the properties inherited from his father and uncle Jörgen. Royal munificence of the magnitude bestowed upon such a nonconformist as Tycho could not help but deepen the already considerable animosity harbored for him by his rivals at court; but so long as Frederick was alive Tycho had no cause to be uneasy. Frederick II was not the type of monarch who took away with one hand what he had given with the other. Moreover, one cannot help but wonder whether subconsciously the king felt a special responsibility toward the foster-son of the man who had saved his life, indirectly forfeiting his own in the process. Whatever the reasons may have been for Frederick's generosity, Tycho, who was only thirty years old in 1576 when the island of Hveen became his estate for life, now possessed all the material wealth required to help him fulfill his great promise as a scientist of the first rank.

Tycho's new domain, which he called the island of Venus, or Hveen, rises like a giant table out of the sea. It is surrounded by white cliffs which leveled off to form the 2000-acre estate of Tycho's day. During the sixteenth century, Hveen's 40 peasant farmers and their families lived in a single village, called Tuna, toward the north end of the island. Nearby were the common fields divided into the three traditional medieval categories of spring planting, fallow, and fall planting. To the south, beyond the fields, lay open meadowland and a small hazelnut grove. For the sportsman there was game aplenty in the form of deer, rabbit, and partridge, while the waters surrounding Hveen abounded in numerous varieties of fish. Even though the island rises to a height of over 150 feet in the midst of salt-water, it has several fresh springs and streams, a number of which Tycho had damned up to form stock ponds for his fish and to provide energy for his paper mill. Tycho drew his own water from a well that is still functional after 400 years. He demonstrated his technical expertise by designing a system of hydraulic pumps through which the water could be lifted into any room of the main house. The method apparently worked so well, one wonders why it was not copied for use on other estates in Denmark.

Although its advantages certainly outweighed its drawbacks, Hveen proved something less than a complete paradise for the serious observational astronomer. It is cold, damp, and cloudy so much of the year, that during Tycho's two decades on the island literally hundreds of nights passed without a single observation being made. Even on clear nights Tycho, like Copernicus, saw less of the heavens than he might have wished. Particularly disturbing was the fact that Mercury frequently lay below the southern horizon out of the astronomer's field of view. Still, the island seemed almost ideal for Tycho's purposes: When conditions were favorable celestial observations could be undertaken with ease, and, because of its seagirt location about halfway between Copenhagen and Elsinore, Hveen protected its lord from considerable numbers of undesired visitors. But what most attracted Tycho to Hveen was not something that the island possessed, but rather what it did not have—a manor house or castle.

The idea of founding an observatory for the purpose of carrying out regular observation, as had been done by Landgrave Wilhelm IV, was Tycho's major ambition; and he was far more interested in designing his own facility from the ground up than in undertaking the extensive modification of an already existing structure. Hveen, whether consciously or not, became Tycho's terrestrial miniature of the all-encompassing celestial universe. Therefore, it is hardly surprising that in choosing a location for his new residence and observatory he selected a site nearly in the center of the island, 160 feet above the sea. Here he would erect his famous Uraniborg, named after Urania, the beautiful mythological muse of astronomy. The cornerstone of the building was laid on August 8, 1576 by Tycho's friend and French envoy to the Danish court, Charles Dancey. The distinguished visitor advised Tycho: "Build something out of the ordinary," and the astronomer proceeded to do just that. Two of the era's most famous architects were employed: Hans van Paschen, a Dutchman, and Hans van Steinwinckel of Antwerp. Tycho got along well with both men, neither of whom seem to have minded his constant alternation of their basic design. Since neither man had ever been called upon to plan and execute the construction of a combination residence and observatory, they were apparently content to let Tycho's imprimatur guide their work.

The finished building was a palatial structure of red brick with sandstone ornamentation, in the so-called Dutch Renaissance style. Only about 37 foot high, the main structure had sides about 49 foot square. On the north and south sides were round towers 18 feet in diameter; each had a platform on top surmounted by a pyramid-shaped roof, forming an observatory. Numerous smaller observation platforms were located along all four of Uraniborg's sides, so that Tycho's pupils could observe the stars from any angle their master desired. The building's spacious rooms were filled with finely carved furniture and elaborately decorated with tapestries and pictures. Native workmen were drawn from all over Denmark, while foreign craftsmen came from as far away as Italy. Tycho's tenants, ostensibly protected against abuse by the terms of the document under which Hveen had been granted to their lord, were forced to neglect their land for months on end while the construction went on. Thus, they received an early indoctrination into the harsh methods to be employed by their master over the next twenty years. Nor could they have taken much comfort from the fact that Uraniborg's grounds contained a prison for those who fell from Tycho's favor while its basement held a large number of alchemical implements, which were believed by them to be the tools of wizzards. Obviously, it did not take very long before the peasants of Hveen began to wish for a return to the more idyllic days of relatively carefree independence they had known before the coming of the imperious Dane.

The house was sufficiently complete to enable Tycho to take up residence sometime late in 1578 or early 1579. During the next year he had his tenants construct a massive enclosure around the main building formed by earthen walls, about 18 feet high and 16 feet thick at the base, with the corners facing the four points of the compass. At the middle of each wall, which ran some 248 feet from corner to corner, was a semicircular bend, over 70 feet in diameter, with an enclosed arbor in each. Entry to the grounds was made through massive stone gates located at the east and west corners of the walls. Mastiffs were kenneled above each entrance to announce, by their barking, the arrival of all visitors. The remaining grounds were laid out with flower gardens, shrubs, and orchards. Located a short distance away were several other buildings including a residence for Tycho's assistants, a private printing press, and a shop for the manufacture of scientific instruments. At the time of Uraniborg's completion in 1580, no astronomer, before or since, has ever enjoyed a more perfect combination of physical facilities and independence from institutional controls than did Tycho Brahe.*

From the founding of the first Babylonian observatories thousands of years ago and continuing down to the present day, the scientific value of these institutions has rested on two fundamental components: the quality of the available observational equipment and the skill of the individuals who have used it. Uraniborg stands at the top of the list in both categories, because as an instrument maker and naked-eye astronomer Tycho excelled all who had lived before him. In reality the instruments with which his observations were made did not differ in general principle from those of the ancients. But in the details of construction and in the quality of the mechanical work, Tycho made numerous and significant improvements. In the first place, most of the instruments he designed, like the Augsburg quadrant, were much larger than the devices employed by Hipparchus, Ptolemy, and the other Greek astronomers. This in turn permitted the use of bigger scales, so that errors in marking the divisions between the major points on the sights were of less consequence than on the smaller scales of more modest instruments. Furthermore, Tycho employed a new method of dividing his scales so that they could be read accurately to fractions of a division. His method involved the use of transversals, or diagonal lines drawn between the divisions on a scale. He cleverly adapted the concept to the curved scales on his astronomical instruments with the result that he was able to take readings accurate to a tenth part of a division—or

---

* In researching this chapter, the thought occurred to me that a definite parallel seems to exist between the type of estate constructed on Hveen and Monticello built on the Virginia frontier by Thomas Jefferson. I soon learned that I was far from the first to make this comparison. See Charles D. Humberd, "Tycho Brahe's Island," *Popular Astronomy,* Vol. XIV, No. 3 (March, 1937), pp. 124-5.

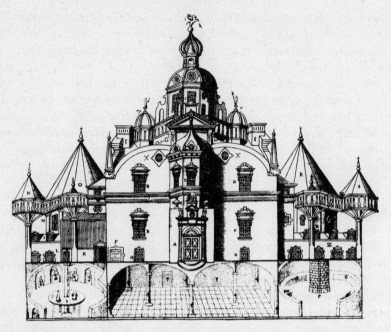

FIGURE 9.   An engraving of the facade of Uraniborg, from Tycho's *Astrono-miae Instauratae Mechanica*. Note the large number of domes and balconies from which observations could be made in all directions. *(Courtesy of the Lilly Library, Bloomington, Indiana.)*

one minute of arc. He even claimed to have designed a steel quadrant, unfortunately lost, that could be read to ten seconds of arc. If so, this means that his observations were something in the neighborhood of ten times more accurate than those of any previous astronomer. Although this claim cannot be substantiated today, we do know that Tycho, even on his worst night, was at least two to six times more accurate in the data he collected than his nearest rivals.

Another of Tycho's technical innovations, and one that greatly increased the quality of his observations, was the substitution of the adjustable eyepiece for the customary peephole. It contained slits that opened to admit more light from a faint star and closed to permit the better observation stars of greater magnitude. And whenever possible he demanded that metal rather than wood be used to fashion his instruments. Wood warps in a damp climate like that of Hveen and, if used to construct large observational devices like Tycho's, it tends to bend under its own weight, distorting the astronomer's measurements. When this happens any effort to achieve accuracy in celestial observation is doomed from the start.

One of the most important instruments at Uraniborg was Tycho's great semicircular quadrant, over six feet in radius, mounted on a wall,

FIGURE 10. The observatory and grounds of Tycho Brahe's Uraniborg located on the then Danish island of Hveen, from the *Astronomiae Instauratae Mechanica.* (*Courtesy of the Lilly Library, Bloomington, Indiana.*)

and used to measure the altitude of celestial bodies as they crossed the north-south meridian. It has been made famous by a painting (see Figure 11) in which it is presented in full action under Tycho's direction with his assistants pointing, looking, and noting, while his faithful dog lies at his feet. Tycho is also pictured as a seated observer in the foreground. After a delay of several years Tycho's Augsburg globe finally arrived, and he had it mounted on a pedestal in the main library. The great brass-covered sphere, some five feet in diameter, was engraved with two circles, one of which represented the equator, the other the zodiac. Over the years, the individual stars and constellations were filled in as their exact positions were determined from scores of observations, just as Tycho had planned. When not in use a large cloth of silk, suspended from the library ceiling, could be lowered over the globe to protect it from dust. Before his death, Tycho had brought the total number of stars engraved on the globe to an even 1000. Another type of instrument was the equatorial armillary used for the measurement of angular coordinates in relation to the equator of the sky. It was also employed to determine the exact time because Tycho's clocks were both primitive and inaccurate, a fact of which their owner was well aware.

One of his most highly prized groups of instruments consisted of several sextants used for the measurement of angular distances between two celestial bodies. With these Tycho pioneered the technique of accurate celestial triangulation.

The integrity of his instruments was not the only reason why Tycho's

FIGURE 11.   Mural quadrant at Uraniborg. (*Courtesy of the Lilly Library, Bloomington, Indiana.*)

observations far excelled those of other astronomers. Either he or his assistants would follow a planet or star for months, sometimes years on end, literally taking dozens of sightings before arriving at a final determination of its exact position. He was also the first astronomer to take into consideration the amount of error caused by the refraction of light through the earth's atmosphere. He even calculated the degree of error built into his instruments, and whenever an observation was made he compensated by an equal amount. And even though he used a number of portable devices, Tycho much preferred those which were set on permanent mountings. He knew from experience what many a careless amateur astronomer is only too well aware of today—that even the slightest blow at a critical point on a delicate observational instrument can permanently impair its effectiveness.

Over the years, as the number of Tycho's assistants increased and he constructed one large instrument after another, Uraniborg simply became too crowded to house them all. So in the 1580s he built another observatory on a small hill just outside the walls of the main palace. Tycho christened it Stjerneborg—the Star Castle—which consisted of five underground rooms with ceilings flush with the ground above. In each room, interconnected by a series of subterranean passages, Tycho placed a large instrument mounted on a block of stone located in the center of a round pit with terraced sides, like those of a miniature Greek theatre. All that protruded above ground were five small conical roofs, built so that they could be rotated when observations were being made. The main purpose behind the underground design was to protect the large instruments from the sway caused by the heavy sea breezes, which constantly buffeted Uraniborg's elevated observatories as they swept across Hveen's exposed plateau. The use of two observation centers also proved to be well suited to Tycho's method of teaching, whereby the skills of one student were pitted directly against those of another. He commonly gave two assistants the same problem, and by assigning them to separate observatories they were prevented from comparing notes before presenting their data to Tycho for analysis. Pity the poor student who erred in his calculations, for the wrath of the master was his only reward. In a study located in the center of Stjerneborg, Tycho had the portraits of eight astronomers hung, all in reclining positions: Timocharis, Hipparchus, Ptolemy, Albattani, King Alphonso, Copernicus, Tycho, and his hoped for but yet unborn successor, Tychonides.

As Tycho's fame spread throughout Europe, an increasing number of pupils became eager to make the pilgrimage to Hveen to study under the great astronomer. Some of them were trained at Frederick's expense, others were subsidized by institutions of higher learning or by cities, while several were maintained by Tycho himself. Except for a small handful of Tycho's most gifted students, we know little more than the names of the young men who lived at Uraniborg for periods ranging from a few weeks to several years. Of all his pupils Christen Sörensen Longberg—Longomontanus—one of the early seventeenth century's best astronomers, is probably the most well known. The child of an impoverished widow, Longberg came to Hveen in 1589 on the recommendation of his professors at the University of Copenhagen and remained until 1597. Another of Tycho's early assistants, and perhaps the more promising of the dozens who studied under him, was the German Paul Wittich, whose early death robbed astronomy of a potential genius. While they were learning astronomy from the master, Tycho's assistants were also introduced to the practice of alchemy in Uraniborg's basement laboratory. Though we have no specific knowledge of the particular direction of Tycho's research in this area, we do know that ever since his extended stay at Heridsvad Abbey his interest in chemistry ran deep and was doubtless tied to his belief in the unity of all things, both heavenly and terrestrial. Tycho's assistants also helped in the construction of astronomical instruments, and in the manufacture of paper in the mill Tycho had built on Hveen. One of them, Elias Olsen, was largely responsible for seeing the first book produced at Uraniborg, an astrological and meteorological diary, through the various complicated stages of printing. Those who displayed less talent were frequently used as combination messenger–purchasing agents and travelled all over Europe on Tycho's behalf.

But for the student gifted enough to keep up with Tycho's demanding intellectual and physical pace, the months spent at Uraniborg must have been among the most stimulating of his entire life. On a clear night observations commenced at dusk and continued until dawn, after which each student's data was subjected to Tycho's careful scrutiny. Then, off to bed for a few hour's sleep before an afternoon conference during which the next night's work was parceled out. Food was often taken on the run, and so long as the sky remained clear, the cycle repeated itself until everyone lost track of time, as if anyone still cared. A week or more of this and prayers for rain must have passed the lips (silently, of course) of many a haggard student, including even those most dedicated to the pursuit of the stars. Tycho, on the other hand, thrived on this type of intensive activity, and how he maintained his health under such strenuous conditions still remains a minor mystery.

At this point it seems fitting that we ask what, in the end, does all this add up to? And wherein lay the magnitude of the man responsible for it? Tycho, unlike Copernicus, was no permanent innovator of astronomical concepts. Instead he was responsible for instituting immense changes in the techniques of celestial observation by which the accuracy of the data collected put astronomy on an entirely new foundation. Tycho, rather than being an interpreter of Nature, was above all an observer and recorder. He thus lay the groundwork for the new precision upon which astronomy and all the other modern sciences rest. T. S. Kuhn has written this of his achievement:

> In his own lifetime he and the observers he trained freed European astronomy from its dependence on ancient data and eliminated a whole series of apparent astronomical problems which had derived from bad data. His observations provided a new statement of the problem of the planets, and that new statement was a prerequisite to the problem's solution.[8]

So reliable were Tycho's instruments and observations that it is difficult to see how a much greater accuracy could have been achieved by succeeding generations had not the telescope been invented a few years after his death. Despite the fact that Tycho himself did not accept the Copernican system, his accurate and systematic observations gave Johannes Kepler a place on which to stand. Aided by Tycho's invaluable legacy of empirical data, Kepler developed the first laws of planetary motion in Copernican terms, and by doing so he moved the world.

Headstrong, domineering, and frequently contemptuous of those with less education and wealth, Tycho was nonetheless a great scientist. This is because, like all great men, he cherished and respected an idea bigger than himself. For just as he was the lord of Hveen, master of its simple peasant farmers, he too served a master greater than himself— Urania, muse of astronomy. By gathering so much reliable information about the movements of the celestial bodies, he opened the path which has enabled modern man to understand why these same bodies act as they do. Without Tycho, the work of Kepler and Newton could not have been accomplished, nor could the revolution in thought begun by Copernicus have been brought to a successful conclusion.

The Tychonic System

Beginning with Copernicus and continuing until the development of Newtonian physics, Western man's understanding of the universe went through a period of intense debate and deepening crisis. Whereas the Aristotelian–Ptolemaic system had been accepted almost without reservation by nearly all educated Europeans before 1450, major contradictions in the old system had begun to accumulate to the point that, by the time the Renaissance had reached full flower, they demanded explanation, as never before. Copernicus responded by designing a new heliocentric model of the heavens, even going so far as to hint at the possibility of an infinite universe. Unfortunately, the Canon's imagination was not matched by the required proof and, although he set the scientific community buzzing, many scholars clung tenaciously to the teachings of Aristotle and his earth-centered universe. To a number of other individuals, however, neither the old geocentric model nor the new helio-centric system could adequately account for all the disparate phenomena associated with planetary movement. It was to this latter group that Tycho Brahe belonged.

Tycho was nothing if not the most proficient astronomer of the six-teenth century. Thus his lasting place in the history of science rests on the extensive, up-to-date, and reliable data he contributed to the eventual solution of the problem of planetary motion. Yet he was also conscious of the fact that the compilation of scientific data alone—no matter how accurate and sophisticated—is a useless exercise unless it can be utilized to lay bare nature's secrets. Almost from the time he determined to make observational astronomy his lifework, Tycho dreamed of one day using what he had discovered to construct a model of the universe superior, both in design and operation, to those postulated by the ancient and medieval thinkers who preceded him. Though he was to fail in his efforts as a cosmographer, Tycho's model of the universe moved the scholarly world a few crucial steps farther away from Aristotle and Ptolemy and, although he himself did not realize it, just that much closer to a full embrace of the Copernican system.

Tycho had always been a great admirer of Copernicus, for he could not ignore the mathematical harmonies that *De revolutionibus* had intro-duced into the science of astronomy. Yet neither could he accept certain aspects of the Copernican universe—the most important being the idea of a moving earth. As a great scientific observer, the thing that most im-pressed Tycho about the heavens is that the patterns of the constellations remain unchanged from season to season. During the sixteenth century this absence of stellar parallax could be satisfactorily explained in only one of two ways: either the earth is stationary, as Aristotle taught and Tycho believed, or by comparison with the distance of even the very nearest stars, the diameter of the earth's orbit is little more than a geo-metric point, as Copernicus had postulated. In fact, Copernicus had esti-mated that in December the earth is about 100 million miles away from where it is in June (the actual displacement is about 184 million miles), meaning that the whole immensity of our solar system shrinks to prac-tically nothing when compared with the distance of the stars from the earth. But since, as we know, Copernicus had no absolute proof to sup-port his bold conjecture, while Tycho had compiled reams of evidence, none of which supported the theory of stellar parallax, he was quite correct to reject the Copernican belief in terrestrial movement. The facts appeared incontrovertible, and no scientist in Tycho's position could afford to overlook them.

From Tycho's point of view, there was yet a further objection to the heliocentric system besides the question of the earth's movement. It had to do with the size and distribution of the fixed stars. During Tycho's time, it was thought that each of the stars had a specific diameter based upon its degree of brightness: stars of the first magnitude, for example, were believed to be about the same size as the planets Jupiter and Venus. This meant that if the stars were truly as far away from earth as Copernicus believed them to be, they could not possibly be visible because their light would be swallowed up by the immense intervening space.

Only, according to Tycho's calculations, if the stars were so large they filled the equivalent of the entire space circumscribed by the earth's orbit round the sun would they be visible from this planet. And who could reasonably be expected to accept this as an article of faith? The brain was simply not yet conditioned to think in such staggering terms, particularly when the eye told it a completely different story. Then, even as now, man had to learn that the difficulty of visual conception is not a valid argument against reality, something that Galileo spent the better part of his adult life trying to teach the Aristotelians of his day.*

In addition to the convincing physical arguments mustered by Tycho against the heliocentric system, there remains the question of his religious convictions, and whether they may not have played an important part in keeping him from joining the growing ranks of the Copernicans. Given Tycho's supreme dedication to observational astronomy, it seems much more probable to me that he found the scientific arguments against Copernicus more appealing than those based on theological considerations. It is true that by the time Tycho had settled on Hveen, the significance of the Copernican hypothesis had dawned upon the spiritual leaders of Europe, whether reactionary or progressive. And, as a Lutheran, he was probably even more sensitive to the theological arguments against the new cosmology than were the Catholic astronomers of his time. After all, Luther and Melanchthon were the first major theologians to openly attack the Copernican system. On the other hand, Tycho, while a supporter of his church, had the good fortune to be ruled by one of sixteenth-century Europe's most tolerant monarchs, a man who appreciated divergent points of view in all areas of serious scholarship. Moreover, Tycho's residence on the island of Hveen isolated him from much of the religious and social criticism leveled against the new scientific discoveries. Thus while he gave the usual lip service to those Biblical passages which heliocentric cosmology seemingly violated, his primary objections were derived from the ample scientific evidence he and his assistants had painstakingly col-

---

* Tycho, as it later developed, was correct to reject the concept of stars with diameters as great as the earth's orbit. When Galileo turned the telescope on the heavens he found that while the number of visible stars greatly increased, their apparent size remained the same. The reason: "The angular diameter of stars had been immensely overestimated by naked-eye observations, an error now explained as a consequence of atmospheric turbulence which blurs the images of the stars and spreads them over a wider area in the eyes than would be covered by their undisturbed images alone." (Even so, the stars are far larger than Tycho ever believed.) T. S. Kuhn, *The Copernican Revolution*, p. 221.

lected at Uraniborg. That this evidence did not contradict the Scriptures was something Tycho might well have been thankful for; but it was far from being the decisive underlying factor in his rejection of the heliocentric universe.

While Tycho could not bring himself to an unqualified acceptance of the Copernican system, neither was he able to give complete support to Ptolemy's geocentric model of the universe. For just as with Copernican cosmology, the data collected during a lifetime of observation simply would not permit it. Of primary concern to him was the information he had gathered on the nova of 1572 and the comet of 1577. Tycho was able to prove that because the new star did not move, it could not possibly have been located in the sublunar region. The comet also inhabited the celestial realm and, by its steady movement through the planetary orbits, would have smashed the crystalline spheres of ancient astronomy, had they existed. For those willing to give credence to the new science, the appearance of such superlunar phenomena as these clearly destroyed the time-honored belief in celestial immutability.

Tycho now faced the dilemma of developing a working model of the universe by which the anomalies of both the Copernican and Aristotelian system could be reconciled with his own scientific findings. The initial breakthrough occurred on February 13, 1578. On that day Tycho drew the first sketches of what is commonly called the "Tychonic system." Fortunately for posterity, these drawings were later bound into the back of his copy of Copernicus' *De revolutionibus*, now located in the Vatican Library.* Like Ptolemy, Tycho placed the immovable earth in the center of the universe. The moon circles the earth and, at twenty times the moon's distance, so does the sun. But the five planets, instead of moving in orbits centered on the earth, orbit the sun and are carried round the earth with it. Outside Saturn's orbit, the finite sphere of the fixed stars rotates once every twenty-four hours (see Figure 12). Notice, too, that the orbits of Mercury, Venus, and Mars are intersected by the sun's orbit about the earth, which would have been impossible under the Aristotelian doctrine of the inviolable crystalline spheres. In the diagram the orbits are simply represented by circles, though in a detailed drawing of the Tychonic system epicycles, eccentrics, and even equants are employed to account for the variable rates of orbital motion.

What Tycho had done, and what makes his system so ingenious (if not very original), was to fashion a compromise between the Copernican universe and the traditional one. His system avoided the most serious

---

* Although the outlines of the Tychonic system have been well known for nearly 400 years, the precise steps leading up to its discovery were discovered quite by accident only a few years ago. For the fascinating story of how Tycho's working manuscript was located, see Owen Gingerich, "Copernicus and Tycho," *Scientific American*, Vol. 229, No. 6 (December, 1973), pp. 87–101.

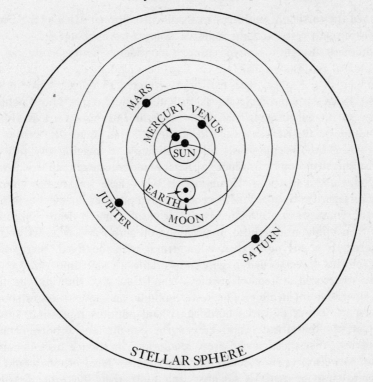

FIGURE 12.   The Tychonic system.

criticisms to which Copernican cosmology had been subjected, but pre-served all its mathematical advantages. In fact, the two systems are identi-cal mathematically, while Tycho's model had the additional advantage of reconciling the anomalies discovered in the Aristotelian system with Scripture, something Copernicus had been unable to do. Thus as the Aristotelian–Ptolemaic system became increasingly indefensible, the Tychonic system automatically recommended itself to those who were reluctant to antagonize the church and academic science, yet wanted to "save the phenomena," as the astronomers of ancient times had strug-gled to do. It is little wonder, then, that Tycho's planetary system en-joyed its greatest success among the Jesuits, who used it against Galileo when he openly advocated the heliocentric hypothesis before being silenced by the Roman Inquisition. It was almost as if those who had become dissatisfied with Aristotle also felt themselves cheated by Coper-nicus, for the Canon had gone too far too fast. In the absence of more convincing proof, the acceptance of his system still required a great leap of faith which few were willing to make. As a conservative, Tycho re-traced Copernicus' steps to the point where he had placed the earth in the heavens and expanded the universe to immense proportions. Tycho returned the earth and with it man to their traditional place at the center of creation and restricted the celestial boundaries once again.

Although Tycho's system enjoyed considerable appeal among the non-Copernican astronomers of his day, he failed to convert those few individuals, like Kepler, who had been attracted to the Copernican system by its great symmetry. As T. S. Kuhn points out, "The Tychonic system has incongruities all its own: most of the planets are badly off center; the geometric center of the universe is no longer the center for most of the celestial motions; and it is difficult to imagine any physical mechanism that could produce planetary motions even approximately like Brahe's." [9] Perhaps the main reason why these weaknesses were so easily disregarded by many of Tycho's contemporaries is that they shared his basic conservatism, which was reflective of a much deeper historical crisis then confronting those in positions of authority, whether secular or sacred. It seems more than coincidental that the fixed, hierarchical order of the universe had begun to crumble at the same time as the rigidly structured system of medieval feudalism was entering the final stage of its fateful decline. The restive middle and lower classes sought to burst the bonds of social conformity and political repression, just as the questing Renaissance mind craved the enlightenment born of new discoveries. Since most of the great revolutions in science have occurred during periods of major social and intellectual ferment, it would be wrong to suppose that the scientist, any more than the professional in other fields, is insulated from what takes place in the larger society of which he is a part. Tycho, to his credit, at least sought the middle ground between the old and the new, as he attempted to reconcile his scientific discoveries with the legacy of the ancients whom he deeply revered.

As a final comment on the Tychonic system, it is worth noting that the attempt to reach an accommodation between two radically divergent schools of thought, without completely rejecting one or the other, has also played a significant part in scientific revolutions other than the one in astronomy. For example, a number of nineteenth-century naturalists rejected certain aspects of the Biblical interpretation of the world and creation; yet neither could they bring themselves to accept Darwin's theory of evolution. Instead, they compromised by settling on a doctrine called catastrophism, which held that the planet earth has been subjected to a series of violent and cataclysmic natural disasters interspersed by long periods of calm. Life was supposedly created anew after each such upheaval, which conveniently explained the existence of ancient fossil remains but at the same time denied the possibility of biological evolution through natural selection over vast periods of time. According to the catastrophists, man made his appearance only after the last of nature's great upheavals; therefore, he could not possibly be the product of evolution from simpler life forms. Thus the general outlines of the Biblical account of divine creation were preserved despite a growing body of knowledge concerning long extinct forms of plant and animal life. Yet even as they sought to strike some sort of balance between Darwinian principles and the Bible, the catastrophists, like Tycho Brahe and his followers, gradually moved the scientific community toward a complete acceptance of the very revolutionary theory they themselves rejected. As he lay dying, Tycho pathetically begged Kepler, his associate, not to let his system be forgotten. Ever the gentleman and diplomat, Kepler promised the great astronomer that he would do everything in his power to accomplish what was asked of him yet, in his own mind, Kepler was already convinced of the truth of the Copernican hypothesis, and it is indeed one of the great ironies of modern science that he used Tycho's own data to help demonstrate the essential validity of the heliocentric system.

The Exile

During the two decades Tycho lived at Uraniborg, life there was hardly the quiet, routine affair that one might expect of a scholar and his family, but rather more that of a Renaissance prince and his court. Brahe was not only a giant of methodical observation but a man of many appetites, ranging from his interminable poetic outpourings to alchemy, instrument making, fish breeding, agriculture, paper manufacturing, and printing. Distinguished visitors from many countries frequently crossed the Sound to the little island, which had suddenly become famous by reason of its renowned occupant. Though Tycho had chosen this spot for his observatory specifically for the purpose of carrying on his work undisturbed, he displayed all the vanity of genius and grew increasingly fond of presiding over a steady succession of hard-drinking banquets, during which he held forth on the significance of his latest observations. Beneath the table, at his feet, sat one of Uraniborg's most interesting residents, Tycho's jester or fool, a dwarf named Jeppe. Dressed in the traditional buffoon's costume of cap and bells, he was barely three feet tall and chatted incessantly, except when devouring an occassional tidbit from his master's hand. Jeppe had been trained to perform only a single but to Tycho very important task, that of cutting the astronomer's rapidly thinning hair. According to Longomontanus, Tycho's former assistant, the dwarf was also thought by almost everyone to be gifted with the power of foretelling coming events, something Tycho's students apparently used to their advantage. When the astronomer was away from the island his assistants frequently took an unscheduled holiday: Jeppe would warn them of Tycho's return by yelling, "Junker paa Landet!"—The Squire is on the land!" [10] And if anyone became seriously ill, Jeppe was called in immediately to predict whether the individual was going to live or die; he supposedly proved correct in every instance.

In 1588 Frederick II, Tycho's protector and great benefactor, died unexpectedly, much to the astronomer's sorrow and dismay. Frederick's son Christian IV, Prince of Denmark, was elected his successor, although the boy was still too young to take over the vacant throne. Tycho, for obvious reasons, was most anxious to meet with the young man, and in 1592 he got his chance. Christian received permission from his mother Queen Sophia, who had herself been a guest at Uraniborg on at least two occasions, to visit the famous observatory. The importance of making a favorable impression on the future king and his three powerful regents was not lost on Tycho. After a pleasant day of sightseeing and amiable, wide-ranging conversation, Tycho presented Christian with a brass globe that, by means of cog-wheels, simulated the daily rotation of the heavens. The young Prince immediately returned the favor by taking a gold chain, engraved with his own likeness, from around his neck and placing it around his host's. At this moment, neither lord nor vassal imagined that one day the memory of this warm gesture would be permanently effaced by bitterness and misunderstanding that ran so deep that Tycho went into self-imposed exile rather than submit to a reduction of the many generous perquisites granted him by King Frederick.

Life and work at Hveen continued as usual for a number of years after Christian's visit; but gradually the practical consideration of economics and a nobleman's duties to his king and tenants became major impediments to Tycho's extravagant manner of living. Over the years, he had become so preoccupied with being king of his tiny island domain he forgot that a far more powerful monarch resided only a few miles distant across the Sound. Christian IV was no tyrant, but neither did his generosity in financial affairs match that of his father—not at least when it came to the Lord of Hveen. Particularly disturbing to the new king was Tycho's unconscionable neglect of his none too burdensome duties as a holder of several major fiefs. Tycho had been involved in repeated conflicts with his tenants, who submitted an increasing number of petitions requesting that their grievances be examined by the crown. Most common among them was the extraction of forced labor and goods to which Tycho was not legally entitled; if the peasant farmers refused to give in to his demands, they were arbitrarily imprisoned and sometimes subjected to corporal punishment. Furthermore, as Canon of Roskilde, Tycho failed to keep the church in proper repair, the major requirement of his office. The roof leaked so badly that the weather threatened to ruin the church's beautiful interior. Christian wrote letter after letter to Tycho asking that the condition be corrected, but to no avail. Finally, he threatened to revoke Tycho's official position as canon if the damage was not repaired at once. Similarly, Tycho had neglected to keep the lighthouse at Kullen in proper working order, with the result that several boats were lost by crashing against the dangerous rocks along the jagged coast. Tycho acted at last, but by this time both Christian's patience with and confidence in him were exhausted. To make matters worse, he now paid a heavy price for the proud and haughty conduct he had displayed as a young man. His old enemies at court had a long-standing score to settle and, sensing his weakened position in the King's eyes, they did everything in their power to hasten Tycho's downfall. Though no direct steps were taken against him by Christian, a number of his inflated prebends were either abolished or reduced to more reasonable proportions. A less proud and egotistical man might have licked his wounds, brought his standard of living into line with his income, and, sufficiently chastened, gone on about his business. But even though he knew that most of the charges brought against him were true, Tycho felt aggrieved and in 1597 decided to "punish" an ungrateful king and his court by leaving Denmark. There are also some indications from his correspondence and conversations with close friends that he was growing restless with his life on Hveen, and perhaps the King's actions provided the very pretext he sought to enable him to take up his European studies once again.

Tycho had all but his largest instruments dismantled and packed for shipping, as well as his printing press, furniture, library, and paintings. Together with a suite of twenty, including his family, assistants, servants, and Jeppe, he left Hveen for Copenhagen and southern Europe with the hope of carving out a new life for himself under the patronage of a more appreciative and enlightened sovereign, although who it might be he did not yet know. But no sooner had Tycho reached Copenhagen than he apparently underwent a change of heart and began to wish that the king would request that he stay in Denmark, just as his father had done over twenty years before. Surely it was not too much to hope that a worried Christian, fearful of losing his most famous subject, would admit his mistake and ask Tycho's forgiveness. Yet the longer Tycho waited in the capital without hearing from the king, the more embarrassing his situation became. His bluff had been called, and Tycho responded in characteristic Brahe fashion by moving his entourage on to Rostock, outside Danish territory. From there he made a final attempt to salvage the situation by drafting an impertinent letter to Christian in which he explained his willingness to return to Denmark, if it could be done on "fair conditions and without injury to myself." Christian wrote a stinging reply stating that, "Now we will graciously answer you that if you wish to serve as a mathematician and do as you are told, then you should first begin by offering your services and by asking about them as a servant should." [11] Tycho was stunned; for once he had met his match! But for a man of his character and disposition there was no choice but to continue on; he simply could not tolerate the embarrassment of returning to his beloved island, tail between his legs, even though Hveen was to remain his for as long as he lived.

Considering the unfortunate circumstances under which he left home, there is something quite saddening about Tycho's voluntary abandonment of Hveen and his beloved Uraniborg, which he was destined never to see again. It was as if Jefferson, because of a misunderstanding with the Governor of Virginia, had chosen to leave Monticello forever. Tycho had not only ruled supreme over his island estate but he had personally superintended nearly every detail of the design and construction of Uraniborg, Europe's first great astronomical observatory. In it were reflected his artistic tastes, cultural values, and, most important, his peculiar genius as the world's greatest naked-eye astronomer. When we consider how little Tycho accomplished during the remainder of his life, one wonders whether his already magnificent contribution to science might not have been even greater had he chosen to remain in the more productive environment of Hveen. On the other hand, had he not left Denmark it seems quite doubtful that he would have ever made the personal acquaintance of Johannes Kepler. It is almost as if Tycho sensed the need to find a more creative and unfettered mind than his own, one that could use his precious observational data for a higher purpose than he himself was capable of achieving.

After nearly two years of uneventful wandering, Tycho sought and obtained the patronage of the Holy Roman Emperor Rudolph II, a man deeply interested in all matters pertaining to science. He appointed Tycho to the prestigious post of Imperial Mathematician, gave him a castle of his own choosing, and an annual salary of 3000 gold florins. Tycho arrived in Prague in June 1599 and shortly thereafter entered into negotiations with Kepler, who was then about twenty-eight years of age and a professor of mathematics at Gratz. Kepler had already created quite a stir with the publication of his highly mystical and speculative work *Mysterium Cosmographicum (The Cosmographical Mystery)* and, impressed by Tycho's reputation as a great astronomer, agreed to join him within a few months. Kepler departed Gratz for his fateful meeting with Tycho on January 1, 1600, and one can think of no more auspicious way for the first full century of genius in modern science to have begun. Though Tycho had less than two years left to live, he provided Kepler, a man of audacious and fertile intellect, with the cumulative result of some thirty years of strenuous and undaunted labor. With it Kepler solved the problem of planetary motion by inventing the law of ellipses. In a single, powerful stroke of inspired genius the circular dogma that had haunted astronomers from Thales to Copernicus was dealt a death-blow.

Tycho Brahe died on October 24, 1601 from a brief but painful illness which illustrates that his appetite for the excesses of life remained undiminished until the end. He had gone to a friend's home for dinner where, as usual, he drank quite heavily and, in respect for the strict etiquette of his day, did not leave the table to relieve himself. The tension on his bladder became so great that the prostate gland was damaged to the point where the doctors could do nothing for him. Tycho was still in the prime of intellectual life, only two months short of his fifty-fifth birthday. His last days were spent in a state of delirium during which he kept repeating the same words: "Let me not be thought to have lived in vain!" Emperor Rudolph, deeply distressed by Tycho's passing, ordered a magnificent funeral for his Imperial Mathematician, who was buried in a suit of armor in Prague's Teyn Church. In 1604 Christine Brahe, his ever faithful wife, was laid to rest at his side. During the years following Tycho's death, the island of Hveen passed through several unappreciative hands, including those of King Christian's mistress. Uraniborg gradually fell into ruin, and in 1623 its red bricks were sold to a contractor for building materials. Finally, in 1658, the Danish government was forced to cede the island to Sweden as part of the humiliating Treaty of Roskilde, and it has been a Swedish possession ever since. Somehow it seems fitting that during the brief period of its existence Uraniborg truly had only one lord and master.

So here's to you, Tycho! Braggart and tyrant though you were, you made the work of Kepler and Newton possible through your unwavering dedication to scientific accuracy. You are a refreshing exception to the somber, retiring Copernicus and the neurotic, insecure Kepler. Without both your methodical observations and flamboyant personality, the history of modern astronomy would be very much the poorer.

6.   Johannes Kepler: The Mystical Lawgiver

The Misfit

IN ATTEMPTING TO CHOOSE BETWEEN THE TWO, it is difficult to say whether, in the centuries since their death, Nicolas Copernicus or Johannes Kepler has had the least enthusiastic historical press of the five major astronomers discussed in the latter half of this book. Tycho and Galileo, both outgoing and self-confident to the point of insufferability, were perfectly capable of defending their own work against contemporary critics, and in doing so they successfully laid the foundations for a highly favorable analysis of their major scientific contributions by future generations. Isaac Newton, a mathematical genius and synthesizer without peer, quite simply stands in a category by himself. But the quiet and reserved Copernicus has too often erroneously been passed over as little more than an Aristotelian revisionist, while Kepler, equally shy and self-critical, once bore the undeserved reputation of a medieval mystic, who, despite his discovery of the three laws of planetary motion, could not truly appreciate their revolutionary significance. Certainly, until recently, Kepler has been the most neglected of the five, and doubtless the least understood as well.

Johannes Kepler was born in an age when the pursuit of science still bore little resemblance to our current understanding of natural events and the methods by which the modern scientist reveals the secrets of nature through observation and experimentation. The concept of natural mathematical laws that explain causal relationships between phenomena had not yet been established, nor had learned men mastered the inductive method, whereby consequences are drawn from a hypothesis and tested by experience. Kepler's was an age when alchemy, astrology, and witchcraft still remained problems to be seriously argued, while the Roman Catholic Church, reeling from the success of the Protestant heresy, sought to tighten its tenuous grip on a moribund, anxiety-ridden medieval world. For all the unanswered questions raised against the cosmology of medieval Christendom, virtually everyone—Catholic and Protestant alike—maintained that man still occupied a privileged, central position in a closed, finite world.

Even though it is in the nature of modern science that its contributions, because of the special language of mathematics, can be universalized in a more objective manner than those of the nonscientific disciplines, we can see that a great individual scientist like Kepler (if his work is studied closely enough) left his signature on all that he discovered by virtue of the method he used to arrive at his conclusions. Still, we are erroneously led to believe that every step leading up to a great discovery is rigorous, logical, and necessary, as the scientist moves inexorably and dispassionately toward his goal. Intuition, guesswork, and luck seemingly play no part at all. Yet at no time in his brilliant career did Kepler adhere to our modern antiseptic view, that the process of scientific discovery be removed from speculation, passion, and intuition. In fact, he believed just the opposite. Historian of science Gerald Holton writes that "Kepler's embarrassing candor and intense emotional involvement force him to give us a detailed account of his tortuous progress. He allows himself to be so overwhelmed by the beauty and variety of the world that he cannot yet persistently limit his attention to the main problems which can in fact be solved." [1] In every one of his major works, we see a continuous fluctuation between daring, sometimes wild, speculation and precise attention to scientific detail. Kepler fashioned many of his ideas about the universe—both right and wrong—out of an inspired Christian–Pythagorean mysticism. Always the scientist–mathematician is ready to yield to the philosopher–mystic, and vice versa. When one role fails him, he temporarily abandons it and is sustained by assuming the other. Thus, in this single man of genius, the soaring, speculative mysticism of a Giordano Bruno is perfectly melded with the uncompromising rationalism of a Galileo. Indeed, Kepler stands like a lonely giant between the medieval animistic view of the closed world and the modern mechanistic concept of an infinite universe. Because of this, his magnificent contributions to science have either been unjustifiably ignored or foolishly undervalued by men of both worlds.

There is little, if anything, in the family background and early childhood of Johannes Kepler to suggest that this first child of a ne'er-do-well mercenary soldier and an innkeeper's daughter, who was nearly burned at the stake as a witch, would one day become a central figure in the scientific revolution of the seventeenth century. Johannes was born on December 27, 1571, in Weil-der-Stadt, a small Swabian Free City located in southwestern Germany. He lived in the crowded, turbulent cottage home of his paternal grandfather, Sebaldus Kepler, along with aunts, uncles, and numerous brothers and sisters, a number of whom suffered from major physical and mental disorders. Johannes himself was not entirely immune to the family curse of physical infirmity, for he was bowlegged, frequently covered with large boils, and suffered from congenital myopia and multiple vision. This latter affliction must have been particularly distressing to one whose love of the heavens ultimately defined his career.

Even though we know that children of the sixteenth century were indulged far less than are their counterparts in contemporary Western society, one might have expected that Johannes, as the oldest child and a son, would have been the apple of his parents' eye, particularly since he was a seven-months' baby and entered the world in very delicate health. This was not the case, however. His father, Heinrich, by all surviving accounts, was a man of vile temper whose only sustained passion seems to have been a perpetual longing for the adventure associated with military life. When Johannes was about twenty-five he compiled a genealogical horoscope of his family in which he describes Heinrich Kepler as a "vicious, inflexible, quarrelsome" man "doomed to a bad end." His mother, Katherine, seems also to have been a highly restless, contentious woman of equally unstable character. She had been brought up by a mysterious maternal aunt whose days were painfully foreshortened when the local authorities burned her alive at the stake as a witch. Katherine apparently inherited her ill-fated aunt's penchant for dabbling in magic, for she collected herbs, concocted potions and, in later life, dispensed free medical advice to anyone who would take the time to listen. On the face of it, such activities were certainly harmless in and of themselves and, for the times, far from out of the ordinary; but when the European witch-craze of the early seventeenth century swept across Germany, the merest suspicion of trafficking in the occult could and frequently did prove disastrous for both the guilty and the innocent. In her old age, Katherine Kepler, as we shall see, would have almost certainly shared her hapless aunt's unenviable fate had not her son, the most gifted scientist of the times, carried on a prolonged and dogged defense on her behalf.

The marriage of Heinrich and Katherine, as one would expect, proved anything but idyllic. Three years after Johannes' birth, following the arrival of a second son, Heinrich gave in to his passion for adventure by marching off to fight against the insurgent Protestants in the Low Countries, and this despite the fact that the Keplers had been among the first and most respected converts to the new religion. A year later Katherine, who could neither live in peace with her husband nor stand to be without him, left to join Heinrich in the Netherlands, leaving her young sons in the care of their paternal grandparents. Because of their advanced age, the elder Keplers chafed under this unwelcome burden and responded by treating little Johannes, who had unfortunately inherited their ungrateful daughter-in-law's dark complexion and diminutive stature, as an outsider. To make matters worse, the boy suffered a prolonged attack of the smallpox from which he nearly died during his parents' absence. Within a year, Heinrich and Katherine returned, after which Heinrich renounced his right to citizenship in Weil. He then purchased a home in the nearby town of Leonberg, but his plans to begin a new and more settled life came to nothing, for in 1577 he left once more for Holland, where he joined the mercenary army supporting the terror-ridden regime of the Duke of Alba. During his return home from a brief and lackluster campaign, Heinrich was arrested in Belgium for some unknown crime and nearly lost his life on the gallows. His financial resources depleted, he sold his house in Leonberg and quit the community in 1580. He next turned up in the small town of Ellmendingen in Baden, where he assumed the proprietorship of a local tavern. Meanwhile Johannes' education, begun only a few years earlier, was interrupted so that he could wash dishes in his father's place of business. Once again, however, Heinrich Kepler became infested by the lust for adventure, and in 1588 he abandoned his wife and family, now numbering six children, for the last time. He is alleged to have enlisted in the Neapolitan fleet as a captain and is thought to have died on his way home, supposedly somewhere in the neighborhood of Augsburg. In any case, he was neither seen nor heard from again. If Johannes harbored any fond memories of his father, he left no record of them. Certainly, as his astrological diary indicates, Johannes' harsh treatment at the hands of the tyrannical Heinrich left a permanently bitter impression on the quiet, sensitive youth. Neither did his unloving mother and grandparents fill this deep, emotional void; his only resort was to seek consolation by entrusting himself to an all-powerful and merciful God, whom he looked to for guidance in all his affairs, including questions of science. From this time onward, Johannes Kepler suffered from a deep-seated sense of insecurity, which left permanent and readily noticeable scars on his delicate psychic apparatus. Not only did he consider himself unwanted by others, he became a misfit in his own eyes, the lifelong victim of a cruel inferiority complex.

Fortunately the major physical and domestic difficulties that so deeply affected the future astronomer's youth were to a considerable extent mitigated by his special intellectual endowment, signs of which had already become apparent by the time he was old enough to enter school. The Keplers belonged to the Lutheran minority for whom its protectors, the Dukes of Württemberg, had created an excellent educational system in the towns and villages under their control. These so-called "Latin Schools" had the responsibility for preparing a new generation of talented Germans for service in the government and the Lutheran clergy. The Dukes, in their wisdom, also recognized that intellectual ability is not a monopoly of the upper class and provided a system of grants and scholarships for the children of poor but faithful Christian families. It seems very doubtful that Johannes' parents would have financed his formal education without such outside support, which their son received when his talents became known to the authorities.

As a boy of seven, he entered the Latin School at Leonberg, where the family was then living. The curriculum bore marked similarities to the studies undertaken by Tycho Brahe at the University of Copenhagen over 20 years earlier. Latin, the language of scholars and the agent of high culture, was the major subject of study. The boys were relentlessly drilled in grammar, composition, and public speaking. And regulations, as at Copenhagen, demanded that the students speak only Latin to one another. Kepler, as his great scientific treatises so richly demonstrate, completely mastered the ancient language, while his more popular works and private correspondence, mostly written in German, testify that the vernacular was little stressed in the new schools. In contrast to his fluent Latin style, his German is often uneven, cramped, and at times awkward, giving the lie to the then current belief that through the mastery of Latin a student's native language would automatically improve.

The precocious youth could have easily completed this level of schooling in the prescribed three years, rather than in the five it took him, had not his parents moved to Ellmendingen, where the boy was hired out as a part-time agricultural laborer and also used as a dishwasher in his father's rented tavern. Kepler later recalled two pleasant events that transpired during this otherwise unhappy period in his life, occurrences which helped to awaken his first tentative interest in astronomy. In 1577 his mother walked with him to the top of a hill near their village, where she pointed out the great comet whose movements across the heavens wrought havoc among the ranks of the Aristotelians. Three years later, at the age of nine, Kepler recounts that "I was called outdoors by my parents especially, to look at the eclipse of the moon." Although both celestial phenomena made a vivid impression on his inquisitive mind, there was no instant conversion to the life of a star-gazer, as in the case of the youthful Tycho. The young Kepler, with his parents' consent, had already decided upon a clerical career, provided, of course, that he could win further scholarships. It would take considerably more than a comet and lunar eclipse, no matter how spectacular, to permanently alter his carefully formulated plans.

Kepler's brilliance guaranteed his smooth progress from the Latin School in Leonberg to the next rung on the educational ladder, the Convent Grammar School of Adelberg, which he entered in 1584. The curriculum was more broadly based than at Leonberg; in addition to Latin the study of Greek, rhetoric, mathematics, music, and dialectics received considerable scholarly attention. Discipline was rigid: classes in the winter began at five o'clock in the morning, in the summer at four; the students wore drab, short-sleeved cloaks reaching below their knees;

and even the slightest violation of the strict rules brought instant reprisal from those in charge. Kepler's introspective nature, in combination with an almost fanatical compulsion to please his superiors, quickly set him apart from his fellow students, who were obviously envious of his intellectual gifts but also contemptuous of the youth's exaggerated piety and obsequious manner. This overwhelming sense of alienation from his peers prompted the future astronomer to compose a highly revealing self-portrait, which he entered in his horoscope record during his student days. In it Kepler compares himself to a servile dog:

> His appearance is that of a little lap dog. His body is fragile, wiry, and well proportioned. Even his appetite were [sic] alike: he liked gnawing bones and dry crusts of bread, and was so greedy that whatever his eyes chanced on he grabbed; yet like a dog, he drinks little and is content with the simplest food. His habits were similar. He continually sought the good will of others . . . ministered to their wishes, never got angry when they repressed him, and was anxious to get back into their favor.[2]

But despite his desire to please, the "dog" also had a nasty bite if strongly provoked: "He tenaciously persecutes wrongdoers—that is, he barks at them. He is malicious and bites people with sarcasms." Yet he quickly repents his ill-considered hostility: "When he committed a wrong, he performed an expiatory rite, hoping it would save him from punishment."

As a student of theology, Kepler pondered much over the questions raised by Luther's splitting away from the Roman Catholic Church, particularly those issues dealing with transubstantiation, predestination, and the fate of individuals who die without having received the sacrament of baptism. His conclusions, so characteristic of introspective genius, frequently conflicted with those sanctioned by religious authorities, thus setting the stage for the endless persecution he suffered throughout most of his adult life at the hands of both religious groups. It is almost as though, from the moment of his premature birth, he had been destined by some malevolent spirit to an ongoing adversary relationship with life. Kepler, too, seems to have recognized that his life would never be very happy, and he gradually reconciled himself to his condition. Years later he chose a verse from Persius, an ancient satirical poet, as his motto: "O cares of man, how much of everything is futile." [3] It is little wonder that this misunderstood genius increasingly sought solace in religion and in the privacy of his scientific studies.

Having completed the second stage of his education with highest honors in 1588, Kepler received a prized scholarship to the University of Tübingen, already renowned as an intellectual center for Protestant theological studies. Johannes, true to form, proved an excellent student in all subjects including theology, but he soon developed a strong appetite for the study of astronomy and mathematics. It was his good fortune to matriculate while Michael Maestlin, one of the sixteenth century's most learned and esteemed astronomers, was a member of the Tübingen faculty. Although Maestlin advocated the geocentric system of planetary motion in his public lectures (the teaching of Copernican theory seems to have been prohibited by the Lutheran faculty of Tübingen), privately he discussed and accepted the heliocentric universe. Apparently his early recognition of Kepler's genius persuaded the noted astronomer to admit his student to that small circle of intimates who shared his views. A spark instantly ignited in Kepler's brain and, much to the consternation of his superiors, he entered into disputations with fellow students in support of the new astronomy. Kepler also learned a good deal of practical astronomy from Maestlin, and he began to grapple with the revolutionary question of how the heavens would appear to an observer standing on the surface of another planet. He knew from studying Copernicus that the earth is moving very rapidly. Yet those who inhabit the planet's surface are unaware of its tremendous motion because they cannot detect it through the use of their senses. He quite logically reasoned that a man standing on the moon, for example, would share an identical experience: he would see the earth change positions because he would not be a participant in its rotation, just as a moonwatcher on earth observes lunar motion in which he does not participate. This realization and the complex issues it raised became the basic theme of a student dissertation written by Kepler in 1593.

Had the Tübingen faculty been more tolerant of the new astronomy, the thesis he presented in the dissertation would have been publicly debated and doubtless long forgotten. However, when the proposal was presented to the authorities for their approval they vetoed the debate. One of Kepler's few close friends and a fellow student, Christoph Besold, who later became a noted professor of law at Tübingen, appealed to his professor and advisor, Vitus Müller, to permit him, rather than Kepler, to uphold the thesis in a public disputation, but Müller refused. The fact that Besold requested to debate Kepler's hypothesis suggests that the authorities were probably well aware of the close Kepler–Maestlin relationship, and that Kepler and his friend considered it more likely that Müller and his colleagues would sanction the debate if it were led by Besold, a law student, rather than by Kepler, an aspiring theologian. Kepler, no doubt disappointed and perhaps even somewhat embittered, wisely decided to hold his manuscript in safekeeping until the time when a more favorable climate of opinion might prevail. He was realistic enough to know that to further protest his fate—when even his highly respected professor of astronomy was condemned to public silence on matters Copernican—would be foolhardy and perhaps permanently damaging to his career. Still, there was sufficient grit to produce a pearl. Twenty years later the neglected manuscript was resurrected when Kepler renewed his study of the questions propounded at Tübingen during his student days. In 1634, four years after his death, the greatly modified work was published under the title of *Somnium* (or *Dream*), the first work of modern science fiction concerning a space voyage to the moon.[4]

Meanwhile, Kepler was graduated from the Faculty of Arts at Tübingen and enrolled in its famous theological seminary, the Tübingen Stift, where he began final academic preparation for his vocation as a Lutheran clergyman. Kepler's clerical studies were due to conclude during the summer of 1594, but a few months prior to this there occurred a decisive turn of events. Georg Stadius, mathematics teacher at the Protestant gymnasium in Gratz, capital of the Austrian province of Styria, died in late 1593. The Governors, as was the custom, turned to the Protestant faculty of Tübingen, requesting that its members recommend a qualified successor. The Tübingen Faculty Senate, after a short deliberation, selected Kepler!

Why, we must ask, would the professors select a replacement so near the completion of his theological studies? Was it because Kepler seemed the best qualified for the position, or were there other important but less obvious considerations; namely, his affinity for Copernican astronomy and his hesitation to accept certain fundamental doctrines of the Church? In deference to both Kepler and his professors, there can be no doubt that the young man was by far the most qualified for the post. His brilliance as a student of mathematics was unquestioned, and the Tübingen authorities were obviously desirous of having the most competent candidate represent the University. But neither can we overlook the fact that even though Kepler was in many ways a model student, some of his previously discussed activities must have raised certain doubts in the minds of at least some of his superiors about his fitness for the clerical life. After all, the new religion had been founded by a rebel, and God forbid that its hard won victory be jeopardized by giving birth to one of its own! Orthodoxy had to be defended at all costs.

Kepler was taken aback and at first could not conceive of abandoning the promise of a pulpit, especially for the lowly position of mathematics teacher in a provincial school. Yet he felt duty bound to consult privately with several of his professors, including Maestlin, all of whom, much to his surprise, urged him to accept the appointment. Finally, after exacting an agreement that he be permitted to complete his theological studies at some future time, Kepler accepted the Faculty Senate's offer. If nothing else, the position would provide him with his first opportunity for financial independence, something he had longed for ever since he had become reliant upon the generosity of others for his subsistence and education. Probably the worst that could happen is that he would be unsuited to the classroom, but at the age of twenty-one there was ample time to retrace his steps. On March 13, 1594, Johannes Kepler walked out of the gates of Tübingen University to begin his journey to Gratz and beyond. Although he did not yet realize it, his cherished plans for a career in the ministry would soon be permanently abandoned in favor of a new passion, science.

Passions of the Heart and Mind

Kepler arrived in the city of Gratz on April 11, 1594, and soon discovered the religious and intellectual climate in his new home to be far less congenial than that to which he had been accustomed in Württemberg. In his homeland, the peasants and bourgeoisie, like their rulers, accepted Luther's teachings without qualification; but in Styria tensions between the Protestant subjects and their unconverted Catholic leaders mounted year by year. These religious differences might have been tolerable had not the Catholic authorities chosen to enforce to the letter the legal provisions of the Religious Peace of Augsburg signed in 1555, which conferred upon the German princes the right to choose between the Catholic religion and the Augsburgian (Lutheran) Creed for their entire domain. During the late 1560s Archduke Charles, under whose rule Gratz had fallen, summoned the Jesuits to the city for the purpose of launching a Counter Reformation in miniature, a policy continued under his son, the Archduke Ferdinand. In 1573, the Jesuits erected a college in the city and shortly thereafter they founded a Latin school, through which they began the conversion of the young. In direct response to this Catholic challenge, the Protestants, in 1574, had founded the Stiftsschule, to which Kepler had just been appointed teacher of mathematics. Had he known of the highly competitive nature of Gratz's educational system, there can be little doubt that he would have declined the appointment in favor of completing his theological studies. The intensity of religious feelings eventually became so great that six years later he was forced to flee the city.

The Stiftsschule at Gratz was separated into two basic divisions. The lower division, composed of three groups of ten boys each, offered the standard curriculum introduced elsewhere by Luther's associate, Philipp Melanchthon. The upper division, to which Kepler was assigned, contained four classes. The major subjects taught were theology, history, metaphysics, logic, rhetoric, the classics, and mathematics. The instructional staff included twelve to fourteen lay teachers and four clergymen. Since he came as a novice, Kepler had to prove himself to both his colleagues and his superiors. After a favorable introductory interview with the administrators of the school he was placed on probation for several months and granted the meager annual salary of 150 gulden. Kepler commenced his lectures with only a few students and a considerably larger number of misgivings. The fact that the young mathematician was totally inexperienced, never having taught a day in his life, coupled with his strong sense of inferiority, put him on the defensive at the outset. He later wrote that his introductory lectures were poorly structured and overenthusiastic. He continually confused his students by introducing "new ways of expressing or proving his point" before making his intentions clear, and he altered "the plan of his lectures or held back what he intended to say." The reason for this, he says, is that he was incapable of focusing his attention on any one point for a prolonged period of time because of the turmoil produced by several "images of thought in his memory" which "he must pour out in his speech. On these grounds his lectures are tiring, or at any rate perplexing and not very intelligible." [5] These same characteristics seem also to have applied to many of the early lectures given by a young scientific genius of the twentieth century, Albert Einstein. Only the most exceptional pupils can benefit from instruction of this kind, and these are few and far between.

As the teacher of mathematics, Kepler faced another major problem, the solution to which lay beyond his personal control. His students were mainly the sons of the landed nobility and rising burgher class. They had received little or no previous training in higher mathematics; neither did they display very much independent scholarly interest in the subject. To make matters worse, mathematics (as opposed to simple arithmetic) was not a required part of the Stiftsschule curriculum, so that during his second year in Gratz Kepler's classes did not draw a single student! This depressing turn of events was anything but the proper medicine for a man of Kepler's hypersensitive ego, and it proved sufficient to cause him to seriously consider leaving Gratz for the security of his beloved Tübingen. Fortunately, however, the directors of the institution were broad-minded enough to objectively consider the realities of the situation; they absolved Kepler from any personal blame by recognizing that "the study of mathematics is not every man's meat." Instead of releasing the shaken young teacher, they decided to make use of his considerable training in the classics by having him offer lectures on Virgil and rhetoric until, in their words, "an opportunity will arise to make use of his knowledge of mathematics." [6] Kepler's intellectual potential was thus appreciated far more than he had realized, because such a major shift in assignments was not a matter which the authorities took lightly, particularly when there was an abundance of qualified candidates available to teach Latin and the classics.

In addition to his duties as a teacher, Kepler also held the title of District Mathematician, a traditional obligation imposed on Styria's official mathematicus. The main responsibilities of the office included the making of local maps and the publication of an annual calendar of astrological forecasts, at an additional and most welcome income of twenty

florins per year. Kepler issued his first calendar in 1595 and continued
the practice during each of the six years he remained in Gratz, although
only those for 1598 and 1599 remain extant. While this is not the place
to go into detail concerning Kepler's attitude regarding astrology, it
must be said that at this early stage in his career he, like Tycho, believed
that the celestial bodies do exert considerable control over the lives of
individual men. Yet he found it very difficult, if not foolish, to prophesy
major events a year in advance. For these calendars were not only
regulators of the months, days and religious activities, they were also
sources of information about the weather, recommendations for planting
and harvesting, forecasts on disease, and predictions on war, political
events, and natural catastrophe. Thus, Kepler had to go far beyond the
accepted practice of casting the horoscopes of specific individuals; he
was required to compile sweeping prognostications for a whole society.
This grated on his conscience because he had already spoken out against
the predictive methods employed by the vast majority of popular astrologers, dismissing their work as nothing less than "necromantic
monkeyshines."

In spite of his grave misgivings about man's ability to reliably predict the future, Kepler seems to have derived a certain secret pleasure
from privately toying with the rules of astrology. Doubtless this attitude
was reinforced to some degree by the public recognition he received
after he accurately prophesied two major events in his first calendar, one
natural, the other man-made. Kepler had predicted that the winter of
1595–96 would be bitterly cold and that Austria would suffer an invasion
by the infidel Turks. Several months later he reported somewhat boast-
fully to Michael Maestlin that "There is unheard-of cold in our land.
In the Alpine farms people die of the cold. It is reliably reported that
when they arrive home and blow their noses, the noses fall off." As for
the Turks: "On January 1 they devastated the whole country from
Vienna to Neustadt, setting everything on fire and carrying off men
and plunder." [7] The fact that both these prophecies came true markedly
enhanced the prestige of the young mathematician, leading to an increase
in his meager salary. Yet even though this was a euphoric moment for
the attention-starved mathematician, he had the good sense not to surrender permanently to such crude practices. Kepler realized that his
success derived more from a combination of common sense and good
luck than from the successful application of the principles of astrology.
Still, he never joined the ranks of the few well-educated individuals of
his time who completely rejected astrology on either theological or
scientific grounds. He gradually developed the intuitive feeling that the
entire universe is a miracle and that astrology operates as the doctrine
of world unity—a kind of animated cosmic soul that breathes life into
the souls of individual men.

We have thus far primarily concerned ourselves with Johannes Kepler's emotional development as a child and adolescent. We must now turn briefly to the intellectual background of his age, particularly as it affected his perception of Copernican cosmology. The few respected astronomers of the sixteenth and seventeenth centuries who took major exception to the Aristotelian system, including Copernicus, Maestlin, and Brahe, had done so largely because they had grown increasingly disenchanted with its proponents' emphasis on qualitative values to the virtual exclusion of mathematical or quantitative data. Although Aristotle had not opposed the study of mathematics, he had certainly minimized its importance as a key for unlocking the secrets of nature. Indeed, he was more content with describing the operation of natural forces than with truly discovering their underlying causes. Conseqeuntly, he advocated that logic, supported by sense experience, must be the natural philosopher's main tool. On the other hand, Plato, his great teacher at the Academy, fell victim to the Pythagorean passion for mathematics—namely, geometry. The Platonists kept alive the fundamental teaching of Pythagoras and his followers, that number is everything and that the study of nature should rest primarily on mathematical principles, which alone provide man with the only certain knowledge he can ever attain. But by virtue of the magnificent scholarly achievements of St. Thomas Aquinas, Aristotelian philosophy, in combination with Christian doctrine, eclipsed the teachings of Plato and Pythagoras in Europe during the thirteenth century, though their work was not entirely forgotten.

As the inconsistencies in Aristotle's qualitative interpretation of the world began to accumulate during the fourteenth and fifteenth centuries, Platonic thought, in which the Pythagorean element was very strong, underwent a major revival, especially in the universities of southern Europe. Many of the early Renaissance humanists openly rejected both the logic and rationalism of popular scholastic philosophy with its strong Aristotelian underpinnings. During the Quattrocento a number of them became attracted to Neo-Platonism, the study of which was strongly encouraged by the great Italian banker and merchant Cosimo de Medici, who founded the Platonic Academy in Florence. The reputation of the Academy quickly spread; its faculty came to include the highly respected Neo-Platonists Marsilio Ficino and Pico della Mirandola. Under their influence number theory enjoyed a major revival as thinking men once more began to look upon the cosmos as a beautiful musical instrument, perfectly attuned by the harmonious relationships associated with numerical regularity.

It was just at the time when the work of Ficino and Mirandola began to influence scholars in the universities south of the Alps that Copernicus matriculated at the University of Bologna. There the most important representative of the new school of thought was none other than Domenico Maria Novara, professor of mathematics and astronomy, and the man with whom the young Copernicus became most closely associated. Novara's influence on Copernicus during his six years residence in Bologna was considerable. As a result of working with the mathematician, Copernicus became a convert to the Pythagorean belief that the universe is a geometrical and harmonious creation. By taking this radical step, Copernicus was later able to modify the heliocentric cosmology of Aristarchus in such a way that he tied all of the orbital movements of the planets together to form an interlocking and interdependent system. The philosophy of the Pythagoreans had thus been vindicated in the eyes of a small but growing number of perceptive scholars because Copernicus had created a more orderly and harmonious system than that left to posterity by Aristotle. And by transferring the earth from the center of the universe to the previously inviolable Aristotelian heavens, Copernicus had made it subject to the same mathematical relationships that govern its sister planets. Aristotle's artificial distinction between the celestial and terrestrial realms had been seriously challenged for the first time.

What, then, does this have to do with the intellectual development of Johannes Kepler? The answer—everything! During the decades following the death of Copernicus, few individuals were courageous enough to openly advocate the belief that mathematics can be used as a major tool for better understanding the universe and man's place in it. However slow to crystalize and assert itself, this belief, in combination with other major elements of Neo-Platonic thought, continued to make headway in the intellectual circles of Europe. Like Novara, Michael Maestlin, Kepler's teacher at Tübingen, formed a strong attachment to the concept of mathematical order in the universe. In fact, he was so taken by the idea that he became, along with English mathematician Thomas Digges, an enthusiastic convert to Copernican cosmology. Kepler, Maestlin's greatest pupil, seized upon the revolutionary doctrine even more quickly and with fewer reservations than his teacher, and it soon became the prime motivational force governing his lifework.

Kepler began teaching during his early twenties, the very age when the majority of mathematicians destined to make a significant contribution to the expansion of knowledge in their field do their most creative work. The sleepy Austrian town of Gratz provided few distractions from his scholarly pursuits, and since his classes were small Kepler had considerable free time to devote to mathematics and astronomy. He had become increasingly preoccupied with the details of Copernican cosmology, many aspects of which he believed needed far greater study and elaboration. In fact, he referred to Copernicus' revolutionary treatise *De revolutionibus* as "an unharvested treasure of truly divine insight."

And consciously or otherwise, his thinking on these matters was connected to the doctrines developed by Pythagoras and Plato.

On July 9, 1595, a date Kepler later recorded as the most memorable of his life, he was drawing a geometric figure on the blackboard for a class when an idea struck him with such force that his mind exploded and took flight, like a star gone supernova. Why, he was thinking to himself, are there exactly six planets? And is there any direct mathematical connection between their orbital distances from one another, or between their distances from the sun and the time required to complete their orbits? The thought suddenly flashed across his mind that in solid geometry one can construct only five regular three-dimensional solids, no more, no less. Between the six planets in the Copernican system (as opposed to the seven in the Ptolemaic) there are five spaces; why not, then, place one of the five so-called "Pythagorean" or "Platonic" solids in each of these spaces! After considerable trial and error he fitted the five solids, each of which can be perfectly inscribed into a sphere (or planetary orbit) so that all of its corners (vertices) lie on the sphere's surface, between the planets in the following order: Saturn–cube–Jupiter–tetrahedron–Mars–dodecahedron–Earth–icosahedron–Venus–octahedron–Mercury. Believing that these five solids truly determine the distances between the planets, Kepler then proceeded to construct an elaborate geometric picture of the entire system, as shown in Figure 14. The innermost figure is the sphere containing Mercury's orbit; the outermost sphere contains the orbit of Saturn and an inscribed cube.

Kepler now decided to put the system to its first true mathematical test, by computing the radii of the circumscribing spheres. Are these, as he theorized, in proportion to each other as are the respective planetary distances from the sun? Alas, his computations proved they were not. Kepler was crestfallen. But then he joyously remembered what Copernicus, Ptolemy and many of the ancient Greek astronomers had believed; that a planet's orbit is not a single circle but a complex combination of circles on circles! Room had to be allowed for the regular deviations of the planetary motions from their circular paths round the sun. Kepler reasoned that the planetary spheres must possess a thickness great enough to permit each planet sufficient room for its epicyclic and eccentric be-

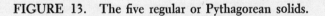

TETRAHEDRON   HEXAHEDRON   OCTAHEDRON   DODECAHEDRON   ICOSAHEDRON

FIGURE 13. The five regular or Pythagorean solids.

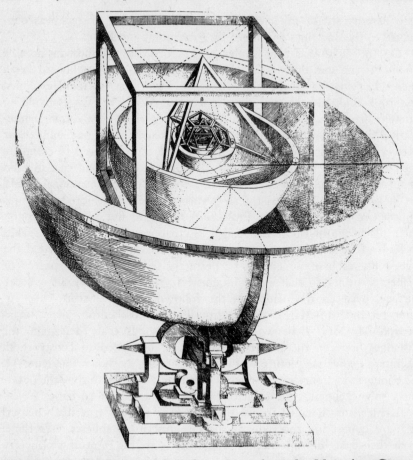

FIGURE 14.   Kepler's model of the universe from the *Mysterium Cosmographicum*. The outermost sphere is Saturn's. (*Courtesy of the Lilly Library, Bloomington, Indiana.*)

havior. By thickening the surface of each circumscribing sphere, its respective planet could perform the various motions peculiar to itself while orbiting the sun. He carefully calculated the new values. They were more satisfying than those computed earlier, but at best they yielded only a rough approximation of what he had hoped to find. He needed much more accurate data on the planetary positions to put his hypothesis to a true test. Still, he was reasonably satisfied that he had made one of the greatest astronomical discoveries of all time.

Kepler's meteoric flight into an exhalted state of Pythagorean euphoria remained the main inspiration of his life, even though he himself later discovered it to have been premature. Like Michelangelo, who wept in disbelief when he gazed upon his just completed masterpiece, *the Pietà*, Kepler shed tears of wonder at having been chosen as God's instrument for revealing a secret previously known only to the Divine Mind. "The delight that I took in my discovery," he later wrote, "I shall never be able to describe in words."

After several months of further refining his newly discovered theory, followed by a visit to his homeland to see his family, Kepler decided to make the results of his work available to scholars through the publication of his first book, *Mysterium Cosmographicum*. He enlisted the aid of Michael Maestlin, who, after obtaining permission from his superiors at Tübingen, generously agreed to edit the text and supervise the printing, the first copies of which reached Kepler in the spring of 1597. The *Cosmographical Mystery* became the first major and unequivocally Copernican treatise to be published since *De revolutionibus* itself, for without the idea of the sun-centered universe the entire rationale of the new book would have been completely indefensible. The work begins with a general outline of the heliocentric system followed by a detailed discussion of Kepler's novel theory that the planetary spheres are separated from each other by the five regular solids.* It was not by chance, Kepler declares, that the Creator constructed a universe of perfect geometric forms. Kepler borrows from Pythagoras the idea that geometric figures are eternal because they are copied directly from the Mind of God, which is itself completely rational and eternal. He consciously rejects the Aristotelian proposition that man learns primarily through the use of his senses in favor of the Pythagorean–Platonic belief that what we know we have always known. Since geometric forms are the only ones that can be totally separated from the confused impressions we receive from the senses, they alone are subject to quantification; therefore, God must have imprinted the geometric data on man's mind at the very beginning. By looking at the world around us, sense impressions simply trigger the mechanism of recall. Or to put it another way, God is a master geometrician who made the human mind in such a way that it can discover truth only through the use of quantification. Kepler's Neo-Pythagoreanism is brilliantly summarized by historian–philosopher E. A. Burtt: "Not only is it true [according to Kepler] that we can discover mathematical relations in all objects presented to the senses: all certain knowledge must be knowledge of the quantitative characteristics, perfect knowledge is always mathematical." [8] Kepler thus rejects Aristotle's world of commonsense experiences because it is nonmathematical and hence devoid of objectivity. Those phenomena which cannot be quantified are of *secondary* and only passing significance, part of a

---

* It would be erroneous to suppose that Kepler truly believed that the five regular solids are actually physically present in space. Neither does he accept the existence of the planetary spheres themselves. Both are geometrical abstractions which will hopefully enable the mathematician–astronomer to compile accurate quantitative data on the planetary movements so that man can better understand God's plan of creation.

lower level of reality. The true scientist must concern himself only with that which can be put into abstract mathematical terms, or what Kepler and Galileo defined as *primary* qualities. This idea, as we shall see, lies at the very heart of the scientific revolution and of the modern obsession with quantification.

Kepler next turned his attention to one of the questions that had most troubled Copernicus: If the Primum Mobile does not drive the planets and stars in their orbits, then what exactly causes the celestial bodies to move? Like Copernicus, Kepler rejects the Thomistic idea that the planets are carried round by celestial intelligences; but he also dismisses Copernicus' traditional explanation that they continually move in circles because it is natural for them to do so. Kepler became the first modern astronomer to seriously question why, if we could travel outward from the sun into space, we would observe that the farther away a planet is from the sun, the slower it moves along its orbit. Some planets obviously have much larger orbits to complete, but this is not the entire answer. Saturn, for example, with an orbit approximately twice as great as Jupiter's, does not complete its circuit in twice the time, as would be the case if it were moving at exactly Jupiter's speed. Instead, Saturn takes about two and a half times as long as Jupiter to circle the sun. Moreover, Kepler took strong exception to Copernicus' designation of an abstract point in space—to one side of the sun rather than the center of the sun itself—as the true center of the universe. A rational God, Kepler believed, was capable of doing better. The sun must be the *true* center of the universe or the demonstrated geometric harmony of the cosmos would collapse.

By focusing his attention on the sun, Kepler singlehandedly launched the modern search for a physics of the solar system. Since God had placed the great globe in the center of His creation, it must somehow hold the key to understanding planetary movements. He remembered that the Egyptians had worshiped the sun as their greatest god because of its lifegiving heat and light. Could not this giant, fiery globe also be the fountain of additional but previously unrecognized powers? Kepler now determined that the sun is also the source of a tremendous propelling force which he called the *anima motrix*, or moving soul. This force field, he hypothesized, spreads outward in all directions like the spokes of a gigantic wheel; as the sun rotates on its axis the energized spokes drive the planets round the sun at its hub. The reason that the outer planets move at a slower rate of speed than those closest to the sun is because the propelling force diminishes in relation to the distance from the center. Or to put it another way: at twice the distance from the sun half as much force from the *anima motrix* would act on a planet. This revolutionary proposition, though later proven false, eventually led Kepler to the discovery of his three laws of planetary motion, and a century later to Newton's discovery of the law of universal gravitation itself. Kepler, to his lasting credit, was the first to seek a universal physical law of celestial mechanics. Through his work, astronomy, however tenuously, was once again united with physics after a separation of some twenty centuries. The foundations of the modern vision of the universe were now beginning to settle into place.

Kepler's quest for a celestial physics was, as we have observed, sustained by his deep commitment to Neo-Platonic metaphysics; indeed the two are indivisible in Kepler's thought. For like Copernicus, who in his cosmology exalted the sun to such a degree that it took on divine associations, Kepler also ascribed great mystical as well as physical powers to this central body. According to Kepler, the sun is the very "heart of the world." By virtue of its dignity, power, and motive force, it alone of all places is worthy of the title: "House of God." Borrowing from the Christian concept of the Holy Trinity, Kepler likens the sun to God the Father; the sphere of the fixed stars to God the Son; and ethereal space, through which the power of the sun is transmitted to the planets and stars, becomes the Holy Ghost. The movement of light and power from the center of the universe outward toward the farthest stars is for Kepler a continuous, symbolic reenactment of divine creation.

Kepler's concept of the universe, like Aristotle's, is hierarchical, though Kepler rejects the Stagirite's concept of an ascending order of elemental purity from the center outward. Aristotle had bestowed upon the Primum Mobile the frictional power to move the celestial spheres. Thus, in the Aristotelian system the closer one moves toward the center of the universe the more physically corrupt it becomes because the distance from the source of divine inspiration increases. Kepler, on the other hand, abolishes the Primum Mobile and transfers the power of movement from the outer surface to the divine inner core of the universe, the sun. He does retain a belief in the Aristotelian concept of a finite universe, however, albeit for different reasons than the Greek philosopher. Since, according to Kepler, the universe is a perfect geometrical construct, it must by definition have limits, no matter how great its boundaries. If such limits do not exist, there can be no structure, and without structure there is no harmony—only chaos. And chaos is anathema to the Divine Mind. Furthermore, the force of the central sun is sufficient for moving only the six planets separated by the five regular solids. Were the universe infinite, the entire system would cease to function because no motivational force is capable of exerting itself across an infinite distance. Even Galileo's use of the telescope did not persuade Kepler to accept infinitist cosmology, although it did force him to admit that the sphere of the fixed stars reaches outward much farther than was previously believed.

Kepler's hope, then, of unifying a classical world picture split into celestial and terrestrial realms rests on both physical and metaphysical principles. There is a direct and crucial connection between Kepler, the metaphysical sun-worshipper, and Kepler, the celestial physicist. Even though he failed in his first work to provide a satisfactory mechanical explanation for the observed motions of the planets, it was primarily because of such metaphysical considerations as the deification of the sun that Kepler supported the Copernican system and continued his quest for a unifying physical principle; one that would vindicate his belief in the existence of an all-pervading mathematical harmony. Never, so far as I know, has a major idea like the one put forth in the *Mysterium Cosmographicum* been so erroneous and at the same time so germinal in molding the future course of scientific development. Michael Matestlin, while he disagreed with certain of Kepler's ideas, immediately sensed the revolutionary implications of the exuberant treatise: "The topic and the ideas are so new," he wrote to a colleague, "that up to now they never entered anybody's mind. For who ever conceived the idea or made such a daring attempt as to demonstrate *a priori* the number, the magnitude and the movements of the celestial spheres and to elicit all this from the secret, unfathomable decrees of Heaven!" [9]

When the first edition of the *Mysterium* appeared in print, early in 1597, Kepler immediately dispatched copies to the leading mathematician–astronomers of his day. He was particularly interested in how Tycho and Galileo would react to his unorthodox hypothesis. In the meantime, Kepler had fallen under the spell of a new passion, one that for a time proved every bit as strong as his zealous commitment to the new astronomy. In December 1595, some friends in Gratz had decided that the young mathematicus was in need of a permanent companion to cheer up his lonely household. And, as luck would have it, they knew just the woman who would make him a proper match! Her name was Barbara Müller, the eldest daughter of a well-to-do mill owner from Gössendorf, not far from Gratz. Although still in her early twenties, Barbara had already been twice married and twice widowed, and she was the mother of a seven-year-old daughter named Regina. Though he picured her as "simple of mind and fat of body," Kepler unfortunately made this observation from the wisdom of hindsight, long after his romantic ardor had cooled to the point of extinction. For the moment, he was blinded by a passion that only first love can kindle, and he immediately entered into a long and difficult negotiation for Barbara's hand with his future father-in-law Jobst Müller, a man who had formed a rather low opinion of Kepler's station and prospects in life. Unable to make headway on his own, Kepler requested that two of his friends plead his case, a physician and Governor of the Stiftsschule, Johannes Oberdorfer, and Heinrich Osius, a retired professor. Meanwhile, Kepler reluctantly left Gratz for an extended visit to Swabia, where his two grandfathers, both aging and in poor health, wanted to see their successful grandson before they died. Since Müller, a hard-headed businessman, had no appreciation for the value of scholarship as compared to the more tangible assets of money and property, the chequered matrimonial negotiations nearly fell through on a number of occasions. Oberdorfer and Osius persevered, however, and the wedding, with Müller's less than enthusiastic consent, took place on April 27, 1597.

Considering Kepler's unhappy past, it would have been too much to hope that married life would alter the situation. Almost immediately Kepler recognized that he had made a serious mistake, but in the society of his time there was no retracing his steps without seriously jeopardizing both his professional and social standing. Barbara Kepler, who like all women of her day was excluded from higher education, neither shared her husband's ideals nor valued his scholarly achievements. Irritable, neurotic, and frequently despondent, she gradually descended into a state of incurable melancholy. To make matters worse, the unhappy marriage produced five children, of which only a boy and a girl survived infancy. Barbara's health declined with each pregnancy until she finally took to bed for long periods during the last years of her life. She died in 1611, fourteen years after her marriage to Kepler, at the age of thirty-seven.

A few months after his marriage, Kepler began to receive the reactions of the scholarly community to the *Mysterium*, which were mixed. The Neo-Platonists, oblivious to its revolutionary character, lauded its author for having greatly advanced the cause of Pythagorean metaphysics, while the more empirically-minded astronomers, also underestimating its modern value, remained highly skeptical. Galileo penned a short letter of gratitude for the copy Kepler had sent to him: "So far I have read only the introduction," he wrote, "but I have learned from it in some measure your intentions and congratulate myself on the good fortune at having found such a man as a companion in the explanation of truth." [10] Pleased that Galileo would respond so quickly but also disappointed with the brevity of his reply, Kepler wrote requesting his opinion of the entire work: "A great desire has taken hold of me," he explained, "to learn of your judgment." Galileo, much to Kepler's disappointment, remained silent. The two astronomers would cross paths again in the future, but only when it suited the great Italian's purpose. Kepler eventually came to realize what many others already knew, that Galileo was a consummate user of men, and he used no one more to his own advantage and with less compunction than Johannes Kepler. Only after his own work came under direct attack by the educational and religious establishment of his day did Galileo, by requesting Kepler's support, reluctantly recognize Kepler's greatness as a scientist in his own right. Despite a series of earlier rebuffs, Kepler came to the beleaguered astronomer's defense. We can be less certain that Galileo would have done the same for Kepler.*

Meanwhile, Tycho Brahe, Europe's greatest naked-eye astronomer, had read Kepler's book with considerable interest. Though Tycho rejected much of Copernican cosmology and many of Kepler's wild speculations, he knew creative genius when he saw it. After three decades of training some of Europe's most gifted young astronomers, Tycho still had not found the person whom he considered worthy of inheriting his massive collection of data. Now the years were rapidly catching up with the homesick nobleman, and his search for a successor had taken on a new urgency. Perhaps Kepler, a young man quite obviously blessed with the conceptual powers which Tycho lacked, was the individual he had been looking for. He initiated a correspondence with Kepler in the hope that he could convince him to come to Prague. Kepler, for his part, also entertained serious thoughts of moving on to a better position. Though his yearly salary at the Stiftsschule had been raised in consideration of his marriage, the pay remained pitifully low and kept him in a constant state of financial difficulty. He also realized that in the absence of more accurate observational data he would never be able to establish the validity of his model of the universe built around the five Pythagorean solids. Tycho's move to Prague meant that the data he craved were now tantalizingly close at hand, provided, of course, he could convince its temperamental owner to share it with him. Moreover, the forebodings of religious intolerance that had haunted Kepler ever since his arrival in Styria had recently taken more substantive form. In 1596, after his accession to power, the Archduke Ferdinand went on a pilgrimage to Loretto where, in a moment of exalted religious fervor, he vowed to root out the Lutheran heresy from his domains. In keeping with that vow, Kepler's school was closed in the summer of 1598, and on September 28 all preachers and teachers at the Lutheran Stiftsschule were ordered to leave the province within eight days. Kepler found it impossible to settle his affairs in such a short time and, like his colleagues, he was forced temporarily to leave his wife and family for the safety of nearby Hungary. However, within a month he received permission to return to Gratz, although his fellow faculty members at the Stiftsschule were denied the same right. The reasons for this special dispensation are not entirely clear, but it may have been because of the considerable respect Kepler had won for his scientific accomplishments among members of the scholarly Jesuit community. The Jesuits also knew of his reservations

---

* It is well to point out that some noted scholars take a significantly different view of the relationship between Kepler and Galileo than the one presented here and elsewhere in this book. See, for example, Stillman Drake, *Galileo Studies, Personality, Tradition, and Revolution* (Ann Arbor: 1970), pp. 123–139.

regarding certain aspects of the Lutheran doctrine, and they doubtless hoped Kepler might be won back into the Catholic fold. Many would have found the promise of peace in return for conversion an offer too tempting to resist, but Kepler, a man of unwavering religious principles, had no intention of compromising his faith. He realized that under these strained circumstances neither he nor his family could expect to be left in peace for very much longer. In desperation, he first turned to Maestlin in the hope that his old teacher might procure him a teaching position at the University of Tübingen. But Maestlin knew that Kepler's religious views coupled with his open advocacy of Copernicanism would make him no more acceptable to the Lutheran authorities than to their counterparts in the Roman Catholic community. Tycho, in the meantime, had been appointed Imperial Mathematician by Rudolph II, and, like it or not, he was now Kepler's last major hope. The beleaguered mathematician finally accepted the Dane's offer to join him in Prague. Kepler left Gratz in early 1600 and arrived in the capital of the Holy Roman Empire on January 26. It proved to be one of the most fateful moments in the history of modern astronomy; for the first time, two giants of the scientific revolution were to come face to face.

A Meeting of Giants

On February 3, 1600, Kepler passed through the gates of Benatky Castle, Tycho's temporary residence located about twenty miles outside Prague. It is difficult to imagine a greater contrast between the character and conduct of two men dedicated to the same branch of science than existed between Tycho Brahe and Johannes Kepler. One was rich, boastful, arrogant, a genius of astronomical observation; the other, poor, emotionally insecure, retiring, a genius of theoretical mathematics. Had they not been so incompatible in temperament, the talents of one would have been the perfect complement of those possessed by the other. They were alike only in their inexhaustible enthusiasm for revealing the wonders of the heavens. This bond alone kept the stormy relationship from a permanent rupture during the remaining eighteen months of Tycho's life.

The situation was made even more unbearable because both men were so beset by major personal problems that neither could fully appreciate the difficulties being experienced by the other. Though he was only fifty-three, Tycho felt his once keen intellectual powers gradually slipping away. To make matters worse, he had no worthy successor to carry on his lifework, and he grew increasingly fearful of what might happen to his treasured observations after his death. And despite his respected position, Tycho remained a stranger in a foreign land. His thoughts turned more and more to Hveen and his beloved Uraniborg, every detail of whose construction and planning he had personally supervised. He longed for the ocean, the salt air, and even the impenetrable fogs, so thick you could almost hear them move as they completely enveloped his tiny island domain. But most of all, he missed the busy stir on clear nights of his assistants reporting to him for their instructions, of making last-minute adjustments on the delicate instruments he himself had so pains-takingly constructed, and of moving from observatory to observatory to make certain that all was proceeding according to plan. Kepler, too, was a man in exile, the object of mindless religious persecution and private misfortune. He arrived at Benatky in poor health and on the verge of poverty, dependent once again upon the largesse of a wealthy bene-factor. The strain of his ordeal in Gratz had also taken its toll in other ways: Kepler's nerves were frayed to the breaking point; for the first time in his life he displayed definite signs of an uncontrollable temper, which, when combined with Tycho's tendency toward frequent emo-tional outbursts, produced an extremely volatile mix. As if the situation was not bad enough, Tycho had an old bone to pick with Kepler over an incident that had taken place several years earlier.

Before Tycho's appointment as Imperial Mathematician, the post was held by one Nicholas Reimers, who called himself Ursus (the Bear). Reimers had started as an illiterate swineherd of eighteen, who by his own unceasing efforts and native talent had taught himself the languages required to master the science of the day. He rose to become a uni-versity professor and finally Imperial Mathematician, only to die in dis-grace and ruin after a long and bitter feud with Tycho, his archrival. In 1584, Ursus, then in the service of the Danish nobleman Eric Lange, visited Tycho's Uraniborg with his employer. At the time, Tycho was at work on his compromise model of the universe, according to which the planets circle the sun while the sun circles the earth. Since he had not yet discussed the system with anyone, the great astronomer was astounded when, four years after the visit, Reimers published his *Funda-mentals of Astronomy*, in which he presented a planetary system very similar to Tycho's. The Lord of Hveen became enraged at what he believed was the theft of his original work; Reimers, Tycho charged, had gone through his private papers like a common thief while everyone was asleep. In truth, the "Tychonic system" was by no means as original as Tycho wished to believe; it had been discovered quite independently by yet a third mathematician, Helisaeus Roeslin. Reimers may well have displayed an injudicious amount of curiosity in an effort to better under-stand the nature of Tycho's work during his visit to Uraniborg, but the system he presented in his book was doubtless as much his as it was Tycho's and Roeslin's. But Tycho refused to believe that anyone else was capable of making the same discovery simultaneously, any more than Newton accepted Leibnitz's claim to the independent discovery of differential calculus. Tycho characteristically vowed to square accounts with Reimers, no matter how long it might take and no matter what the cost.

In 1595, Kepler became unwittingly involved in the growing controversy. At this time he knew neither Brahe nor Reimers personally, but he had read the latter's *Fundamentals of Astronomy* with great enthusiasm. Exuberant over his own newly discovered mathematical theory, Kepler sent an unsolicited letter to Reimers detailing his supposed discovery of a direct connection between the planetary distances and the five regular solids. Kepler, much to his lasting embarrassment, got so carried away that he falsely attributed all that he knew about mathematics to the guidance he had received from Reimers. The Imperial Mathematician did not even bother to pen a reply to the then unknown Kepler. But two years later, after learning of the *Mysterium Cosmographicum*, Reimers, in search of a potential ally, drafted a belated and apologetic reply to Kepler. And then, without Kepler's permission or knowledge, "the Bear" issued his *De astronomicis hypothesibus*, which included a virulent attack on Tycho and incorporated Kepler's highly flattering letter, making it appear as though the young astronomer had indeed taken the side of the Imperial Mathematician against the Danish astronomer. Kepler became deeply upset when the facts came to light. He temporarily overcame his ingrained shyness and made a special visit to the then retired Reimers' home at Prague while on his way to meet with Tycho. Kepler made his feelings known in no uncertain terms by upbraiding the hapless old man for his unethical conduct. Upon arriving at Benatky, Kepler informed Tycho of the visit and personally apologized for any misunderstanding the incident might have caused. Though Tycho passed the matter off as though it had not caused him any great concern, he remained somewhat suspicious and later decided to test Kepler's loyalty by soliciting his aid in composing a rebuttal to Reimers' charges, *A Defense of Tycho against Ursus*. Kepler, by nature, found the undertaking highly distasteful, but he also wanted to assure Tycho of his personal loyalty. He delayed the project in whatever manner possible in the hope that Tycho might undergo a change of heart. Within a short time Reimers died a broken man, but Tycho still refused to let the matter rest. He had instituted a legal action against the rival's estate, hoping to collect damages from his heirs. But these plans came to naught because Tycho himself was dead within a year. Kepler gladly abandoned the *Defense*, which had only served to aggravate the already strained relations between the two astronomers.

For a short time after Kepler's arrival, the old fire blazed brightly in Tycho's eyes. Within a week preliminary arrangements were made with regard to the distribution of work among his greatly reduced staff. Joergen Brahe, Tycho's son, was placed in charge of Benatky's laboratory; Longomontanus, the senior assistant at Uraniborg for eight years, was charged with plotting the orbit of Mars; while Kepler, for the time being, had to be satisfied with the promise that the next planet studied would be his. However, Kepler's eagerness and the difficulties Longomontanus immediately encountered with Mars' seemingly irregular orbit soon led to a significant redistribution of the work. Mars was entrusted to Kepler, while Longomontanus, by no means pleased with this shift in the work load, undertook the further study of lunar theory. Untrained in the arduous techniques of extended celestial observation, but flattered by Tycho's apparent confidence in him, Kepler foolishly boasted that he would solve the problem of the planet's orbit in eight days, and he even made a wager to this effect. If Tycho learned of it, he must have been to some extent amused by Kepler's brashness but also somewhat distressed to think that the man whose talents he had chosen to depend upon was so poorly trained as an astronomer that he would make foolish wagers on the orbits of the planets. Unfortunately, Tycho did not live long enough to witness Kepler's brilliant solution to the complex riddle of Mars' orbit, for the eight days grew into months, and the months into long, tortuous years before the pieces of the puzzle fell into place.

The more time Kepler spent in Tycho's company, the more he came to appreciate what true observational astronomy was all about. Like so many others, he had seriously underestimated both the value and complexity of Tycho's work; the quantity and precision of the great astronomer's raw data was simply overwhelming. But ironically, his growing respect for Tycho's contribution to science only added to Kepler's own deepening feelings of pent-up frustration. Never in his career had Tycho shared the stage with an equal, only with assistants and apprentices whom he had ordered about in the same manner as he did his household servants. Nor was he about to change his habits at the age of fifty-three. Although he was most cordial to Kepler, Tycho shied away from any serious scientific discussions; Kepler literally had to coax and cajole him into answering even his most rudimentary questions. "Tycho gave me no opportunity to share in his experience," Kepler wrote. "He would only in the course of a meal, and in between conversing about other matters, mention, as if in passing, today the figure for the apogee of the planet, tomorrow the nodes of another." It had never entered Kepler's unselfish head that a scientist, for whatever reason, would want to keep secret what he knew; after all, had he not published his own inspired theory of the universe so that all astronomers could share his knowledge? The insecure Kepler wrongly mistook Tycho's characteristic circumspection for a lack of confidence in his abilities. The sensitive young mathematician once again became the victim of his ingrained sense of inferiority whose hopelessly twisted roots reached back into his emotionally thwarted childhood. It suddenly occurred to Kepler that he had made the long journey from Gratz in vain. Tycho would never accept him as an equal but would continue to treat him like an ordinary assistant to be ordered around at will! Moreover, Longomontanus, with the support of Franz Tengnagel, Tycho's assistant and prospective son-in-law, was doing everything in his power to let the newcomer know that his presence at Benatky was not appreciated.

For a time Kepler maintained his usual self-restraint. He had succeeded in obtaining certain valuable information on the orbit of Mars; albeit in niggardly, piecemeal fashion, as though Tycho were passing scraps under the table to Jeppe, his buffoon. Finally, however, Kepler was worn down to the point where he could tolerate the situation no longer. He decided that he could remain with Tycho only if the *grand seigneur* agreed to a number of basic conditions which Kepler set forth in great detail. First, Kepler planned to move his family from Gratz as soon as possible and Tycho must provide suitable lodging either in Benatky Castle itself or in a separate residence nearby. Second, a specific quantity of food and firewood must also be forthcoming in addition to Tycho's pledge that he would seek a guaranteed salary for Kepler from the Emperor. Finally, Tycho must grant Kepler the right to set his own work schedule and the freedom to choose the problems he personally wished to study. The multi-page document was not meant for Tycho's scrutiny; rather its demands were to be handled through formal negotiations conducted by one Johannes Jessenius, a professor of medicine in Wittenberg and a respected acquaintance of both men. But Tycho somehow got hold of the unflattering list of demands which both surprised and angered the imperious nobleman. Yet for all his initial bluster, Tycho took the matter far more gracefully than might have been expected of a man possessed of his hot temper and domineering manner. He offered to assume part of Kepler's expenses in moving his family to Prague and to help defray the cost of their food and housing. He also agreed to a broadening of Kepler's responsibilities and promised to share with him the results of his precious observational data. He also pledged to do his utmost to secure an annual income for Kepler from the Emperor. In the meantime, Tycho asked Kepler to remain at Benatky until Rudolph could be contacted, an invitation Kepler refused. He had already decided to return to Gratz at once for the purpose of resolving his confused family situation. Kepler was also on the verge of physical collapse and in desperate need of a prolonged break from the constant hubub and controversy endemic to the Benatky household. Just before his departure, however, Kepler personally apologized to Tycho for his uncharacteristically agressive behavior, an apology Tycho generously accepted. Only once before had another man gotten the best of the great astronomer, the King of Denmark, Christian IV. Now Tycho had been bested for the second time in his life; not by a nobleman but by a fellow astronomer of common birth—the only man, Tycho realized, who could put his lifework to proper use.

Kepler, in hopes that Styria's intemperate religious climate would moderate in time, thought he might be able to retain his appointment at Gratz and obtain a leave of absence to continue his work with Tycho. But while he was away conditions in Styria had deteriorated. Early in August, 1600, a Roman Catholic Ecclesiastical Commission arrived in Gratz and promptly ordered every Lutheran official to appear before it to publicly state whether or not he would abjure his faith and become a Roman Catholic. Kepler again refused conversion and was immediately discharged from the service of the district. On September 30, he left the strife-torn city with his wife, stepdaughter, and two wagonloads of household goods. Before his departure he had received welcome word from Prague that Rudolph had consented to Tycho's request that Kepler be granted a regular appointment in the Emperor's service at an annual salary of 200 gulden.

During Kepler's absence, Tycho decided to abandon Benatky for new quarters in the capital. By the time the two astronomers took up their studies for a second time, in late 1600, Tycho had only a year to live. He seemed to sense the urgency of the situation and although the two men never became close friends, the longer they worked together the greater became their mutual respect and admiration. Besides his work on Mars, Kepler helped Tycho undertake the compilation of his voluminous data on the stellar and planetary positions with the intention of publishing a new and much more accurate book of astronomical tables. The project advanced so rapidly that they requested a joint audience with the Emperor to obtain his permission to title their work the *Rudolphine Tables* in his honor. Their request was granted, but the project became stalled with Tycho's untimely passing late in 1601. Only after a delay of twenty-seven years was Kepler able to publish the tables, which remained the most accurate available for the next hundred years. But even more important is the fact that Tycho had sufficiently tempered his patrician instincts to the point of accepting Kepler as his associate rather than his subordinate. At last Tycho made him privy to his vast treasure-trove of astronomical data and promised Kepler that after his death it would be his to do with as he saw fit. Thus, both these men of genius, after an unfortunate period of misunderstanding and frequently bitter controversy, obtained from the other what each had sought in the beginning. Kepler, young, bold of imagination, and possessed of great mathematical ability, finally had the information he needed to put his revolutionary theories to the test; while Tycho, aging, methodical of purpose, and without peer as an observational astronomer, died knowing that his unparalleled data were in the best of hands.

In addition to the vast amount of scientific information now at Kepler's disposal, he had also inherited from Tycho an intangible but no less valuable legacy. From his first reading of *Mysterium Cosmographicum* until the day he died, Tycho knew that Kepler's speculative superstition, divorced from hard supportive data, was the young astronomer's greatest weakness. As the first competent mind in modern astronomy to realize the true value of empirical facts, Tycho continually admonished Kepler to submit each conjecture as it arose to rigorous examination. "Do not," Tycho advised him, "build up abstract speculations concerning the system of the world, but rather first lay a solid foundation in observations, and then by ascending from them, strive to come at the cause of things." Kepler, to the lasting benefit of modern science, never forgot what Tycho told him. Though his fertile imagination never ceased to soar, he gradually tempered these speculative odysseys with a determination to find precise formulations in the data left him by Tycho. This insistence on applying theory to observed facts set Kepler apart from his major predecessors. Had this fundamental shift in his thought not taken place, Kepler's discovery of the first physical laws of the heavens would never have occurred. Tycho, in his search for a worthy heir, had indeed chosen and educated him well.

The Lawgiver

By an ironic twist of events, Kepler reached the apex of his career only a matter of months after his humiliating expulsion from Gratz. Tycho died on October 24, 1601; only two days later Johannes Barwitz, an imperial advisor, brought Kepler the good news that the Emperor had decided to appoint him Imperial Mathematician, and to place under his care Tycho's astronomical instruments and incompleted works. At last, Kepler, a few weeks short of his thirtieth birthday, secured the kind of position such as he had only dreamt about. The new appointment inaugurated the most productive period of his life, and he remained in Prague as Imperial Mathematician from 1601 until Rudolph's death in 1612.

When it came to scientific work, Kepler was a man of insatiable appetite. During his eleven years at Prague he laid the foundations of two new sciences: modern optics, whose discovery and development goes beyond the scope of this study, and physical astronomy, by which he formulated the first laws of the heavens. His ultimately successful quest for mathematical principles of physical causation in the universe led to the publication, in 1609, of his greatest and most revolutionary treatise, *Astronomia Nova*, or the *New Astronomy*. The complexity, length, and tortuous detail of the work cannot be fully synthesized here, but we must nevertheless discuss its broad outlines because the book proved as instrumental in revolutionizing the modern science of astronomy as Copernicus' *De revolutionibus* or Newton's *Principia*.

We earlier noted that during Kepler's first meeting with Tycho he was assigned the unenviable task of more fully elaborating the orbital theory of Mars, a problem which neither Tycho nor Longomontanus, his senior assistant, had been able to resolve. With the wisdom of hindsight, we know that all the planets move in elliptical rather than perfectly circular orbits. But these orbits depart only slightly from circularity, and but for Tycho's meticulous attention to detail such relatively minor discrepancies could simply be shrugged off as of little or no importance. With the exception of Mercury, which is close to the sun and therefore difficult to observe, Mars is the planet with the most eccentric orbit, and for that reason it proved the most troublesome to Tycho and his assistant. They simply could not reconcile the ancient belief that the planets move in circles with the observational data thus far collected on Mars. Tycho doubtless hoped that by putting a fresh mind on the problem something worthwhile might turn up, never dreaming at the time just how worthwhile that something would be.

At first, Kepler was no more disposed to question the principle of circularity than were his predecessors. After all, his own cosmological system based on the five regular solids incorporated the doctrine of circular orbits as one of its main components. He thus attacked the problem along traditional lines by literally making dozens of intricate and subtle adjustments of deferents, epicycles, and eccentrics in an attempt to bring Tycho's and his own observational data into line with accepted theory. Each of these intermediate adjustments, it should be noted, were far more accurate in accounting for Mars' movements than the solutions offered by Ptolemy and Copernicus; in fact, Kepler worked out a circular orbit that came within eight minutes of arc of Tycho's observations of the planet—a discrepancy no greater than the thickness of a penny viewed edgewise at arm's length. Yet even this minor deviation was enough to pose a significant problem for Kepler. Indeed, he now faced the most important decision—both symbolically and substantively —of his career: whether to adhere to the ancient–medieval tradition of contenting oneself with approximate mathematical values so long as they supported common-sense experience, or to accept Tycho's data—accurate to within as little as three or four minutes of arc—which strongly hinted at the possibility that the ancient doctrine of circularity was wrong. At this point Kepler almost certainly recalled Tycho's admonition: "Do not build up abstract speculations concerning the system of the world." Rather begin with specific facts and then develop a theory supported by them. Kepler wavered only slightly and then resolutely decided to support the findings of his great teacher. Although he obviously did not fully realize the implications of his decision, by openly professing his faith in Tycho's data he had stepped across the imaginary line dividing classical and modern science. For the first time, the theoretician had enough respect for the accuracy of facts gathered from empirical observation to abandon his prejudices and preconditioning to search for a new hypothesis.

Once he had decided to accept Brahe's calculations as the basis for further research, Kepler gradually came to the conclusion that no system based upon the compounded circles of Ptolemy could solve the problem of Mars' orbit. He began to experiment with a curved orbit which he described as "oviform" or egg-like. It bulges from a semicircle on one end, then flattens while traversing the other. But Kepler, the greatest geometrician of his day, after months of frustrating toil, covering some 900 pages of calculations, at last concluded that the elements of the curve could not be plotted. Exhausted and disconsolate, he sank into one of his periodic states of depression during which he completely isolated himself from both family and friends. When he finally bestirred himself, he returned once again to the abandoned principle of circularity. Perhaps he had made a mistake; perhaps Tycho's observations could be reconciled with the ancient belief in circular orbits after all. Again, however, Kepler experienced nothing but a multitude of dismal failures.

Finally, he decided to establish by even more elaborate calculations exactly how Mars' orbit deviates from a circle. To do this he not only had to plot the planet's movements but also those of the earth on which the observer stands. Suddenly, lightning struck! He hit upon the idea that the orbit of Mars describes not a circle or an oval but an ellipse with the sun at one of its two foci, an hypothesis he later generalized to fit all the planetary orbits. Kepler had discovered what we recognize today as his first law: planets move in ellipses with the sun in one focus. His patient faith in Tycho's observations had borne fruit. The inherited data were sufficiently accurate to render untenable the old hypothesis of epicycles and eccentrics, yet was not accurate enough to reveal the fact that the true orbit of Mars is not a perfect ellipse, but an ellipse distorted by all types of minor variations caused by the gravitational forces exerted by its neighboring planets. Had Tycho's observations been very much more accurate they might have prevented Kepler's momentous discovery. His second law (actually discovered somewhat in advance of the first but supportive of it) states that the radius vector, or line joining sun and planet, sweeps out equal areas of the orbit in equal times. In simple terms, Kepler's first two laws of planetary motion made it possible for astronomers to conceive of a planet's orbit as a single geometric curve whose speed is mathematically computable by a predetermined formula. The astronomer could now predict a planet's position as accurately as he could observe it. At long last the millennial stranglehold maintained by the proponents of the circular doctrine had been broken. One can only imagine the force of the intellectual and emotional impact upon Kepler at the time of discovery. The atmosphere in his study must have pulsated from the dizzying hum of all the great astronomers of the past simultaneously spinning in their graves. Nearly 3000 years of astronomical thought had suddenly collapsed under the onslaught of one man's genius. Astronomy and physics, estranged since Plato's day, were finally reconciled; the ancient dream of the Pythagoreans at last seemed near fulfillment. The fundamental significance of the moment was certainly not lost on Kepler who, in the Introduction to the *New Astronomy*, rightly compared his great iconoclastic quest to the earthshaking voyages of Columbus and Magellan.

During his early work on the *New Astronomy*, Kepler had read and was deeply influenced by William Gilbert's pioneering study on magnetism, published in 1600, titled *De Magnete*. Even before this Kepler had discarded the Aristotelian–Copernican belief that forces exerted between bodies are caused by either their relative positions or their geometrical arrangements. Instead, he was in search of a single physical theory of causation. After reading Gilbert's work, Kepler reached the conclusion that, "Gravitation consists in the mutual bodily striving among related bodies toward union or connection." Thus, according to Kepler, gravitation is no longer a matter of an individual object striving to reach the earth, but rather a case of two bodies, composed of like materials, being attracted to one another. A stone, for example, is not only attracted to the earth, but the earth is also attracted to the stone. This is not only true of terrestrial objects but of certain celestial bodies as well. If the earth and the moon were not restrained "each in its own orbit," Kepler observed, "the earth would move up toward the moon and the moon would come down toward the earth and they would join together." Assuming both bodies to be of the same density, the earth, which he estimated to be fifty-four times as large, would ascend one-fifty-fourth the distance toward its satellite, while the moon would descend the remaining fifty-three parts toward earth. Or, as Kepler wrote: "They would divide the space lying between them in inverse proportion to their weights." Only the motions of their respective orbits round the sun prevents these two bodies from colliding by virtue of their mutual attraction. (Of course, what he did not realize is that these two seemingly irreconcilable phenomena are simply different manifestations of the same principle). Yet how tantalizingly close he had come to the law of universal gravitation! Moreover, Kepler correctly believed that the changing tides on the earth's surface are controlled by this great invisible force, a conclusion that Galileo literally scoffed at in his otherwise brilliant treatise, *Dialogue Concerning the Two Chief World Systems*.

Yet in spite of all that he had learned, Kepler never took the final and crucial step of combining his planetary laws with the concept of gravitation to forge a single universal law of celestial motion. The sun

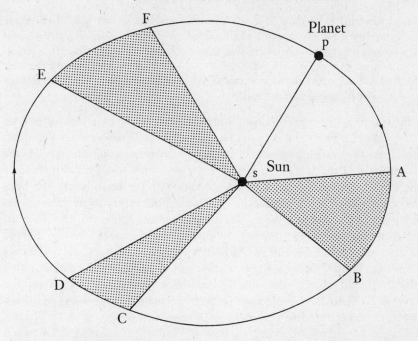

FIGURE 15. Kepler's' first and second laws of planetary motion. 1. Planets move in ellipses with the sun at one focus. 2. A line joining the sun and planet (SP) sweeps out equal areas (SAB, SCD, SEF) in equal times.

in Kepler's cosmology remains the *anima motrix*, the great celestial broom that sweeps the planets along, but not by the force of gravity. Nor did Kepler believe the stars to be terrestrial bodies, each with its own gravitational field. He, like Copernicus, was unable to break cleanly with the ancient belief that the skies are still the ethereal realm, and that as the celestial bodies retreat from our globe they become less amenable to quantitative analysis. Even the outer planets, Kepler believed, do not conform to the geometrician's models as well as those closest to the sun, the most special of all bodies because it is the House of God, the true center of the universe and of mystical powers yet to be satisfactorily explained. There was simply no reconciling this hierarchical view of the cosmos with the formulation of a single, all-governing physical principle. For this reason it was Newton, not Kepler, who took the giant step of unifying gravity and planetary motion into the law of universal gravitation. But it was Kepler, not Galileo or Newton, who was the first to develop mathematical laws that would eventually play a major role in freeing astronomy from the ancient bonds of Aristotelian physics.

In a letter to his friend, Herwart von Hohenburg, which Kepler wrote when the *Astronomia* was almost completed, he describes his ambitious program:

My goal is to show that the heavenly machine is not a kind of divine living being but similar to a clockwork in so far as almost all the manifold motions are taken care of by one single absolutely simple magnetic bodily force, as in a clockwork all motion is taken care of by a simple weight. And indeed I also show how this physical representation can be presented by calculation and geometrically.[11]

This is a prophetic goal indeed; and when we think that Kepler had no predecessor to point the way—as he himself would for Newton—his quest for a single unifying physical concept becomes revolutionary in the truest sense of the word. To Kepler must go a major and belated share of the credit (as well as the grave responsibility) for establishing the basis for what historian of science E. J. Dijksterhuis refers to as "the mechanization of the world picture," the ultimate abandonment of the Thomistic medieval universe animated by diverse intelligences for one moved by inanimate physical forces. Although he was unable to complete this transition himself, it was Kepler's conception of physical causation that ultimately led to the methods of experimentation and research which have since enabled the physical sciences to assume a position of dominance unprecedented in the scope of human affairs.

Ironically, it was the very success of Kepler's mechanistic clockwork system—made famous a century later by Newton—that led, at least until recently, to the major neglect of his work by most members of the scholarly community, both within science and without. For even though the discovery of his planetary laws established the foundations of modern physical science, Kepler committed the unforgivable "sin" of not abandoning his mystical speculations altogether. Unlike the far more popular Newton, he was simply not as committed to the idea that mechanical causation underlays all celestial phenomena. Kepler, for example, would not abandon his model of the five Pythagorean solids on the grounds that it no longer possessed objective value. Rather, he retained the geometric structure by scrapping the idea of circular orbits and inserting slightly elliptical ones in their stead.* Nor did he reject his earlier mystical concept of the Trinity, even though he recognized that the sun is also subject to the influence of such fundamentally physical forces as gravity. And as for the possibility of an infinite universe, he repeatedly declared it "unthinkable!" Thus, throughout his revolutionary search for a celestial physics, he never truly resolved to his own satisfaction the question, in Gerald Holton's words, "of whether the criteria of reality

---

* Since Kepler had previously thickened the walls of the planetary spheres to allow for epicyclic movements (see Figure 14) he could now substitute the elliptical orbits without doing great violence to his geometric model. From an objective analysis of Tycho's data, however, he knew that neither the epicycles nor ellipses truly fit between the five solids. Yet the concept still struck him as too beautiful and harmonious to dismiss.

are to be sought on the *physical* or the *metaphysical* level." [12] He remained, until his death, a man with a split mentality—a theoretical physicist and discoverer of mathematical laws on the one hand, a theocentric Christian mystic on the other. Yet for all this, the great difference which separated Kepler from Copernicus is that Copernicus never questioned the hypothesis of circular planetary motion, while Kepler ultimately rejected it. From this point onward, the history of astronomy—indeed of modern science itself—has been one of a continuous incremental process by which the work of every major individual scientist has built upon and added to the legacy of his predecessors.

Before the Storm

It had taken Kepler six tortuous years to complete the *New Astronomy*, and even then he encountered numerous obstacles to its publication. To begin with, the imperial treasury was perpetually in arrears in the payment of his salary, so that he lacked the necessary funds to cover the basic costs of printing his manuscript. To make matters worse, Kepler had become embroiled in a bitter feud with Tycho's heirs over the disposition of the Dane's observational data, a substantial part of which was indispensable to the *New Astronomy*. It is clear from surveying the historical evidence that Tycho definitely wanted the bulk of his scientific papers placed at Kepler's disposal, though no legal arrangements to this effect had been made as of the time of his death. Quite possibly, Tycho thought that none were needed because neither his children nor any of his other relatives demonstrated the interest or scientific ability to follow in his giant footsteps. On the other hand, he had seriously underestimated the rapacity of the newest addition to the Brahe family, his son-in-law, the Junker Franz Tengnagel, who became determined to cash in on this windfall. At first, an agreement was reached between the Emperor and Tengnagel by which Tycho's scientific instruments and observations were to be purchased for the sum of 20,000 thalers. Unfortunately, the same treasury that proved consistently unable to pay Kepler's modest salary was certainly in no position to expend this considerable sum. Difficulties were compounded by the fact that Kepler, taking no chances after all he had been through with Tycho, took possession of the astronomer's papers immediately after his death. Consequently, he had free access to the entire treasure during the research and writing of the *Astronomia*. Tengnagel, the patrician, had disliked Kepler, the plebeian, ever since their first meeting at Benatky several years earlier. He became enraged by what he perceived as Kepler's highhanded conduct and adamantly refused the astronomer permission to publish any of his father-in-law's work, no matter for what purpose. Kepler, angry and frustrated over this dismal turn of events, only exacerbated the already ugly situation when he compared Tengnagel to Aesop's dog in the manger who eats no straw himself but permits no one else to eat it either. Yet in his private correspondence Kepler confided to a friend that no small share of the responsibility for what had happened rested on his own shoulders:

> The root of the controversies lies in the bad habits and the suspicion of the family but also in my own passion and the pleasure I take in teasing others. Thus Tengnagel found no small reason for suspecting me of bad designs. I had in my possession Tycho's observations and refused to return them to his heirs. Yet Franz was never satisfied with any of my offers to come to an agreement with him by compromise; but he suddenly turned against me with threats as if I were a low slave.[13]

Finally, after a delay of four years and the intercession of Johannes Pistorius, the Emperor's father confessor, a compromise agreement was tentatively reached whereby the printing could proceed provided that Kepler insert a preface by Tengnagel at the beginning of the book. Kepler had no choice but to accept the offer. In the meantime, Emperor Rudolph had instructed the imperial treasurer to grant Kepler a total of 900 gulden to help cover the costs of printing his book, which was finally completed in the summer of 1609.

A highly expectant Kepler found the scholarly community's response to the publication of his latest work most disappointing. The aging Michael Maestlin wrote of the complex issues raised in the *New Astronomy*, "I must confess that your questions were sometimes too subtle for my knowledge and gifts, which are not of the same stature." It was simply a polite way of saying that Kepler's old teacher did not fully grasp the major scientific implications of his former pupil's great treatise. Clergyman and amateur astronomer David Fabricius, one of the men with whom Kepler most frequently corresponded during his years at Prague, bemoaned his friend's abandonment of the circle in favor of the ellipse: "If only you could preserve the perfect circular orbit, and justify your elliptic orbit by another little epicycle, it would be much better." An anguished Kepler replied: "I do not know what to say to your conclusions concerning the theory of Mars . . . Shall I laugh? But you deserve something better by virtue of your extraordinary industry and your unimpeachable conduct." [14] Secretly, Kepler must have felt that he too deserved something better; yet, as befitted his character, here he was, comforting a fellow astronomer who had taken exception to the single most revolutionary proposition contained in the book. As with Copernicus' heliocentric theory, Kepler's discoveries were much better received in England than on the Continent, where his supporters included scientists Thomas Harriot and Jeremiah Horrocks and poet–theologian John Donne. Even in England, however, nearly a century passed before Newton truly grasped the full scientific implications of Kepler's esoteric publication. By and large, Kepler's contemporaries, as had happened in times past and will surely happen again, simply overlooked a seminal work which marked a major turning point in the history of civilization.

Despite this and a number of other difficulties, Kepler had won a position of great respect among the nobility and fellow members of the international scientific community during his residence in Prague. For even though his work, like that of Newton and Einstein, was truly understood by only a few men of his day, Kepler seems to have been surrounded by that indefinable aura that one associates only with true genius. Though his rise in social standing was somewhat less dramatic than Nicholas Reimers,' his predecessor and Tycho's archrival, when one considers the times in which he lived, Kepler had come an extremely long way since his humble beginnings in the small German town of Weil-der-Stadt. Foreign visitors to the imperial court, princes included, eagerly sought his company and advice on a broad range of questions, many of which had little, if anything, to do with scientific matters. Few, it seems, went away disappointed; the many-sidedness of Kepler's knowledge, his penetrating wit, and his facile grasp of complex intellectual issues won him the friendships he had so longed for but had never succeeded in forming during his student days.

When he accepted the position of Imperial Mathematician, one of the things Kepler most feared was the possibility of religious persecution at the hands of certain self-appointed guardians of orthodoxy surrounding the throne, a situation which blessedly failed to materialize. Despite his many weaknesses, Rudolph II was far less concerned with the doctrinal beliefs of his scientists and artists than their creative potential. An inveterate collector of gems, clocks, mechanical instruments, and toys, and a devotee of alchemy and astrology, this small, unwed, eccentric figure would shut himself away from the outside world for days on end while he examined and catalogued his latest acquisitions. This ingrained shyness in combination with an ultimate lack of concern for the affairs of state eventually cost him his throne. Yet while he remained in possession of it, the Emperor created an environment in which men of artistic and intellectual ability thrived as no where else in Europe during the same period.

The knowledge of and respect for Kepler's special abilities, as at Gratz, was not limited to the upper classes. His imperial duties required him to issue annual astrological calendars similar to those published during his residence in Styria. These were highly prized by Rudolph's subjects, many of whom recognized the seemingly omniscient Imperial Mathematician on sight. On one occasion, when a violent thunderstorm swept the city, as predicted by Kepler a few weeks earlier, the frightened populace ran through the streets gesturing skyward and shouting: "It's that Kepler coming." [15]

He was also burdened with the ever growing responsibility of casting the horoscopes of local noblemen and their families. Although the practice provided him with a much-needed source of additional income, owing primarily to the difficulties he encountered in collecting his regular salary, Kepler looked upon this exercise with increased misgivings. Like Tycho Brahe, he accepted without question the age-old proposition that the heavens exert a general influence on nature and man; but this, according to Kepler, has little to do with the ancient art—borrowed from Muslim civilization by medieval Europe—of spying upon the stars to discover the destinies of individual men and women. The soul of each individual, Kepler believed, contains a geometric blueprint of the zodiac at birth. The intermediary light reflected at right angles off the moving planets strikes the triangles, squares, and hexagons inscribed on the soul, thus forming a direct and intimate connection between the planetary movements and man. He rejected altogether the traditional belief that the stars exert major control over human affairs. Furthermore, because each soul has its own specifically selective reactibility to the rays of light reflected off the planets, the relationships between celestial bodies and man are dominated more by the soul of the individual than by the planetary movements themselves. This proposition is based on Kepler's observation, made earlier in the *Mysterium Cosmographicum*, that each planet reacts differently to the light and physical forces emanating from the sun: some planets spin more rapidly than others, some circle the sun at a higher rate of speed, and some reflect more of the sun's light. Thus, it is only logical to assume that individual men will react in different ways to the light bouncing off the planetary surfaces on to their souls. In essence, Kepler believed the human soul was governed by the same yet undiscovered universal law of physical causation as the planets. Man's soul has even taken on a geometric form to match God's geometric model of the universe with its central soul, the sun. Until an ultimate law of universal causation is found, men will never know exactly why they act as they do. Kepler's peculiar and complex conception of astrology won few if any converts, and with Newton's discovery of the law of universal gravitation—totally devoid as it is of any astrological implications—Kepler's theory of astrology was laid to rest once and for all.

On a March day in 1610, the coach of Kepler's good friend and ecclesiastical advisor to Emperor Rudolph, Wackher von Wackenfels, thundered up to the door of his residence. Wackher, in his haste, called for Kepler without getting out of his coach; when the astronomer finally appeared he found his friend in a state of supreme agitation. The news had just arrived at court from Padua that the Italian mathematician, Galileo, with the aid of a new Dutch device called the "perspicillum" or telescope, had discovered four previously unknown planets. For an agonizing instant Kepler's heart nearly stopped! If the report proved factual, it meant that much of his own lifework would be wiped out by a single cruel stroke, particularly his geometric model of the universe.* Indeed Wackher, one of the few known contemporary advocates of the concept of infinity, believed Galileo's discovery truly vindicated the cosmological views of the martyred Giordano Bruno. If there are four "new" planets, then is there no good reason, Wackher argued, to suppose that numerous others also exist? After listening to his friend and recovering his composure, Kepler proposed a somewhat different explanation: the heavenly bodies discovered by Galileo, he postulated, are simply previously invisible moons that circle the other planets as our moon orbits the earth. Reason dictates that the Creator limited the number of planets to six because there are only five perfect solids. As it turned out, Kepler had guessed almost correctly. The four celestial bodies all turned out to be satellites of a single planet, Jupiter, rather than of several different planets. Still, it had been an uncomfortably close call!

Within a month after word of Galileo's telescopic observations reached Prague, the Italian astronomer published a book titled *Sidereus Nuncius* or *The Starry Messenger* informing the world of his discoveries. Compared to Kepler's massive treatise, the *Messenger* was scarcely a pamphlet, but it immediately became a milestone in the history of modern astronomy. Without going into an analysis of the book's contents, something that will be done in the next chapter, it must be said that Galileo's publication struck the final and fatal blow to the Aristotelian concept of celestial immutability. With the aid of the telescope he was able to show that the surface of the moon is rough and mutable; that the Milky Way, thought to be composed of nebulous ethereal clouds, is actually made up of countless thousands of individual stars; and that Jupiter is circled by four sister moons, or, as Galileo named them for propagandistic purposes, the Medicean stars. The ripples caused in conservative academic circles by the publication of this little treatise swelled into a floodtide that swept across Europe from one end to the other. The old-line Aristotelians, their hackles raised and fighting spurs bared, had at last been pushed into a corner. Their only opportunity for escape rested on the successful launching of a direct attack on the credibility of their antagonist's scientific work; hence, Galileo found himself in need of all the scholarly support he could muster. Under the circumstances it is hardly surprising that he turned to Kepler, Europe's most famous astronomer, the man whom he had treated rather shabbily in the past.

---

* This, of course, is eventually what happened but only after Kepler's death. His dedication to the idea of universal harmony was much stronger than his commitment to the laws he had already discovered. He never realized that these laws, rather than his geometric model of a closed universe, would constitute his greatest legacy to science.

Difficult though it is to believe, no convincing evidence has been found to indicate that Galileo ever seriously studied Kepler's laws of planetary motion, even though the German astronomer virtually pleaded with him on several occasions to comment on his work in detail. Galileo's practical mind was of a much different order than the mystical Kepler's; he simply did not possess the patience to follow all of the theoretician's wrong leads and blind alleys in an attempt to uncover the crucial nuggets of proof buried beneath layer upon layer of almost impenetrable, neo-Platonic, mathematical mysticism. Thus, on April 8, 1610, when the Tuscan ambassador to Prague delivered a copy of the *Sidereus Nuncius* to Kepler with Galileo's request that he offer a public opinion on it, the Imperial Mathematician would have been well within his rights to have refused. But, of course, he did not; in fact, Kepler was so thrilled by the new discoveries, he would doubtless have rendered his opinion in an open letter even without a special request from Galileo.*

The prompt delivery of personal correspondence was no easy matter in the seventeenth century, and when Kepler learned that a courier was leaving Prague for Italy less than two weeks after his copy of the *Sidereus Nuncius* was delivered, he immediately suspended his many activities to draft a lengthy reply. Two months later, it was published in Italy under the title *Dissertatio cum Nuncio Sidereo (Conversation with the Starry Messenger)*. In it Kepler heaped such praise on the Italian's accomplishments that he later wrote Fabricius: "Everybody wished that I had been more economical in my praise toward Galileo so that there should be room left for opinions of quite important men who, as one learns, dissent from my views." But Kepler could not be swayed from his steadfast support of his fellow Copernican by the rising tide of criticism leveled at him:

> Never am I a dispiser or concealer of another man's knowledge. Nor do I believe that the Italian Galileo has earned so much of my gratitude, that I, the German, should flatter him for it, by adjusting truth and my deepest conviction to him. None, however, should believe that by my frank agreement with Galileo others should be deprived of the liberty to differ from us.[16]

Kepler's untarnished endorsement of Galileo's research carried more weight by far than the support given him by any of Europe's other dedicated Copernicans. A man of unquestioned integrity, a friend of princes, and the first astronomer of Europe, Kepler graciously overcame his understandable prejudices against Galileo for the sake of scientific

---

* The very fact that Galileo approached Kepler through the ambassador, rather than directly, is itself a strong indication that he felt sheepish about the poor treatment rendered his German colleague in the past. See Edward Rosen, "Galileo and Kepler: Their First Two Contracts," *ISIS*, vol. LVII, part 2 (Summer, 1966), p. 263.

advancement. Besides Galileo's respect, which he richly deserved but never received, he asked for only one thing in return: "You have aroused in me," he wrote, "a passionate desire to see your instruments, so that I at last, like you, might enjoy the great performance in the sky." Since Kepler knew Galileo was sending telescopes crafted in his privately owned workshop to important personages all over Europe, surely it was not too much to hope that his name would be added to Galileo's list of future recipients. After all, Kepler, without actual visual confirmation, had literally staked his professional reputation on the belief that what Galileo had written was true regarding the previously unknown phenomena in the sky. While the Imperial Mathematician never had any real doubts concerning the veracity of Galileo's observations, it was only natural that he would want to view these celestial wonders first hand. Indeed, Galileo promised Kepler a telescope on several occasions, but somehow there was always a powerful skeptic whose claim on the instrument seemed more urgent. Kepler, after all, was already on his side; better, thought the ever-calculating Galileo, to use my limited number of spyglasses to win over my opponents than to reward a staunch supporter. Kepler finally wound up borrowing a telescope from his friend Duke Ernest of Cologne. He became so fascinated with its operation that he founded the new science of refraction by lenses: dioptrics. His book on the subject, *Dioptrice*, published late in 1610, enjoyer a much warmer reception than any of his earlier works.

One would have thought that at some point in this embarrassingly one-sided relationship, Galileo, so continually remiss in his treatment of Kepler, would have publicly acknowledged his great debt to the loyal astronomer. The Italian certainly must have realized that without the latter's support he could not possibly have stemmed the Aristotelian tide which threatened to engulf him during the months immediately following the publication of his *Starry Messenger*. Yet except for a brief letter now and then, Galileo remained silent. Kepler, the eternal optimist if ever one existed, continued to hope that Galileo, whose opinion mattered more to him than anyone else's, would comment on the *New Astronomy*. This desire came to nothing, however. For the moment, at least, the Italian had triumphed, and in the flush of victory old friends are easily forgotten. Europe's two most famous astronomers exchanged letters for the last time late in 1612, although Kepler made a number of fruitless attempts to keep the correspondence, such as it was, alive. Thus, one of the less appealing and little discussed chapters in the history of astronomy had been brought to an unpleasant conclusion. Kepler, ever critical of himself, must have felt that he had somehow failed once again. Yet it was Galileo, the man referred to by Arthur Koestler as being "wholly and frighteningly modern,"[17] who, never having examined Kepler's work in detail, went to his grave some thirty years later—like the contemporary Aristotelians he so despised—still convinced that the planets move through the heavens on circles and epicycles.

Kepler's difficulties with Galileo notwithstanding, the year 1610 had been one of the best of his troubled life. During 1611, however, disaster struck with such force and from so many sides that it nearly toppled Kepler into the murky abyss of total mental collapse. During his stay in Prague, Kepler watched in helpless fascination while his melancholy and half-crazed sovereign became increasingly isolated from his subjects and the affairs of state. Finally, late in 1610, the inevitable happened: an ambitious cousin of the Emperor's, one Leopold, taking advantage of Rudolph's total apathy, raised an army whose capture of Prague and forced abdication of its ruler seemed only a matter of time. But Leopold's boldness aroused the besieged Emperor's brother Matthias, who had himself taken Austria, Hungary, and Moravia from the hapless Rudolph, and he now moved to assert his own claim to the tottering throne. Matthias' superior forces caused Leopold's withdrawal from the outskirts of Prague early in 1611, while Rudolph, his political and military support gone, had no choice but to abdicate. On May 23, Matthias assumed the Bohemian crown and a year later was chosen Emperor in Frankfurt. At the height of his arduous ordeal Rudolph had turned to Kepler in the hope that his Imperial Mathematician (cum psychologist) might, through astrology, offer him some hope, however tenuous, of finding a satisfactory solution to his grave problems. But in a letter written to Rudolph on Easter Sunday, 1611, Kepler observed that, "I am of the opinion that astrology has to be withdrawn not only from the Senate but also from the heads of those who want to advise the Emperor today to the best of their abilities; one must keep astrology from the Emperor's mind." [18] Though Kepler was already aware that Rudolph's cause was hopeless, he was too honest to fire the Emperor's illusions by simply telling him what he wanted to hear.

The turbulent political situation in Prague was matched by widespread unrest and violence elsewhere in Germany, a prelude to the Thirty Years' War. Kepler had already been seriously searching for a new position, and in the wake of Rudolph's expulsion from the throne he intensified his efforts in this direction. As always, he wanted most to obtain a professorship at Tübingen, his old alma mater, and at last it appeared that he might succeed. In 1611, the Duke of Württemberg, anxious to reclaim the talents of this illustrious native son, expressed his willingness to grant Kepler the chair of mathematics at Tübingen, where he would replace his aging mentor Michael Maestlin. However, when the Protestant authorities got wind of their sovereign's intentions they vehemently protested the appointment on the grounds that Kepler's ultraliberal views on the Lord's Supper and related theological matters placed him in the Calvinist rather than the Lutheran camp. His member-

Aristotle, 384-322 B.C. (Kunsthistorisches Museum, Vienna.)

St. Thomas Aquinas, 1225-1274. (Photograph courtesy of Alinari.)

Nicolas Copernicus, 1473-1543. (Yerkes Observatory
photograph.)

Title page from an original edition of Copernicus'
De revolutionibus published in 1543. (Courtesy of
the Lilly Library, Bloomington, Indiana.)

Tycho Brahe, 1546-1601. (Yerkes Observatory
photograph.)

Frontispiece to the 1602 printing of Tycho Brahe's *Astronomiae Instauratae Mechanica* with an engraving of Tycho. (Courtesy of the Lilly Library, Bloomington, Indiana.)

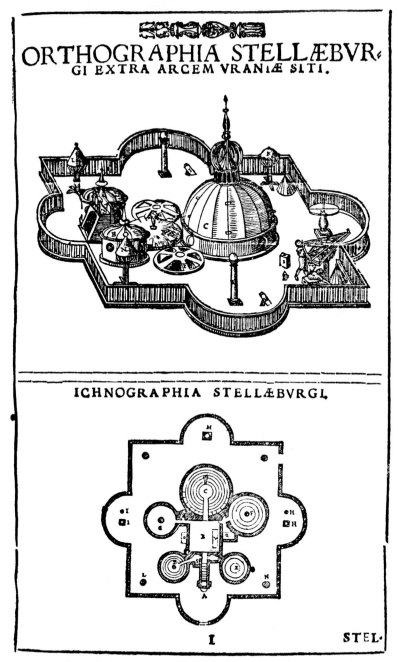

(Top) Engraving of Stjerneborg, Tycho's famous underground observatory which housed his largest and most accurate observational instruments. (Courtesy of the Lilly Library, Bloomington, Indiana.) (Bottom) A view of the underground chambers of Stjerneborg, each of which contained a large instrument such as a quadrant or armillary. From Stjerneborg's domes Tycho made the most accurate observations of the heavens in the history of pretelescopic astronomy. (Courtesy of the Lilly Library, Bloomington, Indiana.)

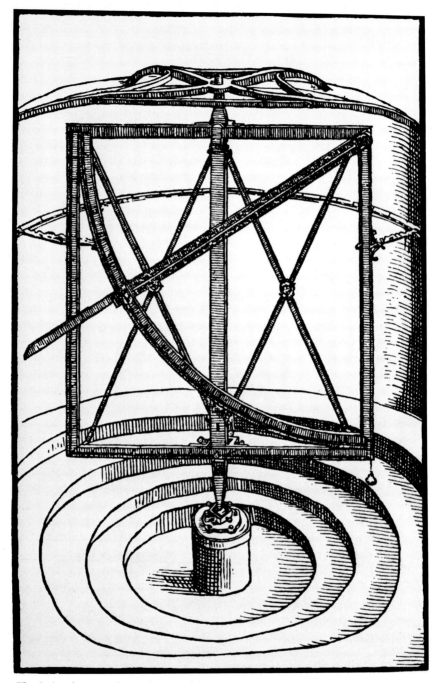

Tycho's giant steel quadrant with a radius of over six feet, from his book *Astronomiae Instauratae Mechanica*. (Courtesy of the Lilly Library, Bloomington, Indiana.)

Tycho's equatorial armillary, from the *Astronomiae Instauratae Mechanica*. (Courtesy of the Lilly Library, Bloomington, Indiana.)

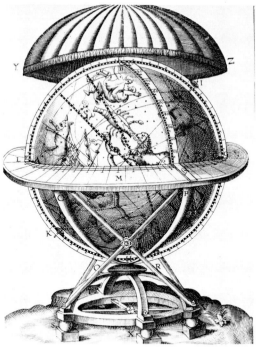

Tycho's giant metal globe on which he charted the positions of about 1,000 stars and the major constellations derived from them. From the *Astronomiae Instauratae Mechanica*. (Courtesy of the Lilly Library, Bloomington, Indiana.)

Johannes Kepler, 1571-1630. (Yerkes Observatory photograph.)

ASTRONOMIA NOVA
ΑΙΤΙΟΛΟΓΗΤΟΣ,
S E V
PHYSICA COELESTIS,
tradita commentariis
DE MOTIBVS STELLÆ
MARTIS,
Ex obfervationibus G. V.
TYCHONIS BRAHE:

Juffu & fumptibus
RVDOLPHI II.
ROMANORVM
IMPERATORIS &c:

Plurium annorum pertinaci ftudio
elaborata Pragæ ,

A S. C. M.ti S. Mathematico
JOANNE KEPLERO,

Cum ejusdem C. M.tis privilegio fpeciali
ANNO æræ Dionyfianæ cIɔ Iɔ c Ix.

Title page from the original 1609 edition of Kepler's *New Astronomy*. (Courtesy of the Lilly Library, Bloomington, Indiana.)

Title page from an original edition of Kepler's *Harmony of the World* published in 1619. (Courtesy of the Lilly Library, Bloomington, Indiana.)

Ioannis Keppleri
HARMONICES
MVNDI
LIBRI V. QVORVM

Primus GEOMETRICVS, De Figurarum Regularium, quæ Proportiones Harmonicas conftituunt, ortu & demonftrationibus.
Secundus ARCHITECTONICVS, feu ex GEOMETRIA FIGVRATA, De Figurarum Regularium Congruentia in plano vel folido:
Tertius propriè HARMONICVS, De Proportionum Harmonicarum ortu ex Figuris; deque Naturâ & Differentiis rerum ad cantum pertinentium, contra Veteres:
Quartus METAPHYSICVS, PSYCHOLOGICVS & ASTROLOGICVS, De Harmoniarum mentali Effentiâ earumque generibus in Mundo; præfertim de Harmonia radiorum, ex corporibus cœleftibus in Terram defcendentibus, eiufque effectu in Natura feu Anima fublunari & Humana:
Quintus ASTRONOMICVS & METAPHYSICVS, De Harmoniis abfolutiffimis motuum cœleftium, ortuque Eccentricitatum ex proportionibus Harmonicis.
Appendix habet comparationem huius Operis cum Harmonices Cl. Ptolemæi libro III. cumque Roberti de Fluctibus, dicti Flud. Medici Oxonienfis fpeculationibus Harmonicis, operi de Macrocofmo & Microcofmo infertis.

ACCESSIT NVNC PROPTER COGNATIONEM MATEriæ eiufdem Authoris liber ante 23. annos editus Tubingæ, cui titulus Prodromus, feu Myfterium Cofmographicum, de caufis Cœlorum Numeri, Proportionis motuumque Periodicorum, ex quinque Corporibus Regularibus.

Cum S.C. Mtis. Priuilegio ad annos XV.

Lincii Auftriæ,
Sumptibus GODOFREDI TAMPACHII Bibl. Francof.
Excudebat IOANNES PLANCVS.

ANNO M. DC. XIX.

Galileo Galilei, 1564-1642. (Yerkes Observatory photograph.)

(Opposite page) Two of Galileo's telescopes now in the Museum of the History of Science in Florence. In the center of the medallion is a lens from a third. (Yerkes Observatory photograph.)

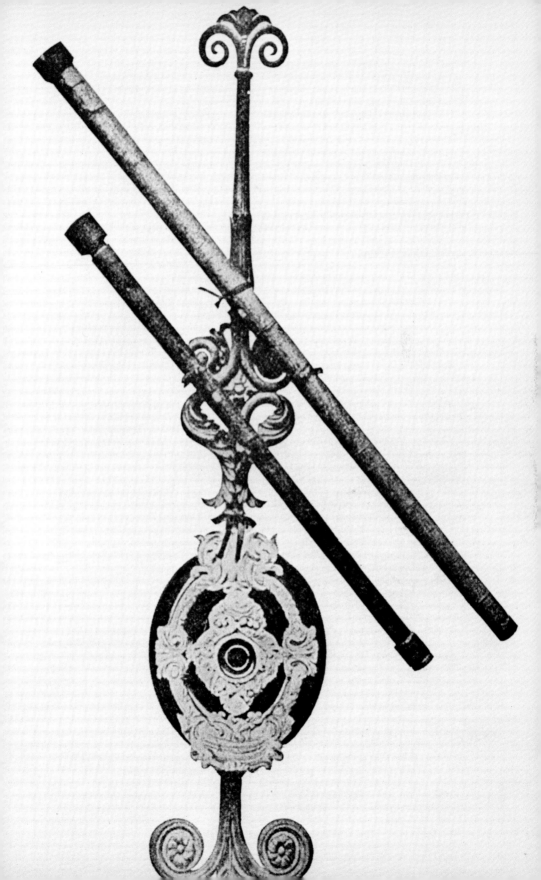

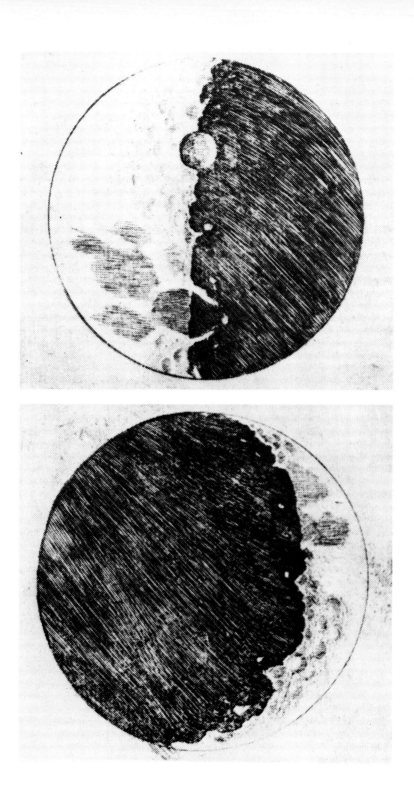

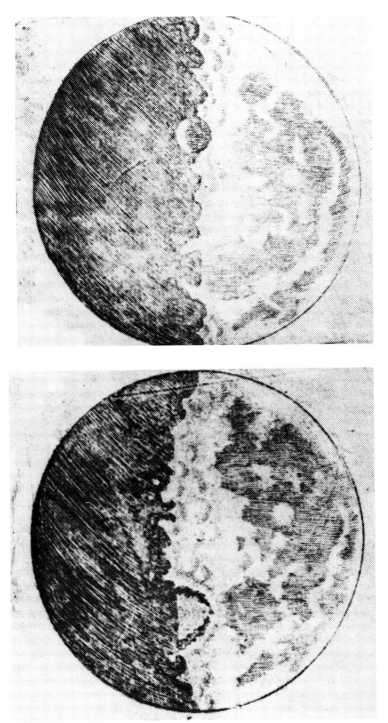

Galileo's drawings of the phases of the moon from *The Starry Messenger*, 1610.

DIALOGO
DI
GALILEO GALILEI LINCEO
MATEMATICO SOPRAORDINARIO
DELLO STVDIO DI PISA.

E Filofofo, e Matematico primario del

SERENISSIMO

GR.DVCA DI TOSCANA.

Doue ne i congreffi di quattro giornate fi difcorre
fopra i due

MASSIMI SISTEMI DEL MONDO
TOLEMAICO, E COPERNICANO;

*Proponendo indeterminatamente le ragioni Filofofiche, e Naturali
tanto per l'vna, quanto per l'altra parte.*

CON PRI — VILEGI.

IN FIORENZA, Per Gio:Batifta Landini MDCXXXII.

CON LICENZA DE' SVPERIORI.

Title page from an original edition of Galileo's *Dialogue Concerning the Two Chief World Systems*. (Courtesy of the Lilly Library, Bloomington, Indiana.)

(Opposite page) Frontispiece to the original edition of Galileo's *Dialogue*. The three figures represent Aristotle, Ptolemy, and Copernicus. (Courtesy of the Lilly Library, Bloomington, Indiana.)

Sir Isaac Newton, 1642-1727. (Yerkes Observatory photograph.)

Title page from Newton's *Principia*. (Courtesy of the Lilly Library, Bloomington, Indiana.)

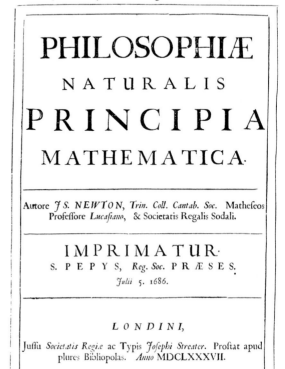

PHILOSOPHIÆ

NATURALIS

PRINCIPIA

MATHEMATICA.

Autore *J S. NEWTON*, *Trin. Coll. Cantab. Soc.* Mathefeos Profeffore *Lucafiano*, & Societatis Regalis Sodali.

IMPRIMATUR·

S. P E P Y S, *Reg. Soc.* P R Æ S E S.

Julii 5. 1686.

L O N D I N I,

Juffu *Societatis Regiæ* ac Typis *Jofephi Streater*. Proftat apud plures Bibliopolas. *Anno* MDCLXXXVII.

ship on the faculty, they argued, would put him in a position to permanently warp the minds of impressionable students which in turn might endanger their chance for eternal salvation. Such arguments seem petty and ridiculous to us today, but at a time when once flexible doctrinal lines were hardening into a state of mindless rigidity the Duke had no other choice but to withdraw the offer of the position Kepler so desperately sought.

In the same year, the Kepler household was visited by disease and death. Frau Barbara became seriously ill with Hungarian fever and epilepsy, slipping back and forth between states of extreme apathy and temporary insanity. She no sooner showed signs of recovering than the three Kepler children were stricken by smallpox, which next to the Black Death was the most feared disease of the time. The oldest and youngest weathered the storm, but Frederick, the six-year-old darling of the family, died on February 19. Kepler, who experienced no greater personal loss in his entire life, wrote of the child and of Barbara's love for him in poetic and deeply moving terms: "In every way one could call him a morning hyacinth of the first spring days whose delicate scent filled the room with an ambrosiac aroma. The boy was so tenderly united with his mother that one could not say that both were 'weak with love' for each other, but rather mad with such love." [19] The death of her son was more than Barbara Kepler, always a woman of delicate constitution, could bear. "Paralyzed by the atrocious deeds of the soldiers and eyewitnesses of the battle in the town; driven to despair for a better future and out of an inextinguishable longing for her lost darling . . . she at last breathed out her soul." [20] The Kepler marriage had never been a very happy one, and as the years passed Johannes began to feel that perhaps the responsibility was more his than his wife's. Partly for this reason, he accepted the position of District Mathematician in Linz just before Barbara's death. In Linz she would be nearer Gratz and her family; perhaps the change in environment would ease the tensions between them which had so frequently led to quarrels and recriminations. Now a widower with two young children, Kepler was even more determined to leave Prague with its all too fresh memories of pain and death. He set out for Linz and his new position early in 1612; the most productive and satisfying period of his brilliant career had come to an undeservedly tragic and bitter end.

Season of the Witch

As the District Mathematician of Linz, capital of Upper Austria, Kepler assumed virtually the same duties as he had during his years at Gratz; namely, teaching school and mapping the locality. However, compared to the stimulating social and intellectual environment of Prague, he found Linz something of a cultural backwater. In Prague he had grown accustomed to mingling with people of importance, and he received great stimulation and encouragement from his numerous friends and fellow scholars who, like himself, loved knowledge and met on a regular basis to discuss the important political and scientific issues of the day. Kepler soon discovered that his new surroundings were narrow to the point of being reactionary, and it did not take long before his superior intellect and incisive wit fostered resentment, if not outright hostility, on the part of local residents.

A few weeks after his arrival, Kepler paid a visit to his new minister, Daniel Hitzler, of whom he requested the sacrament of communion. Hitzler, himself a native of Württemberg, knew all about Kepler's allegedly pro-Calvinist deviations, and he decided to question the recent arrival at length regarding his fundamental religious beliefs. Kepler freely admitted, as he always had, that he could not accept the Lutheran doctrine of the ubiquity (omnipresence) of the body as well as the spirit of Christ at communion. Hitzler's already strong suspicions about this "unhealthy sheep" were thus confirmed; and the minister insisted that Kepler sign a statement in support of the doctrine of ubiquity as defined in the Formula of Concord before he would grant him admission to communion. This, Kepler, on the grounds of conscience, refused to do, whereupon Hitzler banned him from communion and the local congregation. Kepler immediately appealed this action to the Church Council in Württemberg, but to no avail; the discord which had so plagued his footsteps from early childhood had sounded once again.

Despite this unnerving introduction to the less appealing aspects of local society, Kepler remained in Linz for fourteen years, his longest stay at any one place. He was happy to be removed from the turmoil of Prague, and at least the Austrians paid his modest salary on time, something that had never happened during his twelve years of service at the imperial court. He also found some additional compensation for the otherwise disagreeable circumstances of his situation by forming a strong attachment to the small group of scholarly noblemen who had been instrumental in bringing him to Linz in the first place. They deplored the provincialism of their fellow citizens as much as did Kepler, and did everything in their power to make up for the embarrassingly harsh treatment accorded their resident genius.

One would have hardly expected the majority of Linz's residents, poorly educated and ignorant of the new science as they were, to truly appreciate the value of retaining a person of Kepler's unique stature. Somewhat less understandable, however, is the fact that the better educated commissioners of the district apparently considered his services equally expendable. In the autumn of 1616 they collectively decided that the financial outlay for their resident mathematician had become excessive when measured against the services he was rendering to the district. "These expenses," the commissioners curtly observed, "could well be saved by dismissing him." [21] At the time, Kepler knew nothing of the matter which was hotly debated during a number of secret meetings held over a period of several months. His aristocratic supporters, led by the Liechtensteins and the Starhembergs, eventually carried the day, but the bitterness engendered by the controversy exposed the unknowing Kepler to additional chicanery and persecution at the hands of his most ardent detractors.

Fortunately for Kepler, a major event took place not long after he settled in Linz that helped to mitigate the otherwise oppressive later years of his life: the mathematician married for a second time on October 30, 1613. The bride was Susanna Reuttinger, a young woman of twenty-four, who, eighteen years her husband's junior, had been raised in the household of the Baroness Starhemberg following her parents' death while she was still a child. Kepler's ill-considered first marriage left him with the determination not to enter into a second union without first subjecting the field of potential candidates to a very careful screening. As before, he asked his friends to suggest suitable women of marriageable age. With their assistance, Kepler compiled a list of not less than eleven names! True to his scientific training, he attacked the problem as though it were possible to discover a "law of matrimonial infallibility," whereby the assets and liabilities of each maiden could be fed into a sure-fire mathematical formula. Kepler's voluminous personal papers and correspondence, which contain many of the details of his meticulous process of elimination and selection, have been preserved, and even a cursory examination of their contents cannot help but bring a smile to the reader's face. For example, one of the major candidates was a mother with two daughters, both of marriageable age, who were also offered to Kepler as prospective wives. This, he concluded, after giving careful thought to the matter, was "an unfavorable omen," and all three were dropped from further consideration. Another serious candidate, a young maiden, who had apparently caught the mathematician's eye, expressed her willingness to marry him, but Kepler found out, to his considerable dismay, that she had previously given her word to another man. Fickleness was apparently no more appreciated by a suitor of the seventeenth century than by his counterpart today, and Kepler immediately scratched her name from his dwindling list. Yet a third damsel was eliminated because of a lung disease, while a fourth could not measure up because, as Kepler put it, "her features were most abhorrent and her shape ugly even for a man of simple tastes." And so it went. Ironically, Susanna, Kepler's ultimate choice, is mentioned far less in his written deliberations than are the names of her many rivals. However, once the widower finally decided upon her, she seems to have justified his confidence in every way. Their marriage was both happy and peaceful, even though three of their seven children died at an early age. The young woman obviously brought a measure of stability and affection into her husband's life that had been missing ever since his early childhood. Susanna Kepler outlived her illustrious husband by several years, and though Kepler left her in dire financial straits when he died, she remained loyal to his memory to the end.

By the time he settled in Linz, Kepler had already been subjected to persecution enough to last most men several lifetimes, yet he and the members of his family were destined to undergo an even more severe and prolonged ordeal than any they had previously experienced. In 1615, Kepler received word that his mother, Katherine, had been accused of practicing witchcraft and was threatened with being burned alive. To the reader who is unfamiliar with the history of early modern Europe, this may at first seem strange indeed, because ever since the Enlightenment we in the West have been taught to view European history—from the Renaissance onward—as the history of progress built upon the material advancements of modern science and its offshoot, industrial technology. Alas, we have all too comfortably oversimplified the stages of human progress, forgetting, in the process, that there have been many obstacles, both regional and national, to the advancement of the human condition. During the seventeenth century, Germany, as well as many other parts of the Western world, both Old and New, became engulfed in a mass madness such as would not be seen again until the rise of the Third Reich during our own "enlightened" era. The fear of witches was everywhere. Catholic and Protestant, nobleman and commoner, the educated and ignorant, all fell prey to the belief that Satan had forged a worldwide conspiracy whose participants were bent on recovering their master's lost empire. Thousands of women—both old and young—were confessing, under penalty of torture and the threat of death, their secret alliances with the Prince of Darkness. In Kepler's Weil-der-Stadt, populated by no more than a few hundred families, 38 women were condemned to the flames between 1615 and 1629. And in the nearby village of Leonberg, where Katherine Kepler now resided, six luckless females were burned as witches toward the end of 1616.*

It must be admitted that Kepler's mother possessed impeccable credentials to support the trumped-up charges brought against her. She was notorious for her vile temper and generally cantankerous disposition, not to mention the fact that the aunt who cared for her as an orphaned

---

* See Erik Midelfort, *Witch Hunting in Southwestern Germany, 1562–1684* (Stanford, 1972) for a detailed discussion of the subject.

child had herself been burned as a witch. Moreover, Katherine, like so many of her aging contemporaries, employed certain ancient folk nostrums in combination with other mysterious pseudomedical techniques considered pagan by strict Christian standards. The stage for her ordeal was set when Katherine quarreled with a previously close friend, Ursula Reinhold, wife of the village glazier, who accused the old woman of having caused her to become seriously ill. What had really happened is that Ursula, who was apparently somewhat unbalanced and had a bad reputation to begin with, became pregnant by a man other than her husband. In a desperate attempt to hide the side effects of a self-induced abortion, she blamed her weakened physical condition on Kepler's mother, who, Ursula claimed, had given her a drink containing magical properties. Ursula's credulous fellow citizens suddenly began to see the light: Bastian Meyers, a neighbor, reported that his wife had fallen ill and died after drinking one of Ma Kepler's evil concoctions; the wife of Christoph Frick, the local butcher, remembered how her husband felt sharp pains in his thigh whenever the Kepler woman passed him on the street; while the village tailor blamed Katherine for the deaths of his two children, which supposedly occurred after she had leaned over their cradle to offer an unsolicited blessing. Now that the landslide had begun, additional evidence was brought to the attention of the authorities on an almost daily basis. Someone reported having seen the accused, now in her seventies, ride a bucking calf to death, a way of preparing, no doubt, for her regular evening flights to a rendezvous with her cohorts at the infamous witches sabbats held in Germany's Harz mountains. The village sexton suddenly remembered how, several years earlier, Katherine had secretly asked him if it would be possible to have the skull of her father exhumed, so she could have it cast in silver as a gift for her astrologer son. If all this evidence proved insufficient, the local authorities expressed confidence that they would have no difficulty in collecting more.

Katherine's situation was made even more desperate because of a seemingly harmless scholarly undertaking by her famous son. In 1610, Kepler distributed among his friends a few private manuscript copies of the *Somnium* or *Dream*, a greatly revised version of the lunar geography begun when he was a student at the University of Tübingen. The manuscript, published shortly after his death, relates the mythical adventures of a young man, Duracotus, and his equally fictional mother, Fiolxhilde, a "wise woman" who supports her son and herself by selling lucky charms to foreign sailors. The two are taken to the moon by a friendly spirit called the Daemon of Levania, after which Kepler offers a detailed description and analysis of how he believes the lunar geography, flora, and fauna would appear to visitors from the earth. Through a complicated and unfortunate series of events, Kepler lost control of a copy of his manuscript in 1611, and a number of individuals—many of them un-known to Kepler personally—gained access to it, including some of whom the author would not have approved. The *Somnium* had been written for scientists and was therefore little understood, except on the most superficial level, by those lacking a scientific background. Appar-ently the work became the subject of gossip in the *tonstrinae*, the forerunner of the modern coffeehouse. Some of those who knew Kepler and his family (or at least thought they did) discovered sufficient auto-biographical material in the manuscript to further feed the uncontrolled fires of ignorance and superstitution then engulfing Germany. They equated Johannes with Duracotus and made particular note of the simi-larities between the obnoxious Katherine Kepler and Fiolxhilde, the fictional peddler of magic charms and herbs. Especially damning was the description of Fiolxhilde as a "wise woman" in league with celestial spirits. This additional "evidence" of Katherine's alleged satanic activities was dutifully turned over to the magistrates.* Within a short time, official charges were filed, and in 1611 Katherine was placed under arrest. In his attempt to evade a direct confrontation with the Aristotelians by concealing his pro-Copernican work in the guise of a Bohemian myth, Kepler had inadvertently reinforced the position of those whose super-stitious mentality threatened the very destruction of his own mother.

Kepler's reputation as Europe's most famous astronomer by no means served as a guarantee that his mother would escape the fate of thousands of others who had already gone to the fire, many of whom were con-victed on far less substantial evidence than that offered against Katherine. A long, tedious, and extremely taxing legal battle resulted: Only after five years, part of which his mother spent in prison, was the old woman released, but by then irreparable damage had been done. Katherine Kepler died in April 1622 from causes directly attributable to the rigors of her imprisonment and the severe psychological stresses of her trial. The tragedy of his mother's painful ordeal weighed heavily upon her son for the remainder of his life. In addition to his many other burdens, he now considered himself at least partially responsible, however unintentionally, for what had happened. Yet unlike many persons, who, when grief-stricken lose their desire to work, Kepler relied all the more upon intensive study to cauterize his deep emotional wounds. After each of the great personal tragedies or major disappointments of his life, he pro-duced still another brilliant work of science. The period following his mother's trial proved no exception. He soon completed his long-projected *Epitome Astronomiae Copernicanae*, a handbook of Copernican as-tronomy without which his fellow scientists would not have known

* Though the *Somnium* did contain a certain amount of autobiographical informa-tion, Kepler never meant to imply that his mother, like the character Fiolxhilde, was a practicing witch.

how to correlate the laws he had derived from Tycho's observations with the work of Copernicus to arrive at an operable model of the heliocentric system. Another project of major importance was his work, first undertaken with Tycho in 1601, on the periodic position tables of the planets. Finally, after overcoming several major obstacles, the *Rudolphine Tables*, named in honor of Kepler's deceased patron, were brought to press in 1627. Though the great man was now in the twilight of his magnificent career, and though he had suffered much for his fierce dedication to science, the intellectual fires burned just as brightly as they had when he first embarked upon his remarkable quest for scientific truth some thirty years before.

Harmony From Chaos

Within a five-year period, Kepler had been expelled from the Church he loved, suffered the deaths of two infant children, and watched in horror while his mother was caught up in the collective madness of a community witch hunt. To make matters worse, the seeds of religious conflict, sown by the Reformation, were ready to germinate and bring forth a crop of hostilities such as central Europe had never before experienced. In 1618, while Kepler was doing everything in his power to save his mother's life, the Thirty Years' War began in earnest. Before it was over, as many as one-third of the inhabitants of Germany and Bohemia lost their lives to disease, famine, and the senseless attacks of soldiers gone berserk. Devastation was so complete whole regions were converted into veritable deserts, while packs of starving wolves roamed through the charred remains of what were once thriving villages and towns. Within a few years, the indelible stain of the spreading conflict reached the territory surrounding Linz, finally forcing Kepler to leave his home forever. But before this happened, the mathematician had one last opportunity to focus his creative impulse on the question of world harmony as embodied in a physics of the heavens.

Twenty years earlier, Kepler's contemporaries had generally regarded his intention of putting laws of physics into astronomy as a pointless exercise. Even after the publication of the *Astronomia Nova* Michael Maestlin, his respected teacher, bluntly wrote Kepler that, "I do not quite understand this." Disappointed though he was, Kepler never considered abandoning his pursuit of a celestial physics just because others did not grasp the magnitude of his vision. He had, after all, formulated two new physical laws which successfully explained the basic character of the motions of each individual planet. Now he wanted to carry his earlier work to its logical conclusion. Thus, even as he directed Katherine's legal defense, Kepler continued to work on the problem of formulating a structural scheme that would embrace all the planetary orbits at the same time.

In 1619 the masterwork appeared, his *Harmonice Mundi* or *World Harmony*. Kepler considered this work as the further development of his first major treatise, the *Cosmographical Mystery*, in which he contended that both the number of the planets and the size of their orbits could be understood in terms of the relation between the planetary spheres and the five Pythagorean solids. The *Harmonice* contains elaborate drawings and explanations of the regular solids followed by a mystical search for hidden relationships between the planetary distances and times of revolution. He revives the Pythagorean analogy between music and the mathematical harmony of the spheres, and calls Saturn and Jupiter the bass voices, Mars the tenor, Venus the contralto, and Mercury the soprano of the celestial choir. Kepler actually wrote short scores illustrating the various tones which he believed were produced by the planetary movements. And he compares the earth to an animal whose pulsating tides suggest the breathing of a living thing.

Yet for all his continuing preoccupation with the mystical, the *Harmonice* displays a scientific maturity lacking in the *Mysterium*. In point of fact, Kepler was no longer looking for a true music of the spheres such as the ancient Pythagoreans believed could be heard by man and his creator. Rather harmony to Kepler is strictly a mathematical conception, a way of seeking out quantifiable relationships in the heavens. This subtle yet profound shift from a less mystical to a more empirical frame of reference led to Kepler's last major discovery revealed in the fifth Book of the *Harmonice*, his third law of planetary motion which states: the squares of the periods of the revolutions of the planets are to one another as the cubes of their mean distances from the sun. In simple terms, the law rejects the time-honored doctrine of the Pythagoreans that it is the *distances* of the planets from their common center and not their *speed* of revolution which defines the celestial harmony. At last, by a careful study of the comparative rates of speed at which the planets move, Kepler had welded their orbits together into a unified system. The rapture he experienced on discovering his third mathematical law was beyond containment.

> I am free to give myself up to the sacred madness, I am free to taunt mortals with the frank confession that I am stealing the golden vessels of the Egyptians, in order to build of them a temple for my God, far from the territory of Egypt. If you pardon me, I shall rejoice; if you are enraged, I shall bear up. The die is cast, and I am writing the book—whether to be read by my contemporaries or by posterity matters not. Let it await its readers for a hundred years, if God himself has been ready for His contemplator for six thousand.[22]

The die was indeed cast; later in the century Newton, the first person who truly grasped the objective importance of the third law, used it as the first important clue to arrive at his own law of gravity.

In 1626 Kepler was forced from his home in Linz by the war, and he spent the remaining years of his life in a fruitless search for a position that would enable him to pursue his scientific studies in peace. He was accorded temporary sanctuary in Ulm and while there he made the final corrections on the *Rudolphine Tables* before turning them over to his printer. From Ulm the dispossessed astronomer traveled to Prague where he came into contact with the Emperor's famous general, Albrecht Wallenstein, a man whose every military decision was governed by astrological considerations. Kepler had cast Wallenstein's horoscope in 1608 and again in 1624, although on the latter occasion he refused to be pinned down on specific details. Nevertheless, the great landowner offered him an appointment as his private mathematician in the newly acquired Duchy of Sagan. Though he feared, with good reason, that Wallenstein was more interested in him as an astrologer than as a mathematician, Kepler, in debt and in need of a position, found the offer most appealing. Still he wavered and the General responded by offering to finance the printing of Kepler's *Ephemerides*, almanacs of celestial phenomena and weather forecasts covering the years up to 1639. Upon hearing of this pledge of additional financial support, Kepler quickly accepted the appointment and then briefly returned to Linz to settle his affairs and gather his waiting family. When he arrived in Sagan several weeks later, the great man had only two years left to live, most of which were spent making final preparations for the printing of his annual almanacs.

Kepler died in Regensburg on November 15, 1630, from the effects of a severe cold and fever caught while en route to the city in an attempt to collect back interest on two bonds he had purchased several years earlier. He was buried two days later in the cemetery of St. Peter located just outside the city walls. The lengthy epitaph carved on the astronomer's tombstone concluded with the following lines penned a few months earlier by Kepler in obvious anticipation of his death:

> I used to measure the heavens,
>   now I shall measure the shadows of the earth.
> Although my soul was from heaven,
>   the shadow of my body lies here.[23]

Yet even in death the world of man accorded the great mathematician–astronomer no peace and little recognition for his remarkable, indeed, revolutionary scientific accomplishments. Before the Thirty Years' War ended St. Peter's cemetery was decimated, and Kepler's remains were irretrievably scattered by marauding soldiers while, for centuries afterward, the solid core of his work remained submerged beneath a hazy, impenetrable corona of Pythagorean mysticism. One can hardly look back upon the story of Kepler's tragic personal life without recalling the lonely genius' haunting motto: "O cares of man, how much of everything is futile."

7. Galileo Galilei: The Challenge of Reason

A Generational Rebel

FEW WHO HAVE STUDIED THE MAJOR BREAKTHROUGHS in the sciences have concluded that they occur as a result of random chance. Rather, they seem a product of a certain intellectual climate which produces the dynamic tension required to launch a single creative genius on a revolutionary quest. Moreover, as sociologist Lewis S. Feuer has observed, the most innovative of the scientific geniuses seem also to have been among the most unsocialized of the major personalities of their time. "The creative talent at turning points in the history of science has generally the traits of a generational revolutionist." [1] Because of their alienation from the dominant preconceptions of their day, scientists such as Albert Einstein "were still able to observe with naivete, with an innocence that was miraculously unconditioned." [2] Thus being able to see things from a radically different perspective than either their predecessors or their contemporaries, men possessed of great imaginative powers became locked in generational conflict with their tradition-bound peers, who remained steadfastly loyal to the scientific establishment of the times. Though Feuer has limited his study of generational rebellion to the scientific revolutionists of the nineteenth and twentieth centuries, the concept of societal conflict within generations can also tell us much when applied to the European astronomers of the sixteenth and early seventeenth centuries, the period when an activist mode of perceiving and explaining natural phenomena first began to strongly assert itself.

It was not until the latter half of the eighteenth century, in the works of the French *philosophes* and in the tracts published by advocates of rebellion in the American colonies, that the concept of "revolution" gained popular currency. During this period Denis Diderot, editor of that most important of Enlightenment publications, the *Encyclopaedia*, first coined the phrase "scientific revolution" to designate the unprecedented achievements of extraordinary individuals in the natural sciences. The term caught on and is now commonly applied to those major turning points in scientific development associated with the names of Copernicus, Newton, Lavoisier, Darwin, and Einstein. As historian of science T. S. Kuhn has pointed out, the work of each of these men

> necessitated the [scientific] community's rejection of one time-honored scientific theory in favor of another incompatible with it. Each produced a consequent shift in the problems available for scientific scrutiny and in the standards by which the profession determined what should count as an admissible problem or as a legitimate problem–solution. And each transformed the scientific imagination in ways that we shall ultimately need to describe as a transformation of the world within which scientific work was done. Such changes, together with the controversies that almost always accompany them, are the defining character of scientific revolutions.[3]

While both Darwin and Einstein, and to a somewhat lesser extent Newton, realized the truly revolutionary character of their scientific discoveries, it is quite apparent that Renaissance astronomers Nicolas Copernicus, Tycho Brahe, and Johannes Kepler only on occasion thought of themselves as major innovators, and never as generational rebels or revolutionists. Certainly, they harbored no conscious intent of bringing about the rapid overthrow of accepted science for the purpose of substituting an entirely new framework in its stead, although from our historical vantage point we know how each one of them contributed significantly to that very end. Rather, each (with the possible exception of the self-effacing Copernicus) more or less looked upon himself as an important figure of accepted natural philosophy, and all, including Copernicus, considered themselves orthodox representatives of Western Christian culture. Where they favored change at all, they did so as advocates of what political scientists refer to as "the doctrine of conservative revolution"—change brought about with a minimum of disruption, displacement, and passion. Yet despite their strong commitment to orthodox modes of thought, as young men they all manifested certain patterns of behavior that eventually proved inimical to the pursuit of science in the manner familiar to their medieval predecessors.

In terms of outlook, temperament, and life style Nicolas Copernicus seems to have been the most conservative of the group. Reserved by nature, the astronomer was destined at an early age for a clerical career; but even though he remained a loyal servant of the Church throughout his life he was never enslaved, either emotionally or intellectually, by its institutional demands. In fact, he took only the minimal vows required of a canon, never choosing to become an ordained priest. What is more, Copernicus had little to do with the Church, or at least the duties of his appointed office, until the death of his uncle when he was thirty-nine. And even then, his interests were largely bureaucratic as opposed to theological. In the meantime, he had fallen under the spell of the emerging Neo-Platonic academic movement while Novara's student at Bologna. Thus, at an early age he developed a certain mistrust of the unbending scholasticism of his day, a shift in thought sufficiently strong enough to propel him toward a revolutionary mathematical restructuring of the cosmos.

Copernicus seems to have been respected, if little understood, by his fellow canons, and he preferred the isolation of his tower chamber to the company of others. Whether, during those quiet years when he perfected his innovative model of the universe, he experienced a conscious sense of alienation from his surroundings, we do not know; but it is significant that after decades of isolated scholarship he opened his heart and mind to a young Lutheran mathematician, a total stranger, who shared his belief in a sun-centered universe. Rheticus gave him the courage to honor his convictions and the aging canon finally published *De revolutionibus* despite deep misgivings about the reaction it might provoke among the more tradition-bound scholars. Though hardly a rebel in the context in which we apply the word today, Copernicus obviously profited from the fructifying intellectual tension of his cultural milieu, tension that spurred his development of a revolutionary hypothesis on the order of the universe.

The strains of generational rebellion are manifestly more identifiable in Tycho's character and personality than in Copernicus', where they are both subtler and far more obscured by the meager historical information available to scholars. As we have observed, Tycho was openly rebellious by nature, and at an early age went against the traditions of his social class. This lifelong recalcitrance, which ultimately led him into a needless and emotionally painful exile, carried over to his activities as a revolutionary observer of the skies. Tycho's early interest in astronomy had almost nothing to do with the Aristotelian science of his day. He became fascinated—independent of formal instruction—with accurately predicting the periods of celestial phenomena long before he learned that, from the Aristotelian viewpoint, it was not the astronomer's task to observe with great accuracy. Eschewing tradition, Tycho consciously decided to become nothing less than the greatest naked-eye astronomer in the history of the world, a goal he ultimately achieved. This transference of antisocial behavior, which first manifested itself in his stormy relations with family and friends, to the realm of scientific inquiry eventually forced a major revision in the Aristotelian cosmology dominant in the universities of sixteenth-century Europe. Though Tycho did not overthrow the accepted Aristotelian system (or seek to), he became champion of the "progressive conservatives" who adopted his revised model of a still geocentric universe, but one that rejected the principle of celestial immutability. What is more, Tycho, in the splendid isolation of his island domain, fashioned his own loyal circle of potential young revolutionists, who, by their very presence reinforced their master's unorthodox research and then spread word of it throughout Europe. Indeed, as Feuer points out, the support forthcoming from a generational circle, such as Tycho had at Uraniborg, has proven a major factor in sustaining scientific revolutionists in their work. It is partly because both Tycho and Galileo had a strong core of activist supporters that they shared a supreme confidence in the validity of their scientific endeavors. Copernicus and Kepler, on the other hand, had few active supporters, both were chronically preyed upon by the twin nemeses of hesitancy and self-doubt. Copernicus, like the lonely and reserved Charles Darwin, delayed publication of his manuscript until late in his life, while Kepler, like the insecure Freud, developed a number of physical illnesses compounded by neurotic symptoms. By contrast Einstein, who, like Tycho and Galileo, found strong support for his unorthodox discoveries among a close innner circle of students in Zurich and Berne, suffered few, if any, of Freud's, Darwin's, and Kepler's physical or mental disorders.

The origins of Kepler's creative discontent doubtless stem from the deep psychic damage he sustained as a neglected child. He became alienated at an early age from his hostile family environment and compensated by seeking refuge first in religion, later in the mystical, mathematical teachings of the ancient Pythagoreans and Neo-Platonists. Yet even then he could not please those whose respect and affection he most craved. Both Catholics and Protestants considered his views on religion heterodox, while his belief in a rigorously geometrical–mathematical universe flew in the face of traditional Aristotelian doctrine. As in the case of Tycho, Kepler became an exile; but unlike his Danish contemporary he had neither the financial resources nor the scholarly support to undertake a major campaign in favor of the new science. Neither an ancient nor a true modern, Kepler balked at exploiting to the full his three laws of planetary motion whose revolutionary implications gave him nightmares.* His, like Copernicus' and Tycho's, was not a desire to overthrow the scientific system of his day, but the intention to modify certain of its fundamental teachings. Tragically, Kepler died without truly knowing how great his impact on modern science would be. Yet he was indeed a generational revolutionary, though one without a clear concept of the cause for which he was fighting.

In the figure of Galileo we encounter for the first time in the history of modern science a generational rebel possessed of at least a quasi-revolutionary intent, as opposed to Copernicus and Kepler who were largely unconscious revolutionists. So enthralled was Galileo by all facets of the new science that he made the mistake of not limiting himself to

---

* Witness, for example, his tenacious clinging to the five Pythagorean solids inscribed in *perfectly circular* orbits decades after his discovery of a mathematical law to the contrary.

the solution of a particular and clearly defined set of problems. Instead, he attempted to make a clean sweep of everything, including history itself, and this challenge to established authority is what made him so dangerous and eventually led to his downfall. Whereas Kepler wrenched in agony when he discovered the hitherto undreamed of principle of ellipses, Galileo relished almost to the point of perversity the prospect of debunking the shopworn teachings of those who defended the scientific status quo. Whereas Copernicus held back the publication of his manuscript for fear of being publicly ridiculed, Galileo—while under the ban of papal authority—wrote and published a sizzling dialogue in which he attacked the Aristotelians, an open invitation to public censure and worse. Galileo, even more than the gregarious Tycho, was the supreme activist. He derived an instinctive satisfaction from doing battle with others, and thus became the central figure in the first major public conflict between established authority and the search for truth in the history of modern science.

Galileo Galilei, eldest son of Vincenzio and Julia Galilei, was born at Pisa on February 16, 1564—the year that also saw the birth of William Shakespeare and the death of Michelangelo. The Galileis were Florentines, descendents of a noble family that in times past had been among the most powerful and respected households of the great Renaissance city. The original family surname of Buonaiuti (Bonajuti) had been changed to that of Galilei several generations earlier, apparently for the purpose of perpetuating the name of a revered family member of the fourteenth century, a physician named Galileo. Though his parents may have had this illustrious ancestor in mind when they named their child, it seems more probable that Galileo's rather interesting Christian name derives from the Tuscan custom, dating back to Roman times, of duplicating the surname in the first-born son, as, for example, in Romano Romani.*

By the time of Galileo's birth the family fortunes had rather severely declined, at least financially, and it was for this reason that Vincenzio moved to Pisa, where, in addition to pursuing a career as a professional musician, tradition maintains that he supplemented the family income by becoming a cloth merchant. Yet whatever he might have lacked in a material sense, Vincenzio was largely able to compensate for in other ways; he was a man of high culture and considerable talent. A lutist

---

* It is interesting to note that of all the great scientists from the seventeenth century onward, history has chosen to honor only two, Tycho Brahe and Galileo Galilei, by calling them by their first names. (They are even indexed by their Christian names in many books.) The thought never crosses one's mind to refer, in conversation, to Newton as Isaac or Einstein as Albert, nor would one be understood if he referred to Copernicus as Nicolas. Whether the character and temperament of these two astronomers had anything to do with the practice of calling them by their first names I cannot say, but it seems only reasonable that it should have.

and composer, he authored several published works on counterpoint and was also well versed in the classical languages and mathematics. We have an excellent clue to his attitude toward education, indeed toward life itself, from a revealing passage contained in his *Dialogue on Ancient and Modern Music:* "It appears to me," says one of the principal speakers,

> that they who in proof of any assertion rely simply on the weight of authority, without adducing any argument in support of it, act very ab- surdly, I, on the contrary, wish to be allowed freely to question and freely to answer you without any sort of adulation, as well as becomes those who are truly in search of truth.[4]

Statements similar to this are frequently expressed in the writings of the son; in fact, the quotation above might well have been taken from Galileo's *Dialogue Concerning the Two Chief World Systems.* It was doubtless from his financially troubled but highly cultured father that Galileo first developed his love of independent inquiry, not to mention an enduring distrust of and contempt for argument based on the weight of authority alone.

Both Galileo's intellectual and artistic abilities manifested themselves at an early age, so that Vincenzio had intended that his son become a merchant, he soon thought better of the idea. Though he was economi- cally unable to provide the boy with the benefits of a formal education, the elder Galileo—Renaissance man that he was—undertook the task of instructing Galileo himself. The precocious youth learned both the theory and practice of music, and is said to have excelled his own father on the lute, which he continued to play until the time of his death. It was probably from his father that the child also received his first lessons in drawing, for which he demonstrated great natural flair. In later life he often told his friends that had painting not been considered an unde- sirable career for a nobleman he almost certainly would have chosen to become an artist. This assertion hardly seems the product of idle reminiscence, for the artist Ludovico Cardi da Cigoli, who was thought by Galileo to be the finest painter of his day, claimed that he owed to the great scientist all that he knew about perspective. And, like Tycho and Copernicus before him, Galileo tried his hand at poetry, albeit with limited success. Though what remains of his verses is relatively undis- tinguished, his prose works are among the most beautifully written of the time. By whatever standard of measure one might use, it is obvious that Galileo was more favorably influenced by his father's example than were any of the other major astronomers of the sixteenth and seven- teenth centuries. What lasting influence the future scientist's mother, Julia, may have had on her gifted son is not known.

The Galilei family remained at Pisa until 1574, returning to Florence when Galileo was ten. After receiving his early instruction at the hands of his father, the child was sent away for a brief period to be further educated at the Jesuit monastery at Vallombrosa, where he learned Latin, some Greek, philosophy, logic, and, of course, religion. And it was at Vallombrosa that he was probably first introduced to the science of Aristotle. There is some evidence to indicate that while he was a pupil in the monastery school Galileo either actually joined the Jesuit novitiate or at least made plans to do so. Whatever the case, Vincenzio, whose worldly upbringing as a Florentine nobleman included a healthy strain of anticlericalism, removed his son from the monastery to save him for the more practical and economically rewarding profession of medicine. This episode, however, sheds important light on a major trait in Galileo's character—his genuine respect for religion. One can, of course, write this incident off as nothing more than an adolescent's immature fascination with a radically different style of life than any he had previously experienced. But if this is done, then it becomes very difficult to explain Galileo's almost continuous efforts (after he had perfected the telescope) to persuade the Church hierarchy of the religious value of seeking an accommodation with the new science, let alone his submission to papal authority at his trial some fifty years later—a trial he could have easily escaped by settling, at the urging of friends, in the Republic of Venice, or crossing over into Switzerland. His behavior makes sense only if we recognize that Galileo maintained a lifelong allegiance to the institution that eventually silenced his disquieting voice by quite literally bringing him to his knees.[5]

In the autumn of 1581, Galileo, at great financial sacrifice to his family, was matriculated at the University of Pisa as a student of medicine in the Faculty of Arts. Vincenzio had confidence that once he enrolled Galileo's unusual intellectual ability in combination with the still highly respected family name would gain his son one of the forty scholarships annually awarded to poor students. Things might have gone as planned had not the liberal education Galileo received at the hands of his free-thinking father gotten in the way. He soon established a reputation as a brilliant, though disputatious, student—traits which resulted in his being christened "the Wrangler" by his peers. Though he had not as yet become an opponent of Aristotelian cosmology, he displayed an outward contempt for the highly authoritative manner of contemporary university education whose practitioners, Galileo sensed, falsely clothed themselves in the mantle of the Greek master. While the works of Aristotle and other ancient Greeks were deemed the basis of scientific truth, to understand them was hardly considered the primary responsibility of the student; it was sufficient that he be able to recite the authorities by heart. The truly learned man was one capable of quoting from the ancient sources verbatim, not one who could weigh the facts, measure them against his own opinions, and arrive at increased understanding. Since the ancients were believed to have discovered all that it was possible to know, it seemed pointless to foster the independent spirit of investigation. Though orthodox in his faith, when measured against this criterion Galileo was indeed a rebel. As Vincenzio might have anticipated, his son became a thorn in the side of the educational establishment, one who provoked considerable enmity among professors totally unaccustomed to and intolerant of aggressive contradiction in the lecture hall. It is for this reason that Galileo failed to obain a coveted scholarship, and he was eventually forced to return home without the important degree of "Doctor."

In the meantime, two unforeseen events occurred which were to have a profound impact on Galileo's future. The most significant of the two took place during a vacation period in 1583 when Galileo, then aged nineteen, paid a visit to the Grand Ducal Court of Tuscany in residence at Pisa, where mathematician and family friend Ostilio Ricci was teaching geometry to the court pages. By chance Galileo happened in on one of Ricci's lectures, became captivated by what he heard, and afterward, without his father's knowledge, made arrangements with the mathematician to give him private lessons in the subject. Though Vincenzio Galilei also had a scholarly interest in mathematics, it is the one subject he had neglected (perhaps with good reason) to teach his gifted son. For Galileo it was love at first sight; Ricci was literally astounded at the passion and aptitude with which the novice tackled the unfamiliar discipline. Ultimately, the secret of his special instruction could be maintained no longer, and Galileo, with Ricci's moral support, had to face his father. The reaction proved far less severe than Galileo had anticipated; after learning of his son's rapid progress Vincenzio agreed that the lessons should continue, provided they not impinge upon his preparations for a much more lucrative career in medicine. But within a short time the young man had mastered all that Ricci could teach him and was soon carrying on independent research of his own. Vincenzio, seeing there was no hope of holding his determined son back and not wishing to provoke an open confrontation, relented altogether by granting Galileo permission to devote his career to mathematics rather than medicine—and this despite the fact that a professor of medicine, as Vincenzio well knew, could earn twenty to thirty times the wage of a mathematics professor. Perhaps Vincenzio, a high-spirited and questioning man in his own right, realizing that he too was something of a rebel, and liking it, did not have the heart to quash this first strong intimation of intellectual rebelliousness in his first-born child.

Not only was Galileo's introduction to mathematics by Ricci a crucial turning point in his career, and in the history of science itself, of equal importance was Ricci's orientation toward the Greek geometricians, especially the great Syracusan Archimedes. Galileo's biographer Ludovico

Geymonat is correct in pointing out that: "Devotion to Archimedes was to be one of the most precious legacies transmitted to Galileo by Ricci." [6] The reason is quite simple, yet profound in its consequences. Archimedes (287–212 B.C.) was one of the few natural philosophers of the ancient world whose interest in mathematics coincided with an affinity for actual experimentation. He was the first to apply mathematical methods of integration in order to determine areas and volumes, and he won lasting fame for his studies of how bodies behave when immersed in water. It is of more than passing interest that Galileo related in writing the famous story he had read of how Archimedes, at the command of his king, Hiero of Syracuse, detected a jeweler's attempt to cheat Hiero by substituting a quantity of silver for a like amount of gold in the royal crown. Archimedes, so the story goes, weighed the crown and then taking an equal weight of gold he immersed first one then the other in a basin brimming with water. More water spilled over the sides when Archimedes immersed the crown than when he immersed the pure gold, proving that the volume of the crown was greater and had therefore been tampered with by the unscrupulous jeweler, who, we are told, paid for his treachery with his life.

Whether this story is historically true does not concern us here. Rather it illustrates how mathematics can be directly applied to experimentation for the purpose of proving a logical hypothesis. As we shall see, Galileo's fascination with mathematics, from the very beginning, was of a practical, Archimedean nature. He looked upon geometry as man's supreme God-given tool for drawing out nature's most intimate secrets. However, unlike Kepler, who gained his inspiration from the theoretical and mystical properties of numbers when applied to the heavens, Galileo used mathematics for the express purpose of coming to grips with what he perceived as the true (meaning quantifiable) world around him. To Galileo, the more knowledge man acquires through the mathematical–experimental method, the less will he be forced to rely upon the vagaries of superstition which have haunted him since the dawn of time. It is this conscious and radical shift in thinking that lies at the very heart of the scientific revolution in astronomy—indeed of all modern scientific revolutions. So crucial, according to Alexandre Koyré, was the influence of the ancient geometrician from Syracuse on Galileo's thought processes that, "The true forerunner of modern [Galilean] physics is neither Buridan, nor Nicole Oresme, nor even John Philoponos, but Archimedes." [7]

Galileo's conversion to what is commonly called the "experimental method" was both triggered and personally vindicated the year he began the study of mathematics by a second important incident that took place while he attended Mass in the cathedral at Pisa. During the service, Galileo became preoccupied with the gentle swinging of a huge lamp suspended above the cathedral's high altar. As he watched it sway to and fro he observed, by placing a finger on his pulse, that although the amplitude of the lamp's swings diminished the time of oscillation remained the same. In other words, every swing of the lamp lasted the same number of beats. From this (chance?) observation Galileo made his first modern scientific discovery in physics, "the law of isochronism" or "equality of time" which states: The time of oscillation of a pendulum is independent of the size of its swings so long as the swings are small.* Galileo's first "experiment" had probably already been conducted by Arab scientists, and the results may have also been known a century earlier to the great Florentine artist and engineer, Leonardo da Vinci. Still, Galileo made the discovery of isochronism independently, an indication that in his mind there was already developing a keen and restless spirit of observation, soon to be fired by a growing commitment to mathematical demonstration through the use of experiments. Moreover, Galileo applied this discovery to the practical invention of a pendulum suited to measure pulsebeats—the *pulsiogium*—a device quickly adopted for general use by the physicians of his day. Shortly before his death, over fifty years later, Galileo returned to his studies on pendulums—so long interrupted by his campaign on behalf of the Copernican system—and endeavored to construct a pendulum clock, whose design he dictated to an assistant after he became totally blind.†

---

* Though most Galilean scholars rightly approach the account of the incident in the cathedral of Pisa, first related by Galileo's admiring and not very objective contemporary biographer, Vincenzio Viviani (1622–1703), with due caution because of its obvious allegorical overtones, a majority of them concede that the episode did in fact occur about as related above. Galileo also proved his theory of the equal acceleration of bodies by experimenting with pendulums. In his masterpiece titled *Dialogues Concerning the Two New Sciences* he tells of constructing a device whereby two threads of equal length were suspended from a horizontal rod. To these threads he attached two balls, one of lead, the other of cork. Galileo found that despite the great differences in the respective weights of the two "pendulums" they took an equal amount of time to complete equal arcs.

† However, it was the Dutch physicist and astronomer Christian Huygens (1629–1695) who became the key figure in the development of the pendulum clock. Though Galileo may have occasionally employed the pendulum to time some of his many experiments on the movement of different sized bodies, the results could not have been very satisfactory. While clocks were in use when he was born, they were anything but the precise timepieces which today run uniformly day in and day out. The primary purpose of the medieval clock was to divide the day into major parts: dawn, midday, and sunset. They were particularly suited for use in the monastery as devices for regulating the canonical hours of worship. It was not until the late seventeenth and early eighteenth centuries—with the invention of the recoil escapement by Robert Hooke (1660) and the deadbeat escapement by George Graham (1715)—that clocks capable of accurately measuring time in the modern sense were first crafted. The most accurate timepiece of Galileo's day was still the ancient clepsydra or water clock, a modified version of which he employed in many of his experiments.

The Silent Conversion

Impoverished both financially and in the desire to continue his medical studies, Galileo withdrew from the University of Pisa in 1586 and returned to his family in Florence, where he quietly passed the next few years. He was without a profession, and his prolonged presence in the Galilei household, interrupted only by an occasional visit to a neighboring city, must have become something of an embarrassment to his disappointed parents, especially to Vincenzio who had nurtured high hopes for his gifted son. Yet even though our knowledge pertaining to this period in Galileo's life is sketchy to say the least, his biographers consider these years to be very important, for they permitted Galileo the opportunity to further develop his recently discovered mathematical and mechanical skills, as well as to broaden his cultural contacts. He now possessed the maturity to take full advantage of his father's far-reaching intellectual and artistic interests by entering into Florence's many lively cultural groups, of which the senior Galilei was an important and respected member. He impressed a number of influential noblemen with the breadth of his youthful erudition; among them was the Marquis Guidobaldo del Monte, who seems to have taken a special interest in the emerging scientist's early work. Guidobaldo, brother of Francesco Cardinal del Monte, had read Galileo's first Archimedean-inspired scientific essay called *The Little Balance (La Bilancetta)*. It circulated in manuscript in 1586, and provided a detailed description of his recent invention, the hydrostatic balance, a device used to determine the specific gravity of precious substances.[8] On the basis of this and a number of other yet unpublished works, the Marquis recommended Galileo to his brother, the Cardinal, who, in typical Renaissance fashion, recommended him to Ferdinand I, the Grand Duke of Tuscany. Ferdinand was favorably enough impressed by what he heard to appoint Galileo lecturer in mathematics at the University of Pisa. Thus, at age twenty-five, the aspiring mathematician made what must have been to him and to the members of his family a triumphant return, as a professor, to the very institution which had denied him financial support as a student only a few years earlier.

It must be said that Galileo's newly won position was anything but a major windfall, at least from a financial point of view. He received the pitifully meager wage of sixty scudi per year, only 3 percent of the annual 2000 scudi income of Girolamo Mercuriale, Pisa's famous Professor of Medicine. Under the circumstances, one wonders if Galileo was not periodically haunted by unwelcome memories of his father's unfulfilled desire that he obtain a medical degree. Yet his lowly position on the academic ladder notwithstanding, he had at least managed to get his foot, however ill-shodden, inside the door of the educational establishment. Doubtless, Galileo would have much preferred the generous patronage of a wealthy nobleman; but, barring that, he knew that only a regular teaching position, supplemented by income from long hours of private tutoring, could provide him with the economic means indispensable to the continuation of his scientific studies. And with diligence and hard work he could anticipate both a promotion and an increase in salary within the not too distant future. Furthermore, the contacts he was destined to make first at Pisa and later at Padua with certain of his colleagues, but even more importantly among admiring students from influential families, would in the future prove invaluable in his campaign to win educated laymen over to the Copernican system.

Since it had been little more than four years between Galileo's departure from Pisa and the delivery of his inaugural lecture as a professor in November 1589, few major changes had occurred in the composition of the University faculty. Therefore, it was hardly to be expected that the teachers who had denied a scholarship to an inquiring and headstrong student, one who refused to knuckle under to authority, would later welcome his appointment as their colleague. Galileo, for his part, made no discernible attempt to temper the manner in which he displayed his adverse judgment of both the methods and content of Aristotelian education. Like Tycho and many others marked by genius, he had that air of exaggerated superiority which demanded that he always be right against everybody and in everything. He campaigned against the old science, from the very outset, like a man burdened with the unendurable knowledge that only he possesses a special truth—which Galileo obviously did. And in the final analysis, one is strongly tempted to say that the end Galileo achieved justified the means, unpalatable though these means sometimes became during the heat of battle.

Though he had not as yet formulated a clear conception of the fundamental principles of the modern scientific program to which his name has become inextricably linked, Galileo almost immediately began to undermine the authority of the time-honored educational system to which his colleagues were both culturally and philosophically bound. In his zealous quest for the truth, he branded the scientific errors of the Aristotelians as if they had been moral faults; and perhaps to him they were. Because of this unendearing character trait, he not only alienated those whose unwavering support of all propositions contained in the ancient texts—no matter how absurd—went beyond reason and common sense, he also provoked the hostility of the more progressive Aristotelians, who were more puzzled and annoyed by their own inability to explain the mounting anomalies in their system than by the opposition of honest critics. We are told that because of his aggressive irreverence toward the old order Galileo had only one close friend on the faculty at

Pisa, philosopher Jacopo Mazzoni. The rest of his colleagues were either standoffish or openly hostile to the arrogant newcomer.

Professional displeasure with Galileo spread into areas other than questions of natural philosophy. For example, a regulation was in effect requiring professors to wear their togas not only while teaching classes but also during the public pursuit of their nonacademic affairs. Galileo responded by composing a poetic satire, "Against Wearing the Toga," in which he not only ridiculed the ancient custom but praised the benefits of going naked. This witty work naturally found an appreciative audience among the student body, but provoked nothing but increased animosity within the faculty whose exalted status was symbolized by the wearing of academic regalia.

Galileo's frankness also got him into difficulties with his patron, the Grand Duke, a man whose favor he could ill afford to alienate at this early stage in his career. Giovanni de Medici, Governor of Leghorn and nephew of Duke Ferdinand, had designed a machine for dredging the city's harbor. The plans were submitted to Galileo for inspection, and he honestly, though not very diplomatically, replied that if constructed the machine would be virtually useless. This annoyed Giovanni who nevertheless built the apparatus, but Galileo proved to be correct, which annoyed Giovanni even more. This incident simply provided Galileo's enemies with another weapon which they did not hesitate to use against him.

By this time Galileo, whose three-year contract was nearly at an end, realized his position was in considerable jeopardy. Even if he succeeded in obtaining a second contract, which at this point seemed most unlikely given the animosity of his patron and peers, he could expect no improvement in his financial situation. Had he only had himself to consider, Galileo might have tried to salvage his position, but family considerations intervened. Vincenzio Galilei died in 1591; as the oldest son, Galileo inherited the tremendous responsibility of providing for his mother and six brothers and sisters—something his modest salary at Pisa would never permit him to do. So for a second time he turned to his trusted friend the Marquis del Monte, who again came to the rescue. Del Monte warmly recommended the young professor to his friends at the University of Padua in the Republic of Venice, and as a result Galileo was appointed professor of mathematics, in 1592, with a four-year contract. Though his salary was mercifully tripled the additional burden of supporting his large family more than offset the increased compensation. Still, the Venetians were known for their generosity in rewarding true merit, and Galileo had reason to expect regular and substantial salary increases, as had been granted his predecessor, Guiseppe Moletti.

Though the fortunes of the great commercial Republic had begun to decline as a result of the circumnavigation of Africa and the opening up of the New World by the powerful seafaring nations of Western Europe, in Galileo's day the Mediterranean was still the center of the civilized world, and Venice remained the hub. In light of the increasing competition for economic and military supremacy brought on by their major rivals—Spain, Holland, and England—the Venetians were constantly on the lookout for talented men who might further their economic and political interests. News of Galileo's work in fundamental science and practical invention had preceded him, and these considerations, at least in part, inspired the Republic's authorities to take a chance on the controversial young professor. They cared not a whit that he was a rebel, so long as his scientific studies held the promise of furthering the interests of the State. This utilitarian attitude held a strong appeal for Galileo, who found the intellectual climate that surrounded the neighboring cities of Venice and Padua intoxicating. He rubbed shoulders with ambitious men from all walks of life, who, like himself, had come to northern Italy to work in an environment free of the nagging social, political, and religious restraints common to the time. This broad freedom of thought extended equally into the lecture halls of the University, where Galileo applied his sharp and inquiring mind to the many practical and unanswered questions raised by Aristotelian physics. It is little wonder that years later he nostalgically wrote to a friend that the eighteen years he spent teaching and conducting research at Padua were the best he had ever known.

Thus far, little mention has been made of Galileo's scientific interest in or contribution to the field of astronomy, the subject of primary concern to us. This is due to the fact that while he was aware of the major developments in the field, he contributed little of consequence to astronomy until his perfection of the telescope between 1609 and 1610. During the early stages of his career, both at Pisa and Padua, Galileo spent most of his spare time working in the field of kinematics—the study of motion exclusive of the influences of mass and force. Despite a still popular belief to the contrary, it was in kinematics—not astronomy—that he made his greatest contribution to modern science. Moreover, it was from his early work on the mechanics of motion that Galileo first became suspicious of the validity of Aristotelian physics. Once he had proven the Aristotelians wrong to his own satisfaction in this field it seemed only natural to transfer his ingrained skepticism to the study of the heavens. Thus, before we turn our attention to the great Italian's astronomical studies, it is important that a brief general analysis of his major contributions to mechanics be attempted; for along with Kepler's three laws of planetary motion, Galileo's discoveries in terrestrial physics make up one of the two main components of Newton's law of universal gravitation.

Galileo came to mechanics because he had a natural inclination for the subject; as with da Vinci, he was an inventor at heart. Unlike Aristotle, who established separate and impenetrable boundaries between the earth and the heavens, Galileo thought of mechanics as a cosmological science that would eventually demonstrate the existence of a natural link between terrestrial and celestial phenomena. If, as Copernicus and Kepler maintained, the planets run as a machine, then must they not be moved and controlled by mechanical forces, and must not the laws which those forces obey be the laws of mechanics? The truly revolutionary element in Galileo's work in this field was not only the manner in which he answered these questions, but in the way he framed them. To achieve this new approach he had to extricate himself from a scientific and philosophical tradition that had dominated man's concept of the universe for over sixteen centuries.

When writing of Aristotle's view of the world system in chapter two, I attempted to stress, because it has been widely unappreciated, that Aristotelian science is a very reasonable and thoroughly thought out body of both exact and theoretical knowledge. Even today the Stagirite's physics seem, at first sight, to be in full agreement with the way we perceive the universe through our senses; but because we are conditioned from early childhood to operate from a quantitative as opposed to a qualitative frame of reference in the sciences, the ancient philosopher's teachings are branded as obsolete. Yet before Galileo's pioneering studies the concept of obsolescence in the sciences did not even exist. Historian Herbert Butterfield hit the nail directly on the head:

> The supreme paradox of the scientific revolution is the fact that things which we find it easy to instill into boys* at school, because we see that they start off on the right foot—things which would strike us as the ordinary natural way of looking at the universe, the obvious way of regarding the behaviour of falling bodies, for example—defeated the greatest intellects for centuries, defeated Leonardo da Vinci and at the marginal point even Galileo, when their minds were wrestling on the very frontiers of human thought with these very problems.[9]

Quite simply, it took a Galileo to see things from a radically different point of view than generation after generation of his intelligent predecessors—but then, after all, this is what true genius is all about.

On the central question of motion, as with other phenomena, Aristotle

---

* Butterfield is a British historian, and at the time the book quoted from was written, in 1957, "Women's Liberation" had not yet penetrated the male dominated classrooms of Professor Butterfield's England. Indeed as the events related in this study show—events that span centuries—women played no direct part in the scientific revolution in astronomy. With the recent and rather radical shift in thinking regarding the role of women in Western society, we can anticipate that they will be more directly involved in scientific revolutions of the future.

taught that we must seek the reasons in the essence or quality of things. Every material object has a motion of its own corresponding to the materials of which it is composed. The four elements: fire, air, water, and earth, are endowed with special inherent tendencies which in turn determine the nature of their motion—fire and air move upward, earth and water downward. Whereas all earthly elements move in straight lines—either up or down—the motion of the celestial bodies, composed as they are of the quintessential element, ether, is circular and eternal. Centuries after Aristotle's death, Thomas Aquinas specifically declared that man must not think of the material comprising the celestial spheres as anything analogous to the stuff of our earthly bodies. The heavens are therefore incorruptible, the home of all natural motion, while the world below the lunar region is subject to change and decay. Thus, to Aristotle and his medieval proponents straight-line or rectlinear motion, which only occurs on earth, is violent and *unnatural* motion, a contradiction of the celestial principle of circular immutability. Such motion as the throwing of a rock or shooting of an arrow, they argued, occurs only in the terrestrial realm and is strictly dependent upon the operation of a physical mover. Consequently, the Aristotelian doctrine of interia, in contrast to the modern explanation of Galileo and Newton, is a *doctrine of rest*. In speaking of the earth, it was *motion* not *rest* that always required an explanation.

Aristotle and his followers maintained that every transmission of movement requires direct physical contact. A stone or arrow keeps on moving only so long as it is in touch with whatever is propelling it. Once it loses contact with its mover the body falls to the ground, where it again enters into its natural state of rest. The question then arises as to exactly what this moving force is. Certainly, in the case of the rock and arrow, it could not remain the arm of the thrower or the bowstring of the archer because both instantaneously lose physical contact with their respective projectiles, yet neither falls directly to the ground. Aristotle explained the continued movement of a projectile by reasoning that it is pushed in its flight by the continuous force of onrushing air. As the projectile moves forward it displaces the air directly in its path which then rushes round behind to prevent a vacuum from forming. (Aristotelian physics did not permit the existence of a vacuum in either the celestial or the sublunar regions.) The problem with this explanation is twofold: in the first place, if a stone or arrow is being constantly thrust forward by the force of onrushing air, why does a piece of thread tied to one or the other always trail behind rather than move ahead on the propelling air current, as we would have reason to expect? An even more important question is why, given these conditions, does not an object in flight go on forever?

Galileo was not the first to become discontented with these and other major anomalies in Aristotelian mechanics.* As early as the sixth century, John Philoponos, a skeptical commentator on Aristotelian physics, originated an alternative explanation of motion called the "impetus theory." This concept, passed from one generation to another, finally reached its point of greatest popularity among scholars at the University of Paris during the fourteenth and fifteenth centuries. Its two best known advocates among what historians of science call the "Parisian school of physicists" were Jean Buridan and Nicholas Oresme, who are also important for other reasons besides their teaching and written commentaries on impetus.

These thinkers, in contrast to Aristotle, believed that a projectile is carried forward by a force or "impetus" which it acquires from the simple fact of being in motion. Butterfield states that it "was supposed to be a thing inside the body itself—occasionally it was described as an impetuosity that had been imparted to it; occasionally one sees it discussed as though it were itself movement which the body acquired as the result of being in motion." [10] This concept obviously offered a better explanation of how it is possible for the motion of a body to continue after contact with its original mover has been lost, but a major question remained: what would cause the motion of an "impressed" body to cease? The answer, according to Oresme, is that impetus, like the heat given off by a live coal taken from a fire, gradually weakens to the point where it is no longer capable of moving the projectile forward in a straight line. It is much like the rippling effect that goes on long after a rock has been cast into a quiet pond, but which gradually fades away. Buridan, on the other hand, did not advocate Oresme's self-expending view of impetus. Rather, he theorized that a body's impetus is destroyed by the contrary resistance of the medium through which it passes. Once impetus exhausts itself the projectile ceases to move in a straight line and curves round, taking a downward path to the earth. This, of course, could not occur in the heavens because there is no air to impede the perpetual motion of celestial bodies through ethereal space.

The impetus theory, though a departure from original Aristotelian physics, was still more ancient (or perhaps we should say medieval) than modern, for it still held that motion on earth is basically unnatural —nor could it be sustained without the continued application of force. Thus, the impetus theory is very much the counterpart in physics of the modified geocentric system of the universe fashioned by Tycho Brahe. It saved the phenomena but added relatively little that was new.

Little progress in physics had been made since the late fourteenth

---

* It is only fair to point out that Aristotelian mechanics is a considerably more detailed and sophisticated system than can be adequately encompassed in this broad study. Rarely, if ever, did the ancient master expound anything as silly as he is often credited with by his modern detractors.

century, so that matters stood pretty much as outlined above when Galileo undertook his work in the field. Like a true scientist, he began by eliminating first one then the other of the most speculative elements of the ancient doctrine of motion, carefully stripping away the centuries of accumulated patina. His study of mathematics had already led him to the conclusion that a law of nature must contain no element incapable of mathematical verification, no matter how appealing it might be to commonsense experience. He began his studies in mechanics at Pisa by experimenting with bodies of different weight. Aristotle taught that heavy bodies fall faster than light ones: when two objects are dropped at the same time from an equal height they will supposedly fall with speeds proportional to their weight. A ten pound cannonball, for example, would fall ten times faster than a one pound cannonball. While Galileo did not disprove this erroneous proposition by climbing to the top of the Leaning Tower of Pisa, as his biographer Viviani reported, and dropping two different sized cannonballs to the ground, he did something far more valuable.* He measured the time it takes objects of different weight to roll down an inclined plane. By keeping time with a specially designed pendulum, he was able to show that, whatever the angle of slope, a ball rolls in such a way that the total distance it has gone is directly proportional to the square of the time it has been traveling. This is true no matter what the weight of the ball, and therefore proves that objects of different weight fall with the same acceleration, or at least they do so in a vacuum where no air-resistance is encountered. The importance of this proposition is that it is valid for *all* bodies, and consequently acquires a universal character. And since it applies throughout the terrestrial realm, Galileo privately began to wonder if it might not also extend to the celestial region beyond. The revolutionary fires were heating up.

Galileo's second major contribution to the demise of Aristotelian physics occurred while he was at Padua, though a detailed analysis of it was not published until the appearance of the *Two New Sciences* in 1638. After a period of intensive thought and experimentation, he reached the conclusion that once a body is set in motion it will continue to move until some extraneous force causes it to stop, a general approximation of Newton's statement of the law of inertia. Contrary to Aristotle and his followers, Galileo believed that motion is as "natural" as rest. His arrival at this unheard of conclusion required a quantum jump in abstract reason-

---

* Had it simply been a matter of disproving Aristotle's proposition by dropping two objects of different weight from a tower it would have been universally rejected long before Galileo's day. Such an experiment requires a special apparatus which had not yet been invented. For an enlightening account of this stage of Galileo's studies see Lane Cooper, *Aristotle, Galileo, and the Tower of Pisa* (Ithaca: 1935).

ing, and by taking it he forever changed the course of man's thinking about physical causation. For if it was as natural for a body set in motion to continue moving on earth as in the heavens, then there must be little if any difference between the two realms; indeed the universe truly becomes one great expanse unmarked by physical divisions. To quote historian of ideas Leonardo Olschki:

> The young scientist opened up to systematic investigation the unlimited physical space in which a privileged form of motion has no sense. In that space every sort of motion is only a function of "gravity" to which everything in nature is subjected—even fire and air—because gravity is the fundamental cause of motion independent of the shape and the "nature" of the mobile body. In this way celestial mechanics becomes only an aspect of the general rules which explain the fall or swing of bodies, the inclined plane, and the different forms and variations of motion. It is on this basis that the heliocentric system was transformed into a problem of mechanics and could be inserted into the framework of mathematical and experimental science.[11]

While several others might have made Galileo's telescopic discoveries, there was probably no other man of his generation capable of founding the new science of mechanics. He, along with the unsuspecting Kepler—like midwives presiding over the birth of a Promethean giant—had at last succeeded in severing the umbilical cord which had hitherto prevented modern science from progressing beyond the fetal stage.

Though most Galilean scholars agree that it was the Italian's work in kinematics which convinced him of the universal application of physical laws—and of the validity of Copernican cosmology itself—the exact time of his conversion to the heliocentric doctrine remains unknown. It is an undisputed fact, however, that until he began working with the telescope in 1609, Galileo continued to teach the Ptolemaic system even though he had become convinced of its inherent falsity many years earlier. His first declarations in support of Copernicanism were made in private correspondence dating from 1597. In May of that year he confided his belief in the new doctrine to the only close friend he had on the faculty while at the University of Pisa, Jacopo Mazzoni. A few months later, on August 4, he addressed a letter to Kepler in which he thanked the young astronomer for sending him a copy of his recently published *Mysterium Cosmographicum*. As was observed in the previous chapter, Galileo confessed to having "read only the introduction" but was quick to add: "I have learned from it in some measure your intentions and congratulate myself on the good fortune of having found such a man as a companion in the exploration of truth." He then went on to reveal that:

Many years ago I became a convert to the opinions of Copernicus,* and by that theory have succeeded in fully explaining many phenomena, which on the contrary [Aristotelian–Ptolemaic] hypothesis are altogether inexplicable. I have drawn up many arguments and confutations of the opposite opinions, which however I have not hitherto dared to publish, fearful of meeting the same fate as our master Copernicus, who, although he has earned for himself immortal fame amongst a few, yet amongst the greater number appears as only worthy of hooting and derision; so great is the number of fools. I should indeed dare to bring forward my speculations if there were many like you; but since there are not, I shrink from a subject of this description.[12]

We see in Galileo's reply to Kepler an interesting contradiction of character; one which reveals itself on several occasions to anyone who closely examines the early stages of his long career. While he was supremely self-confident and often quite merciless in his criticism of the opinions held by others, he had not yet gained sufficient confidence in his own scientific theories—at least those on astronomy—to submit his views either to the close scrutiny of his scholarly peers or to an examination by the public at large. It is interesting to note that the same year he revealed his pro-Copernican stance to Mazzoni and Kepler, Galileo dedicated his just completed *Treatise on the Spheres, or Cosmography* to the Ptolemaic system. He hung back from a public declaration in favor of the new astronomy, not because he feared religious persecution or, for that matter, scholarly attack, but because, as he wrote Kepler, he was fearful of the same "hooting and derision" that had so haunted the great Copernicus. Galileo wanted more tangible proof of the heliocentric system before he was willing to risk his reputation by taking up the scientific cross on its behalf. Strangely, it was the twenty-six-year-old, insecure Kepler, seven years Galileo's junior, who, in his reply to the Italian felt it necessary to bolster his colleagues lagging spirits: "Be of good cheer, Galileo, and appear in public. If I am not mistaken there are only a few among the distinguished mathematicians of Europe who would dissociate themselves from us. So great is the power of truth." [13] Although several years passed before Kepler's advice was taken, this Galileo eventually decided to do. Armed with the results of his telescopic research, he finally went public in support of the heliocentric doctrine, convinced that he alone could build up a floodtide of opinion capable of sweeping the new science into irreversible acceptance. Unfortunately, his "proof" of Copernican cosmology was far less convincing

---

* If, as Galileo claims, he had been a Copernican several years prior to 1597, his conversion might conceivably date as far back as the period before he obtained his first teaching appointment at the University of Pisa. However from this writer's point of view it seems more likely that he was first drawn to Copernicus' work because of its strong mathematical appeal as a hypothesis and only later, while at Padua, did he accept it as being scientifically true.

to others than to himself. As a scientist, he made no greater mistake than not reading beyond the introductions to Kepler's *Astronomia Nova* and *Harmonice Mundi,* for deep within the pages of these mystical tomes lay the very physical laws of the heavens which, when combined with his own research into terrestrial physics, culminated a generation later in the first scientific revolution.*

The Devil's Scepter

In many ways, the eighteen years Galileo spent at Padua could serve as a nearly perfect example for those who support the contention that the concept of Renaissance man was not merely an ideal, but a living reality. The great physicist had grown up in a humanistic and artistic environment almost totally absent of scientific education, but his intellectual and aesthetic interests were developed far beyond those of the majority of his peers.[14] While lesser men would have doubtless been content to pursue the quiet, respected life of a professional scholar, the entire period of Galileo's residence in the Venetian Republic was one of unceasing industry and cultural enrichment. Not content with limiting his professional and social contacts to the university community alone, he frequently travelled to nearby Venice, where he established close ties with influential families and with clever young aristocrats, who were captivated by the sharp wit and incisive mind of the youthful professor.

Galileo's teaching duties encompassed a broad range of subjects: as-

---

* In both this chapter and the previous one I have argued that Galileo did not delve deeply enough into Kepler's work to truly appreciate the significance of his three laws of planetary motion. It is worth pointing out that there are those who take a different view. Albert Einstein, for example, contended that Galileo indeed knew of Kepler's laws but that "this decisive progress did not leave any trace on Galileo's life work—a grotesque illustration of the fact that certain creative individuals are often not receptive." Galileo Galilei, *Dialogue Concerning the Two Chief World Systems,* transl. by Stillman Drake, foreword by Albert Einstein, 2nd ed. (Berkeley: 1967), p. XV. Essentially the same argument is presented in Edwin Panofsky's superb little monograph *Galileo as a Critic of the Arts* (The Hague: 1954), pp. 22–25. The problem with this position is that when confronted with the pressing need to provide proof of the heliocentric system, other than his telescopic discoveries, Galileo never once mentioned Kepler's laws. Nor can his failure to employ the laws be convincingly explained away by his unwillingness to share the limelight with Kepler, at least after 1630. After all, he had turned to the brilliant German as early as 1610 for support of his telescopic discoveries. Two decades later his need was much greater, and Kepler's laws would have been of critical importance, particularly after the rejection of his erroneous theory of the tides (see p. 310). Lastly, why, if Galileo had known of the details of Kepler's discoveries, did he not once mention them in their surviving correspondence in which Kepler literally pleads with the Italian to comment in detail on his work? It is, of course, possible that even had Galileo been conversant with the principle of ellipses his commitment to the circularity of planetary orbits was too deeply ingrained to be rejected.

tronomy, mechanics, geometry, and fortifications, among others. That he was an exceptional teacher is undisputed; in fact, his lectures became so popular the halls assigned him were not always large enough to accommodate both his students and interested visitors. When this happened he sometimes lectured in the open air. Galileo was a practitioner of the Socratic method and often used the erroneous opinions of his students, which he dismantled piece by piece, to illustrate some major point. And as if in knowing preparation for future verbal jousts with the Peripatetics, he would present an accepted belief in its strongest light; then, after he had his audience nodding in enthusiastic agreement, he proceeded to undermine their seemingly unshakable credulity by refuting the argument point by point. Not only was this an effective method of teaching, it made for excellent theater in the bargain; and no one enjoyed center stage more than Galileo.

We have already observed how, after the death of his father, Galileo had to assume almost the entire burden of maintaining the other members of his family. To augment his rather meager salary, he followed the established custom of offering private lessons to those students who could afford them. Within a relatively short time word of his reputation had grown to the point where not only university students but the sons of foreign noblemen sought the young professor out. In response to these demands Galileo purchased a large residence in which were usually boarded a dozen or two of his private pupils. Though at times the burden must have been quite onerous, Galileo reaped a second reward, equally as important as the financial one, from this experience. For the first time in his life the physicist was continuously surrounded by an intimate group of admiring supporters—just as Tycho had been at his private observatory on Hveen. This nearly ideal situation enabled Galileo to try out new ideas and to conduct experiments, which even in the liberal environment of the University might otherwise have been viewed with a jaundiced eye. As a result of the countless hours which he spent in the company of his eager disciples, Galileo became all the more convinced that in Copernicus and mechanics he had discovered the crucial turning-point between the old science and the new.

Galileo's faith that his merit as a teacher would be both recognized and rewarded by the Venetian authorities proved well founded, for with each renewal of his contract the mathematician's salary significantly increased. His income rose from an initial 180 florins per annum in 1592 to 1000 in 1609, the year he unveiled his first model of the telescope. When he decided to resign his position at the University a short time later, his employers offered him even more money, but he refused to reverse his decision. Still, his income never came near to matching his expenditures during this period. Besides basic living expenses, there were always additional family demands on his ever dwindling purse. The most onerous of many such burdens was the handsome dowry Galileo provided his sister Virginia upon her marriage to Benedetto Landucci in 1591. It took years to pay off the money he had borrowed to meet this obligation, and then, in 1601, he was called upon to begin the process all over again, when his younger sister Livia wed Taddeo Galletti. This time he was forced to seek a two years' advance on his salary from the Republic, and the situation was made even more difficult because Michelangelo Galileo, his irresponsible younger brother, defaulted on the payment of his share of the dowry contract, which he had jointly signed with Galileo. Within a short time Michelangelo also married and repeatedly turned to his hardpressed brother for "loans" that were never repaid.

The generosity with which Galileo met the unceasing monetary demands of his close relatives, coupled with the great scientific and social issues that were then taking shape in his mind, had a decidedly adverse effect on the scientist's personal life. A few years after his arrival in Padua he had formed a liaison with a young Venetian woman, Maria Gamba. Maria, at Galileo's request, came to live in Padua, but not in his home. Although Galileo's affection for his mistress was genuine, as was her's for him, the couple never married, a not uncommon or necessarily frowned-upon occurrence of the times. The relationship lasted for ten years, during which Maria bore Galileo three children, two daughters and a son: Virginia (1600), Livia (1601), and Vincenzio (1606). During the remaining years of Galileo's residence at Padua the children were entrusted to their mother's care. However, when their father returned to Florence in 1610 he decided to take Virginia and Livia with him. Galileo soon regretted this action, however, for in addition to the many other problems confronting him he simply did not wish to assume responsibility for raising two young girls. Neither could he return them to their mother because in 1613 Gamba married a Venetian, Giovanni Bertoluzzi, after which seven-year-old Vincenzio also became Galileo's personal responsibility. Galileo now made and executed a fateful decision which most of his biographers look upon as a discreditable performance at best. He sent Virginia and Livia to the impoverished Convent of San Matteo at Arcetri, a short distance from Florence, where in 1616 and 1617, respectively, the Galilei daughters assumed the vows of nuns. Virginia, who took the name of Sister Maria Celeste, ultimately reconciled herself to her fate and became a source of comfort and affection to her father after his shattering encounter with the Inquisition. But Livia, the younger daughter, who became Sister Arcangela, remained dispirited and deeply embittered by her enforced confinement. By the time her aging father recognized his error, all chance of repairing their seriously damaged relationship had been lost. In this episode is revealed that irrepressible egotism in Galileo's character, which caused him to sacrifice the happiness of his own offspring for what he considered the higher calling of his profession. Family members who made financial demands from a distance could be tolerated, even indulged, but when the possibility developed that his own children might interrupt his work by their continued presence, it was more than he was willing to risk.

In addition to his other sources of income, Galileo also attempted to turn his natural talent for invention into a money-making proposition. He set up a small workshop near the University, where he employed a close friend and excellent craftsman, Marcantonio Mazzoleni, who helped fashion the marketable instruments which Galileo so expertly designed. Among these were compasses, surveying tools, a glass apparatus—sometimes called the thermobaroscope—for measuring the expansion of liquids, much like a thermometer, the delicate hydrostatic balance discussed earlier, and, of course, the first models of his famous telescope. Except for his excellence as a teacher, there is no activity which could have pleased his employers more, for as Jacob Bronowski has observed, "This was sound, commercial science as the Venetians admired it." [15]

It was over a dispute relating to one of the instruments sold by the scientist that the typical "Galilean style" of scholarly combativeness first publicly asserted itself. Though the incident has proven of no significance to the history of science, it might well have served as a warning to those who in the future were to greatly underestimate the brilliance, wit, and resourcefulness of their worthy opponent. Galileo had developed a method of greatly improving an old instrument called the "military compass," which was used primarily by engineers to solve mathematical problems. It had become one of the few truly profitable items sold in his workshop, and in 1605 he composed a handbook, in Italian, to guide those who purchased the handy device. A year later, a Paduan teacher of mathematics and colleague of Galileo's named Baldassar Capra published a much-too-similar work in Latin. That plagiarism had occurred was obvious, and Galileo successfully brought suit against Capra in the university courts. The latter's work was duly confiscated and the copies burned. Unfortunately, Capra's public disgrace did not satisfy Galileo, whose fighting instincts had been deeply roused. He overreacted by publishing a broadside, in 1607, in which he called the mathematician everything from "a venom-spitting basilisque" to a "greedy vulture," and worse. As Koestler has pointed out, "Not even Tycho and Ursus [Nicholas Reimers] had sunk to such fishwife language; yet they had fought for the authorship of a system of the universe, not of a gadget for military engineers." [16]

In the meantime, Galileo's scientific interests were gravitating more in the direction of astronomy. This process was accelerated by the appearance of a "new" star in 1604, still called "Kepler's nova," which, as we observed earlier, had attracted the attention of astronomers and educated laymen throughout Europe. It is known that Galileo lectured on the unusual celestial phenomenon to large audiences on at least three separate occasions. Unfortunately, the text of these lectures has not survived, but from information derived from other contemporary accounts scholars have concluded that the physicist took an anti-Aristotelian position on the subject. After a series of careful observations Galileo reached the same conclusion about the nova of 1604 that Tycho had regarding the new star of 1572: it was located in the celestial realm, not in the sublunar region, and therefore contradicted the timeworn argument for celestial immutability. Doubtless Galileo's already privately stated commitment to Copernican theory was further strengthened by the nova's timely appearance. At first attracted to Copernicus' work because of its mathematical appeal, he had now discovered corroborative evidence through actual observation. Yet Tycho had used evidence of a similar nature to support his own revised model of the geocentric universe—something of which Galileo was perfectly aware. The time had not yet arrived when the truth of Copernicanism could be satisfactorily demonstrated, and Galileo wisely chose to maintain public silence on the matter pending further developments.

It was five years later, in 1609, that the forty-five-year-old scientist first received news of the telescope. Apart from the regional reputation he had established as a teacher–inventor, Galileo was not particularly well known. Certainly knowledge of his scientific work came nowhere near rivaling the reputation of the younger Kepler, who had become Imperial Mathematician at the age of 30. This is not to suggest that Galileo's research in mechanics did not compare favorably with Kepler's mystical–mathematical foray into celestial physics, for it did. It was simply that, contrary to the pattern of publication followed by Kepler and most important scientists since, Galileo, with more than half his life already behind him, was yet to publish a major work. But this set of circumstances changed far more quickly than even he dared dream. Within a few short months Galileo's name had eclipsed even that of Kepler, for the Italian published a little treatise which instantly elevated his name into the ranks of the immortals in his profession.

No one can be certain who first invented the telescope, although the man most often credited with this achievement is Hans Lippershey, a Dutch spectacle-maker who applied for a patent in 1608. Rumors about the invention had reached Galileo in Venice a few months later, but at first he seemed little interested. However when the French nobleman Jacob Badovere, one of his scientific correspondents, wrote him from Paris confirming the new discovery, Galileo immediately set his mind to the problem of recreating the recent invention. Within a few days he succeeded in working out the design of the new instrument and then proceeded to construct the first of many Galilean telescopes, a process whose steps he recorded for posterity.

> First I prepared a tube of lead, at the ends of which I fitted two glass lenses, both plane on one side while on the other side one was spherically convex, and the other concave. Then placing my eye near the concave lens I perceived objects satisfactorily large and near, for they appeared three times closer and nine times larger than when seen with the naked eye alone. Next I constructed another one, more accurate, which represented objects as enlarged more than sixty times. Finally, sparing neither labor nor expense, I succeeded in constructing for myself so excellent an instrument that objects seen by means of it appeared nearly one thousand times larger and over thirty times closer than when regarded with our natural vision.[17]

Galileo grasped the commercial potential of the telescope as quickly as he had its technical ramifications, and he sent word to friends in Venice that he was about to unveil a revolutionary mechanical device of great importance to the future economic well-being of the Republic. He arrived in the city on August 24, and the following day presented a telescope to the Doge and Senate during a public ceremony attended by many of Venice's most influential citizens. Their reaction was enthusiastic to say the least, and who could blame them? These were capitalists who made their living by financing the movement of precious goods on the open sea; now they possessed an instrument that would enable their crews to identify ships several miles distant and two hours' sailing time away. What better protection against piracy or wrecking on some unknown shore than this? The authorities generously responded to Galileo's gift by raising his pay to 1000 florins a year and conferring on him life tenureship of his chair of mathematics at the University of Padua. This proved the high-water mark of Galileo's eighteen years' service to the Republic and constituted the ultimate vindication for those who had supported hiring the young rebel in 1592.*

Though Galileo was not the inventor of the telescope, in the final analysis the question which matters far more is: How was the instrument used? So far as is known, Galileo became the first person to direct his spyglass to the heavens, and the result proved astounding. It was almost

---

* Though Galileo reinvented the telescope and significantly improved its optical components, he has frequently been charged with dishonestly presenting the instrument as his own original invention. If he did so, which seems doubtful, this charade could have lasted only a short while because he does not claim to have been its inventor in *The Starry Messenger*. After all, if he had word of the Dutch "spyglass," as he called the telescope, so did others. That he opportunistically sought the glory and monetary benefits deriving from the instrument is hardly debatable, however. For a detailed discussion of this question see Edward Rosen, "Did Galileo Claim He Invented the Telescope?," *Proceedings of the American Philosophical Society*, Vol. 98, no. 5 (Oct. 15, 1954), pp. 304–312.

as if Satan, in a diabolical attempt to even a long overdue account with the Creator, had placed a magic wand in the hands of the ever-questioning Job, which enabled him to reveal the most intimate secrets of the cosmos. Each day between September of 1609 and March of 1610 Galileo impatiently waited for the sun to dip below the distant horizon, the signal that his nightly vigil could commence. When clouds obscured his vision of the moon and stars it was pure agony, for there was so much to see and so little time for a man chasing the illusive promise of immortality. He knew that it would not take long for others to publish the results of their observations, perhaps before he could make public his own. This fear was finally put to rest with the publication, in 1610, of his splendid little book *Sidereus Nuncius (The Starry Messenger)*, which gives a graphic account of one of the most momentous developments in the history of astronomy.

Galileo was first struck by the radically altered appearance of the moon, which "is not robed in a smooth and polished surface but is in fact rough and uneven, covered everywhere, just like the earth's surface, with huge prominences, deep valleys, and chasms." [18] Countless stars and nebulous objects also appeared where once only empty space had been. Our Milky Way "is, in fact, nothing but a congeries of innumerable stars grouped together in clusters. Upon whatever part of it the telescope is directed, a vast crowd of stars is immediately presented to view. Many of them are rather large and quite bright, while the number of smaller ones is quite beyond calculation." [19] This was also true of other patches of whitish clouds: "If the telescope is turned upon any of these it confronts us with a tight mass of stars." In the constellation of Orion, for example, Galileo counted more than 500 new stars within only one or two degrees of arc, besides those already visible to the naked eye. To the three stars in the Belt of Orion and the six in the Sword he observed that, "I have added eighty adjacent stars discovered recently, preserving the intervals between them as exactly as I could." [20] Suddenly the vast expanses of the universe, indeed the possibility of infinity itself, seemed far more reasonable than all but a daring few had even dreamed. Giordano Bruno's mystical vision of the limitless cosmos had been instantaneously transformed into the new reality.

Galileo saved his worst news for the Aristotelians until last. When the physicist cum astronomer pointed his telescope at Jupiter, he discovered what he termed "three starlets," small, but very bright. At first he thought them to be fixed stars, but his curiosity was aroused because they were all in a straight line parallel to the planet's orbital path. He began to closely monitor the movements of these bodies and found that they significantly changed positions every day. Then, on January 13, 1610, a fourth starlet appeared. Only one plausible explanation presented itself; Jupiter's companions were actually satellites like the moon. And like the earth's celestial companion they appeared, vanished, and then reappeared depending upon the relative position of their respective orbits round the mother planet. There, far out in space, the *same* natural process was being played out as in the terrestrial realm.

Many other anti-Aristotelian arguments, including the phases of Venus and the famous sunspots, were subsequently derived from Galileo's telescopic observations. Collectively, they signalled an end to the reign of Aristotelian–Ptolemaic cosmology, which had dominated Western man's conception of the universe for 1000 years and more. The sudden appearance of previously invisible stars made a mockery of the notion that they had been created for man's pleasure alone—unless, of course, God intended that all men should go about at night carrying an optical machine. The telescope also battered away at the hierarchical concept of the universe as conceived by Aristotle and Christianized by St. Thomas. The ethereal realm suddenly vanished, Galileo could neither see the gates of God's Empyrean Heaven nor the celestial intelligences whose responsibility it was to move the planets in their orbits. Along with the disappearance of the angels, the telescope destroyed the artificial designation between higher and lower forms of matter. The jagged, mountainous lunar surface was composed of the same elements as exist on earth, which demolished Aristotle's concept of celestial immutability, and with it the idea of a two-sphere universe. Did Galileo perhaps dream of what the photographs brought back by the lunar astronauts of the twentieth century so spectacularly illustrate—that man is in heaven, not below it, floating through space on a beautiful blue and white ball? Probably not, but he had reached the conclusion that if there is a geometrical center of creation, man and the planet he occupies are not a part of it. Thus Galileo, a man truly possessed of an overweening ego, dashed the medieval Christian belief in the egocentric universe. It is little wonder that many of his most ardent detractors looked upon his magnifying tube as nothing less than a tool of the Devil.

Campaigning for Copernicus

Instrument of the Devil or not, Galileo now possessed what he believed to be irrefutably factual evidence of the heliocentric system's existence. And if this evidence, garnered from several months of intensive telescopic observation, was proof enough for Galileo, he believed it should also be proof enough for other reasonable men. Yet he also realized that the opponents of Copernicanism would not give up without mounting some type of major offensive directed against either his revolutionary discoveries or his interpretation of them, perhaps both. With this thought in mind, the astronomer decided to launch his own program of public education on behalf of the new science, and to do so before his rivals could gather their dazed forces.

Whether knowingly or not, Galileo had already taken the first major step in this direction when he presented Venice's political and commercial leaders with a model of the telescope in August of 1609. Though he had not as yet charted the heavens with the instrument, he immediately won powerful allies who, mesmerized by the practical commercial implications of the recent invention, were more than willing to testify to its yet untold value. Galileo's next step was both carefully planned and dualistic in purpose. During past summer vacations he had arranged to give private lessons in mathematics to the future Grand Duke of Tuscany, Cosimo II de Medici. It so happened that Cosimo came to power not long before Galileo published *The Starry Messenger*. The scientist cleverly hit upon the idea of honoring the Grand Duke by giving his illustrious family name to the four moons of Jupiter, which now became the "Medicean stars." Galileo, in the Dedication of the little treatise, heaped such praise upon the youthful and yet untested Cosimo that it bordered on the obsequious:

> Your virtue alone, most worthy Sire, can confer upon these stars an immortal fame. No one can doubt that you will fulfill those expectations, high though they are, which you have aroused by the auspicious beginning of your reign, and will not only meet but surpass them. Thus when you have conquered your equals you may still vie for yourself, and you and your greatness will become greater every day.[21]

Galileo's flair for turning almost any situation to his personal advantage was never better demonstrated than in this instance. By tying the name of the powerful House of Medici to his yet most important astronomical discovery, he had skillfully dealt a solid blow to the opponents of the Copernican system before they even had an opportunity to look through a telescope, let alone formulate an alternative explanation of the newly discovered satellites. Galileo further ingratiated himself with the Duke by sending him one of his best telescopes with the promise that another, even better model, would be forthcoming.

Besides seeking a powerful ally in the Medici, Galileo had an additional motive for so lavishly singing the praises of his former pupil. During the period preceding his telescopic discoveries, he had grown increasingly disenchanted with the dual responsibilities of classroom teaching and tutoring a large number of students on the side. He wanted both the freedom and economic security to pursue his research in physics and astronomy before old age robbed him of his physical and intellectual vitality. And he also wanted to return to Florence, the city of his family and youth. Thus, in praising the name of Cosimo and his ancestors, Galileo launched a second campaign whose objective was to secure for himself the position of Court Mathematician to the Grand Duke. With the help of friends, including a strong written recommendation from the ever loyal Kepler, he obtained the coveted appointment four months after publishing *The Starry Messenger*.

Needless to say, the reaction triggered in Venice by his resignation from the University of Padua proved something less than enthusiastic. Was it not the Venetians who had offered him a respected teaching position after he had fallen from favor at Pisa? Had they not treated him well? The answer to these and similar questions was of course affirmative, as Galileo himself well knew. Still, we can sympathize with the scientist's desire for the leisure of experimenting and writing, coupled with his deep yearning to return to the city of his ancestors. Galileo knew that while at Padua he had returned value for value received. The bitterness engendered by his imminent departure proved that much if nothing else. In July, 1610, Galileo left the service of Venice for that of Florence; he never laid eyes on the great commercial center again. Ironically, Venice was the only state in Italy that had successfully challenged the authority of the papacy of Rome. In 1606 the powerful Society of Jesus (Jesuits) had been banished from the Republic, an act which Galileo viewed with grave misgivings. Little did he dream that certain influential members of the Society would one day play a major role in forcing him to recant his fervent belief in the Copernican system. But this event was years into the undiscernible future; for the present, Galileo was pleased to be able to count the Jesuits among his staunchest supporters as he undertook his campaign to single-handedly create the first scientific revolution of modern times.

The Starry Messenger was rushed to print so that its author could establish undisputed priority to his claim of being the first astronomer to employ the telescope. During the months following the book's publication, he continued his astronomical research through which the arguments for the Copernican system were multiplied both in his own mind and in the minds of his followers. Most important among these new observations was his discovery of the phases of Venus in January, 1611. If Venus revolves about the sun, as Copernicus had said, then it must show evidence of phases like the moon. But neither Copernicus nor any other naked-eye astronomer could test this hypothesis because of the planet's proximity to the sun. To the unaided eye, Venus appears as a brilliant, shapeless point, but with Galileo's telescope the phases were clearly revealed for the first time. Thus another theory had been transformed into a part of the new reality, or so it seemed.

It was during this same period that Galileo first observed Saturn's unusual shape. When viewed through his telescope the planet appeared to be what the astronomer described as "tricorporate" or three-bodied rather than a perfect sphere, a discovery which he considered to be of major importance. He died without knowing that he had mistakenly taken Saturn's rings for part of the planet's surface, because none of the telescopes in his possession had the power to bring them into clear view.* Galileo, true to his earlier pattern of behavior, immediately communicated word of the discovery of Venus' phases to friend and nobleman Giuliano de Medici, while notice of Saturn's three-bodied appearance was sent to Kepler in Prague. But since these discoveries were announced in the form of anagrams, neither correspondent knew what Galileo had found. In this manner, the astronomer could be certain of establishing the priority of his find without revealing its content to anyone, even to his most trusted colleagues and friends. Should another astronomer subsequently claim one or both of the discoveries as his own, the anagrams could then be translated and Galileo's priority established by authoritative witnesses. It was doubtless in the back of the great scientist's mind that one conflict of the type he had engaged in with Capra was enough.

Within a matter of months, Galileo Galilei, the professor of mathematics with a respected regional reputation, became Galileo Galilei, Europe's most famous scientist—perhaps even its most famous man. Thus, an opportunity unique in the history of science presented itself to him, and he was determined to utilize it to the full. Shortly after his arrival in Florence, he received permission from Cosimo to journey to Rome for the purpose of publicizing and seeking official clerical acceptance of his astronomical discoveries. Galileo knew the mission would be made easier because of a letter he had already received from Father Christopher Clavius, chief professor of mathematics at the Jesuit Roman College, confirming the existence of the celestial phenomena as described in the *Messenger*. In fact, Galileo knew he had far less to fear from the Church hierarchy at this juncture than from the academics who had staked their professional reputations on the validity of Aristotelian doctrine. This belief was confirmed during his visit to Rome, where the astronomer was received with the greatest honor by Pope Paul v and the College of Cardinals, including Maffeo Barberini, the future Urban viii. At the request of Cardinal Robert Bellarmine, head of the Jesuit Roman College, four of its members, Clavius among them, gave their scholarly opinion on what they had seen through the telescope. To a man they fully confirmed Galileo's observations, an endorsement tantamount to an official sanction by the Church itself.

It is of utmost importance, however, that we keep in mind exactly what the Jesuit confirmation of Galileo's observations did and did not do, something Galileo himself either lost sight of or, more than likely,

* The eventual discovery of Saturn's rings was made by Christian Huygens in 1653.

unwisely chose to ignore. The astronomical studies conducted by the Jesuits supported Galileo's rejection of the traditional belief in celestial immutability; they also challenged the planetary system of Ptolemy because the phases of Venus proved that the planet revolves round the sun and not the earth. On the other hand, these findings did not force Bellarmine, Clavius, and their well-trained colleagues to concur with Galileo's belief that Copernicus had been proven factually correct. Indeed, the telescope proved nothing of the kind. Rather, it offered reasonable men a choice between the system devised by Canon Nicolas Copernicus of Frauenburg and the modified geocentric system advanced by the Danish nobleman Tycho Brahe. (Of course, those who refused to listen to reason or look through the telescope, and there were many, rejected both.) Since the proponents of either system could justifiably claim equal support for their respective positions from the phenomena revealed through the telescope, the settlement of this momentous question, at least for the time being, turned on other considerations. Most important among these was the powerful psychological momentum advocates of the Tychonic system derived from their close ties to traditional Aristotelian–Ptolemaic cosmology. Though Brahe had moved halfway toward Copernicus, he had retained the most important element in Aristotelian cosmology, the geocentric universe. This had tremendous appeal to those who fought to retain the old ways, especially if they had not been deeply influenced by the Neo-Platonic revival in mathematics, as had Copernicus, Kepler, and Galileo. Furthermore, it did not take a Jesuit to understand the grave theological implications involved in embracing the heliocentric system. To do so would destroy the beautiful Thomistic synthesis forged from ancient Greek science and the Christian commitment to an anthropocentric creation. The Jesuits' friend Galileo, along with his fellow Copernicans (certain Jesuits included), were free to advocate it as a working hypothesis; but because it seemed contrary to the teachings of the Bible, it must not be presented as established truth. Only if incontrovertible proof could be produced on behalf of the heliocentric doctrine, were the Jesuits willing to reconsider. Under the circumstances, this position was both reasonable and correct. The ultimate veracity of heliocentric cosmology rested not on the telescopic observations of Galileo, but on the working out of its revolutionary mathematical laws by Kepler and Newton. Although Galileo did not realize it, his own lack of appreciation for Kepler's laws of planetary motion would cruelly mock him for the rest of his days. And though he popularized astronomy through the use of the telescope, he could offer no conclusive proof that the system for which he so ardently campaigned was true beyond a reasonable doubt.

The many honors Galileo received during his triumphant visit to Rome were by no means limited to those conferred upon him by the clergy. He was dined and feted by some of the city's best educated and most influential laymen, including the young Roman Prince Federico Cesi. It was Cesi who in 1603 had founded the *Accademia dei Lincei,* the Academy of the Lynx-eyed, into whose exclusive ranks Galileo now entered. The Lincean Academy was the first formal society in Europe that specialized in the pursuit of science and philosophy. It offered inquisitive men an alternative to the conservatism and pedantry of the universities, and its example was later copied by educated laymen from many other nations. England's Royal Society, over which Newton presided, and the American Philosophical Society, founded by colonist Benjamin Franklin, are today just two of its highly respected spiritual heirs. Galileo himself became only the sixth Lincean, an honor he obviously valued. From 1611 on, most of his important scientific correspondence bore the bold signature—Galileo Galilei Linceo.

Before setting out for Rome, Galileo made the decision to concentrate as much as possible on winning the support of educated laymen for the Copernican doctrine, and he devoted the major part of his time between 1611 and 1615 to that end. While the Church hierarchy had not accepted the heliocentric system as being factually true, it had certainly treated him with cordiality and respect. In time it might be completely won over—or so Galileo hoped. The major opposition to the Copernican interpretation of what he saw in the heavens came, as he anticipated, from vested academic interests. Various Aristotelians had already accused him of everything from falsely claiming credit for the telescope's invention to playing tricks on those he was able to entice into looking at the heavens through his optical tube. An excellent case in point is that of the Aristotelian professor Cesare Cremonini, who, despite major philosophical differences with Galileo, had been a close friend of the scientist during his employment at the University of Padua. Not only did Cremonini reject Galileo's interpretation of the latter's telescopic discoveries, he adamantly refused to use the instrument, claiming that looking through those spectacles would give him a headache. Furthermore, Cremonini argued, had God meant men to use such an instrument to better understand His universe He would have endowed them with telescopic vision at birth. One cannot help but smile at what seems an absurd argument today, but it was anything but a laughing matter to the men who made it in Galileo's time. Man's senses, they believed, were perfect as created by God. Experimentation of the type advocated by Galileo could only degrade man by breeding irreverence and contempt for noble tradition.

A number of other Aristotelians gave lectures and published books in which the existence of the celestial phenomena, including the Medicean stars, were altogether refuted on logical grounds: since Aristotle had made no mention of them, they simply could not exist. The acknowl-

edged leader of this faction was University of Bologna professor Giovanni Antonio Magini. Magini had earlier corresponded with Tycho
and Kepler and looked upon himself as an authority in their field.
When Galileo visited Bologna in April, 1610, he attempted to counter
this particular argument by showing Jupiter's moons to Magini and
several of his colleagues. He could have spared himself the effort, however. They unanimously agreed that nothing could be seen of the
planet's satellites. To make matters worse, one Martin Horky, a close
friend of Magini's, published a libelous polemic against Galileo which
bore the title: *A very brief excursion against the Starry Messenger*.
Galileo suspected, with good reason, that Horky had been persuaded to
undertake the attack at Magini's suggestion.

Whether Magini and his associates saw nothing of the Medicean stars
because of their ingrained anti-Copernican bias or because of technical
difficulties encountered with the telescope, we will never know. But in
fairness to Galileo's opponents, it must be pointed out that the first
telescopes were extremely primitive by present day standards: the lenses
were poorly ground and contained numerous imperfections, the images
produced were irregular and surrounded by a multicolored corona, and
when held in the hand the instrument vibrated, making the new stars
virtually invisible. It even took Kepler several days before he learned
the proper techniques of telescopic observation on a model borrowed
from a friend. (What Galileo would have given to have had even the
cheapest model of the telescope sold by the thousands in almost all of
today's countless discount stores.) It was only through tremendous concentration and extreme care that the secrets of the heavens were gradually
revealed. Add to this the unyielding defense of an obsolete doctrine
whose closed-minded proponents were still powerfully entrenched in the
day's educational establishment, and our appreciation for Galileo's efforts
on behalf of the new astronomy can only increase. Copernicus, it will
be recalled, had neither the heart nor the stomach for a battle of this
magnitude. He thus addressed himself "to mathematicians only." Galileo,
in a foreshadowing of the Enlightenment, sought to extend the benefits
of scientific knowledge to those outside the university community. As
historian Giorgio de Santillana has observed, "He believed . . . that in
all walks of life, from the highest to the lowest, there arise men who can
think by themselves and who are the natural elite." [22] In a well-planned
attempt to broaden the base of his support, he chose to abandon the
time-honored tradition of writing in the classical Latin. He would by-pass
the academic world and address himself in the vernacular to members
of the intelligent lay public. He was not—as has so frequently been
asserted—writing on behalf of the masses, for whom his regard as a
nobleman was no greater than that of a Montesquieu or a Voltaire.

Rather, it was to the natural aristocracy of the mind that he made his direct public appeal on behalf of Copernican astronomy.

Galileo returned to Florence in the early summer of 1611 in a buoyant mood. As the first great publicist of modern science he had neither completely silenced his Aristotelian critics nor won the unqualified support of the Church Fathers for Copernican doctrine. He had, on the other hand, captured the imagination of the educated public with his numerous demonstrations of the telescope, and was received with great cordiality by influential members of the upper clergy. Given the momentous importance and highly sensitive nature of the issues involved, he had accomplished far more than could have been reasonably expected. A man of lesser vision and commitment would have doubtless been more than satisfied with the result, but in Galileo's case the promising nature of recent events only served to strengthen his all-consuming desire to achieve a total victory. To him Rome was not the end of the struggle, merely a satisfying beginning.

The astronomer had barely returned home before the first of several important skirmishes preceding the main battle over Copernicanism took place. It involved not a matter of astronomy but of physics, yet it would have a profound effect on Galileo's future relations with the Aristotelians. Galileo, along with a number of other distinguished guests, had been invited to a dinner hosted by the Grand Duke. The night was very warm and ice had been provided to cool certain of the specially prepared dishes. During the course of the evening's conversation, the question arose as to why it is that ice floats when it appears to be heavier than water. One of the guests, an Aristotelian doctor, argued that ice, like any other object, floats or sinks because of its shape, a proposition taken directly from the teachings of the ancient master. Galileo, who had experimented on the subject since his student days, correctly maintained that it is not the shape of a body which causes it to float, but whether or not its specific gravity is less than that of water. He cited Archimedes' famous experiment discussed earlier and supported it with examples drawn from his own work with the hydrostatic balance. Among those present were two cardinals visiting from Rome. One of them, Maffeo Barberini, took Galileo's side, and the two men soon became warm friends. It was this man who, as Pope Urban VIII more than twenty years later, issued orders for Galileo's condemnation by the Inquisition of Rome, proof indeed that history takes ironic twists.

The Grand Duke, impressed by the evening's discourse, requested Galileo to publish his scientific views on the arguments presented by both sides. Galileo obliged and within a few months his *Discourse on Floating Bodies* was published. Since he wrote the book in Italian it could be read and understood by almost any literate layman. The scientific arguments alone were devastating to the Aristotelian position, but Galileo could not resist making his opponents appear even more ludicrous by injecting liberal doses of his vitriolic wit into the dialogue. He not only rebutted all the arguments made against him, he added and answered several others that his opponents had not even thought about. Nor could the Aristotelians fall back on the argument that what Galileo's experiments showed was open to conflicting interpretation. The hydrostatic balance was cheap and simple to construct and unlike the telescope, free of optical illusions. Nonetheless, they felt duty bound to respond to his open challenge, and during the next six months published four books in an unsuccessful effort to refute the Galilean position. It so happened that their leader was Ludovico delle Colombe, whose last name means pigeon or dove. Galileo, in a flash of mischievous insight, dubbed Colombe and his colleagues the "Pigeon League," a satirical rubric quickly adopted by Galileo's followers to identify the Aristotelian opposition. While we can appreciate the humor of the moment, such tactics did nothing to raise the level of the increasingly heated debate.

A second and far more serious scholarly controversy developed a few months later, early in 1612. Because of it Galileo came into direct conflict with an important member of the clergy for the first time. The affair, however, did not involve basic differences regarding fundamental theological issues, as one would have reason to expect, but centered on the question of who was entitled to priority in the discovery of a major celestial phenomenon. Shortly before he departed Padua in 1610, Galileo made his first observations of sunspots, whose appearance had been noted by various scholars dating at least as far back as the ancient Roman poet Virgil. He was preoccupied with other matters and, though he spoke of the spots with his friends, he did not bother to investigate them in detail. As was bound to happen sooner or later, given the large number of individuals now using the telescope, another astronomer claimed the discovery (or to be more precise the telescopic confirmation of it) as his own. This was the German Jesuit Father Christopher Scheiner, professor of mathematics and Hebrew at the University of Ingolstadt. Scheiner was eager to bring word of his discovery to print but was denied permission to publish in his own name by a superior, who feared it might be misinterpreted as a Jesuit endorsement of the Copernican system.*

---

* Indeed, there is sufficient evidence to show that Scheiner, who became one of Galileo's most ardent detractors, was, like many of his fellow Jesuits, a Copernican at heart. One of the most unfortunate aspects of this and similar episodes is that questions of science, no matter how well argued, frequently took a back seat to conflicts of personality and temperament. Furthermore, the honor of first publishing a treatise on sunspots belongs neither to Scheiner nor Galileo but to Johann Fabricius of Wittenberg, whose booklet appeared in 1611. Another early observer of sunspots was the English mathematician–astronomer Thomas Harriot.

Scheiner decided upon an alternative, if less satisfactory, method of making his discovery known. He drafted a detailed letter to Mark Welser, a wealthy Augsburg merchant and patron of Kepler, in which he revealed the appearance of "several black drops" on the face of the sun. Welser, with Scheiner's permission, immediately had the letter published in booklet form using the pseudonym "Apelles." A copy was then forwarded to Galileo, asking his opinion on the matter. Though he would not learn Scheiner's identity for another year, based on the nature of the scientific arguments presented Galileo concluded that it had been composed by a Jesuit. Still, he not only took umbrage at Scheiner's claim of observational priority, he brusquely rejected the Jesuit's interpretation of the phenomenon. Scheiner, in an attempt to skirt the theological implications of the discovery, accounted for sunspots by presenting them as small hitherto invisible planets which move across the sun's surface, periodically obstructing the astronomer's vision. Of course the very existence of the spots conflicted with the perfection of the celestial region, but the Jesuits, through Father Clavius, had already rejected at least this much of Aristotelian cosmology. By equating the sunspots with satellites, however, they, like the moons of Jupiter, could be fitted into the context of Tychonic astronomy.

Galileo had no time for Tycho's cosmology, and he had long since broken free of the philosophical scruples which still bound the Jesuits to a geocentric interpretation of the universe. He therefore placed the dark spots on the sun's surface, or at least as close to the sun as the clouds are to earth. In fact, he equated sunspots with this very terrestrial phenomenon, an argument diametrically opposed to Aristotelian cosmology. He further observed that the apparent motion of the spots across the sun's disk is additional proof that the sun, like the planets themselves, revolves on its axis, as Copernicus maintained. In all, Galileo wrote three major letters on the subject to Welser. These were subsequently published in their original in Rome under the auspices of the Lincean Academy in 1613. Scheiner countered by drafting a reply to Galileo's *Letters on Sunspots* [23] entitled *A More Accurate Discussion of Sunspots and the Stars which Move around Jupiter*. For the first time in print Galileo openly endorsed the Copernican system as a reality rather than a promising hypothesis; and he did so in the vernacular. In his mind's eye he may well have envisioned this act as taking him beyond the point of no return. That he had taken this step while at the same time claiming priority for the discovery of a long-known celestial phenomenon and antagonizing a powerful Jesuit in the bargain seemed to bother Galileo not at all. His science, he believed, was quite sound. What he did not know is that his failure to temper the manner of its presentation had already set in motion that long series of events which, in the end, turned both the Jesuits and the Church against him.

Despite his open advocacy of the Copernican system, Galileo encountered no formal clerical opposition to his latest book. The work won great popular acclaim, and he even received letters of admiration from Cardinals Boromeo and Barberini. This, indeed, was part of the problem. Galileo's combination of scientific genius and polemical brilliance set him in a class apart from other men. An adoring public reveled in his seemingly effortless capacity to show up any and all opponents, no matter what arguments they might employ against him. Had Galileo's intellectual and literary gifts been less profound, it is reasonable to assume that the aura of righteousness and invincibility with which he imbued his cause would not have gotten so far out of hand. It was only with the deepest reluctance and as a last resort that the Church moved against him. Even its warning of 1616 was, as we shall see, of a highly restrained and cordial nature. With these facts in mind, it becomes somewhat easier to understand how this truly great scientist could be lulled into a dangerously false sense of security when even his closest friends quite literally begged him to temper the rasher aspects of his abrasive behavior.

The Call of 1616

Thus far Galileo had succeeded in handling his critics with little or no difficulty. However, now that he had openly endorsed the Copernican system in his *Letters on Sunspots*, the controversy entered a new and far more serious stage. Up until this time—1613—Galileo's main opposition came almost exclusively from the university-based Aristotelians, his dispute with Father Scheiner over the discovery and explanation of sunspots notwithstanding. Still, it seemed only a matter of time before the thirteenth-century synthesis of Aristotelian science and Christian doctrine would prompt the more conservative elements within the clergy to join forces with their counterparts in the academic world. Galileo might have anticipated this strategy sooner than he did had he taken more seriously an incident which occurred toward the end of 1612. While a guest in the Florentine villa of his good friend Fillippo Salviati—a main figure in his two great *Dialogues*—he received news that an elderly Dominican Father named Niccolo Lorini had attacked his scientific views from the pulpit during a sermon on All Souls' Day. Galileo promptly sent Lorini a note asking for an explanation. The embarrassed Dominican, unnerved by the unexpected communication from the famous scientist, hastily drafted a reply disclaiming any desire to become involved in such matters. "It is indeed true," Lorini wrote, "that I did say a few words just to show I was alive. I said, as I still say, that this opinion of Ipernicus—or whatever his name is—would appear to be hostile to the divine Scriptures. But," he added, "it is of little consequence to me, for I have other things to do." [24] Because Lorini obviously knew little or nothing about Copernican astronomy, Galileo was satisfied with this explanation. He and his friends had a good laugh over the Dominican's naive ignorance and the matter was quickly forgotten.

What Galileo did not know is that Lorini—while certainly no expert on scientific matters—appeared far less harmful than was actually the case. Though he posed no direct threat in his own right, the Dominican had close friends both among the opposition in the academic community and within his own powerful religious order. The Dominicans, in contrast to the Jesuits, constituted one of the most conservative elements of Roman Catholicism, particularly when it came to the matter of Christian cosmology. It was, after all, the Order's two greatest scholars, Albertus Magnus and his brilliant student St. Thomas, who had made the Aristotelian universe plausible to Christian theologians some three centuries earlier. This synthesis had since been elevated to the level of dogmatic principle with the result that if someone seriously questioned Aristotle on an important matter of science, his religious outlook was likely to be suspect as well. The Dominican aversion to all matters Copernican was further sustained by the lingering memories of the apostate monk Giordano Bruno, himself originally a member of the Order. Bruno, an unrepentant heretic, had been burned at the stake for doctrinal reasons which had nothing to do with heliocentric astronomy. Nevertheless, while in exile in England, he had drafted a number of rabidly pro-Copernican polemics, which in the new light of telescopic observation suddenly appeared to have a certain basis in fact. Thus the stigma of Bruno's iconoclastic mysticism clung like the odor of dead fish to the Dominican cowl, serving as a continual source of embarrassment and an ever-present reminder that even the hottest flames cannot always be counted upon to cleanse all. Add to this the long-smoldering Dominican antipathy toward the more liberal Jesuits, many of whom supported certain major aspects of Copernican doctrine, coupled with the determined efforts of the Aristotelians to hold their remaining ground, and all the ingredients necessary for combustion were present in more than ample amounts.

On December 14, 1613, the Grand Duke, Cosimo II, invited a number of learned men to a banquet in Pisa, where the court was in residence for the winter. Galileo could not attend because of illness, but his faithful friend and former pupil, the Benedictine monk Benedetto Castelli, was among the guests. Castelli, a confirmed Copernican, had been recently appointed chief mathematician at the University of Pisa, having received strong backing from his famous mentor. Among the others present were the Duke's mother, the Duchess Christina, his wife Madeleine of Austria, and Dr. Cosimo Boscaglia, a professor of philosophy. The Duchess dominated the conversation which was primarily concerned with the new astronomy. She especially wanted to know all about the Medicean stars, whether they were actually in the heavens and, if so, what were their positions. Both Castelli and Boscaglia assured her that their existence was indeed genuine. Castelli left a short time later, but no sooner had he gone out of the palace when Madame Christina's porter overtook him with the message that the Duchess wished to continue her discourse with the monk. Castelli immediately became suspicious. In his written account of the affair sent to Galileo he related how Dr. Boscaglia, an Aristotelian, who had the ear of the Duchess during most of the banquet, conceded "as true all the new things you have discovered in the sky" but one: "He said that only the motion of the earth had something incredible in it and could not take place, in particular because the Holy Scripture was obviously contrary to this view." [25] Castelli's thoughts must have flashed back to a recent incident involving Galileo's old nemesis Ludovico delle Colombe. The leader of the "League" had authored a widely circulated manuscript titled *Against the Motion of the Earth*, in which he cited several texts from Scripture, all in seeming contradiction to the proposition of terrestrial motion. The most graphic of these passages was taken from the book of Joshua (10: 12–13): "Then Joshua spoke to the Lord in the day when the Lord gave the Amorites over to the men of Israel; 'Sun, stand thou still at Gibeon, and thou Moon in the valley of Aijalon.' And the sun stood still, and the moon stayed, until the nation took vengeance on their enemies." Galileo was well aware of the danger involved if he allowed himself to be pulled into the quagmire of scriptural exegesis. He discreetly refused to rise to Colombe's tempting bait. Colombe had scored a major point, however. When Castelli, Galileo's pupil, took up his teaching duties at Pisa, he was forbidden by the overseer of the university, Arturo d' Elci, to discuss the possibility of the earth's motion, even in private.

When Castelli returned to the palace he was ushered into the chambers of the Grand Duchess. There he encountered Christina, her son, the Grand Duke, and several of the other banquet guests including Dr. Boscaglia. By this time Castelli was truly on his guard.

> Madame began, after some questions about myself, to argue the Holy Scripture against me. Thereupon, after having made suitable disclaimers, I commenced to play the theologian with such assurance and dignity that it would have done you good to hear me. . . . I carried things off like a paladin. I quite won over the Grand Duke and his Archduchess, while Don Paolo came to my assistance with a very apt quotation from the Scripture. Only Madame Christina remained against me, but from her manner I judged that she did this only to hear my replies. Professor Boscaglia said never a word. [26]

The limits to Galileo's self-restraint had always been extremely narrow. He had maintained silence on this matter once before, but only by exercising a supreme effort of the will, something he could not bring himself to do a second time. The time had come, he decided, to meet the scriptural challenge head on, just as he had directly confronted Lorini in 1612. He drafted a long letter to Castelli offering his opinions on the proper relationship between science and religion, and included a possible Copernican interpretation of the events recorded in the miracle of Joshua. Galileo conjectured that when the ancient Israelite commanded the sun to stand still, what had in fact happened is that it ceased its rotation, and the earth in consequence its daily and annual motion. This was obviously an undignified grasping at semantic straws, something his scholastic opponents had become famous for. Moreover, it detracted from the main theme of the letter, to be discussed in greater detail in the next section of this chapter, that not all statements in the Bible should be taken literally because they were written in the language of the common people, who are not schooled in the terminology and methods of science. Copies of the letter were freely circulated, and for a time it appeared that Galileo's critics had been silenced once again. It proved a deceptive calm, however. A storm of major proportions had been in the making for a long time.

In Florence, on December 20, 1614, Father Tommaso Caccini, a Dominican Friar, preached a sermon from the pulpit of Santa Maria Novella using for his opening text a quote from the Acts of the Apostles (1: 2): "Ye men of Galilee, why stand ye gazing up into the heavens?" Caccini branded mathematics a tool of the Devil and recommended that mathematicians be expelled from Christian lands. Especially dangerous were mathematicians who believed in the Copernican doctrine of the moving earth. Heresy, though not openly charged, was strongly implied. Certainly the friar left no doubt in anyone's mind at whom his vitriolic attack was directed. Santillana has documented Caccini's close ties to Colombe and the League, thus presenting a convincing case for conspiracy. It was apparently agreed beforehand that Caccini would stir up a public controversy centered on Colombe's earlier argument that Galilean astronomy directly violated Holy Scripture. This it was hoped would compel the Christian leadership in Rome, which was still reluctant to become involved in the matter, to step in for reasons of preserving order and authority. "The name of 'provocateur' had not yet been invented," writes Santillana, "but the dodge was as old as the world." [27]

By now, Galileo, cautioned by his friends, suspected a conspiracy, and for the time being held his peace. Meanwhile old Father Lorini, during a visit to Pisa, was shown a copy of Galileo's *Letter to Castelli*. The conciliatory attitude adopted by the Dominican toward the natural philosopher some three years earlier was suddenly forgotten. On his return to the Convent of St. Mark in Florence, Lorini conferred with his fellow monks regarding the contents of the *Letter*. They shared his deep concern and agreed that the document should be forwarded at once to the Holy Office of the Inquisition in Rome for further examination. On February 17, 1615, Lorini sent a copy of the *Letter* to Paolo Cardinal Sfrondrati, one of the Inquisitors-General. Galileo, apprised by his allies of Lorini's action, became deeply troubled. The *Letter to Castelli* had been hastily drafted, and he knew that even a slight change in its phrasing could place his scriptural analysis in a very bad light indeed. He did not know exactly what form Lorini's letter to the Holy Office had taken, but wisely he expected the worst. Lorini, emboldened by the support of his fellow monks, and genuinely aggrieved over Galileo's approach to scriptural interpretation, was filled with a zealot's desire to stamp out the growth of what appeared to him as incipient heresy. His letter to Cardinal Sfrondrati in Rome confirms this:

> All our Fathers of this devout convent of St. Mark are of the opinion that the letter [to Castelli] contains many propositions which appear to be suspicious or presumptuous, as when it asserts that the language of Holy Scripture does not mean what it seems to mean . . . . Ever mindful of our vow to be the 'black and white hounds' of the Holy Office, when I saw that they [the Galileists] expounded the Holy Scriptures according to their private lights and in a manner different from that of the common interpretation of the Fathers of the Church; that they strove to defend an opinion which appeared to be quite contrary to the sacred text; that they spoke in slighting terms of the ancient Fathers and of St. Thomas Aquinas; that they were treading underfoot the entire philosophy of Aristotle which has been of such service to Scholastic theology . . . I made up my mind to acquaint your Lordship with the state of affairs, that in your bold zeal of the Faith may, in conjunction with your illustrious colleagues, provide such remedies as will appear advisable.[28]

Lorini closed with an expression of priestly charity by stating that "those who call themselves Galileists are orderly and good Christians all, though a little too wise and conceited." Yet there can be no doubt that his letter to the Inquisition was open to only a single interpretation—the Dominicans wanted Galileo and his followers silenced once and for all.

Cardinal Sfrondrati set the machinery of the Holy Office in motion by turning the matter over to an Inquisitor for a detailed examination. When Galileo heard of this, he immediately dispatched an authentic copy of his *Letter to Castelli* to his friend Archbishop Piero Dini in Rome. He wanted to be very certain he would not be credited with any statements he had not made. He also informed Dini that the *Letter* had been composed in haste, and that he was now revising and expanding it for purposes of clarification. He completed this task in June, 1615, and titled the little treatise *Letter to the Grand Duchess Christina*. Meanwhile, Galileo's friends made enquiries and wrote letters on his behalf to the Holy Office. Dini forwarded the copy of the *Letter* sent him by Galileo to the authorities. Though its contents were viewed as a matter of some concern, it was not generally thought to deviate seriously enough from Catholic doctrine to necessitate further action. Thus the Holy Office dismissed the case at this point. It was clearly a defeat for the conservatives and a source of personal embarrassment to Lorini and his faction of the League in particular. Still the reactionaries continued their pursuit of Galileo in full cry.

The struggle becomes increasingly complex at this point, so that we shall be able to trace only the highlights of what followed. It should be kept in mind, however, that the refusal of the Holy Office to further pursue Lorini's denunciation did not mean that it had passed favorably on the new astronomy. It meant only that the charges did not warrant a formal hearing at this time. Copernican astronomy was still a moot point in Rome, where the Church hierarchy adopted a wait and see attitude.

It was still not too late for Galileo to have turned away from the controversy forced upon him by his enemies, on the grounds that as a layman he was no more equipped to deal with theology than were most of his adversaries with science. Instead, he decided to fight and would have gone to Rome at once had he not been detained by a prolonged illness. Galileo believed that if only he could personally place his scientific arguments for Copernicanism before the members of the Holy Office, he could win their approval. In his own mind, he was not attacking the Church, something most of his detractors accused him of, but desperately attempting to prevent this great institution from turning its back on the rational product of human inquiry. Though there was nothing in the geocentric views of the Old and New Testaments that had anticipated the heliocentric system, Galileo could not understand how any intelligent and well-educated citizen of his day, clerical or secular, could deny its validity. Yet the fact remains that the only way one could assimilate the new cosmology was to abandon the entire medieval outlook and enter into a new expansion of reason as he had done. He could never bring himself to accept the idea that the Church hierarchy was far more interested in order and authority than in scientific truth. Nor did the Church truly understand Galileo. They were, as Santillana has pointed out, like two ships passing in the night.

A few days after the Holy Office dismissed Lorini's complaint, Galileo received a second and equally welcome piece of news. Prince Cesi, founder and President of the Lincean Academy, wrote Galileo from Rome that a Carmelite friar, Paulo Antonio Foscarini, had just published a book in defense of the Copernican system. Furthermore, the author had sent a copy to Cardinal Robert Bellarmine, the seventy-three-year-old general of the Jesuit Order and Consultor of the Holy Office, asking his opinion of it. Bellarmine, a theological conservative, was the most respected theologian in Rome, Pope Paul v notwithstanding. A specialist in applied theology, he authored the Catholic catechism in its present form and led the battle against all deviationists including the Lutherans, Calvinists, and Anglicans. An opponent of any and all attempts to compromise orthodoxy and papal supremacy, he was nevertheless a very reasonable and gentle man. He was also a personal admirer of Galileo, whom he had met and entertained during the scientist's triumphant visit to Rome in 1611.

Bellarmine, despite an innate conservatism, made every attempt to reach a working accommodation between the new science and traditional theological doctrine, a most difficult and unenviable undertaking. On April 12, 1615, he drafted a reply to Foscarini's inquiry in which the unofficial but quite authoritative position of the Church was carefully set forth. Bellarmine expressly mentioned Galileo by name to show that the response was addressed to him as well. Unaware that the preface to *De revolutionibus* was composed by Osiander rather than Copernicus, the Cardinal expressed his view that Copernicus spoke of the heliocentric system in hypothetical terms only, and urged Foscarini and Galileo to follow suit. Informed by the Jesuit astronomers that the new system, in light of their recent telescopic observations, saved the appearances better than did the Ptolemaic model of the universe, he was willing to admit this much. However, to go further by attempting to establish Copernicanism as fact would be asking for trouble.

> To say that the assumption that the Earth moves and the Sun stands still saves all the celestial appearances better than do eccentrics and epicycles is to speak with excellent good sense and to run no such risk whatever. But to want to affirm that the Sun, in very truth, is at the center of the universe and only rotates on its axis without travelling from east to west, and that the Earth is situated in the third sphere and revolves very swiftly around the Sun, is a very dangerous attitude and one calculated not only to arouse all Scholastic philosophers and theologians but also to injure our holy faith by contradicting the Scriptures.[29]

Bellarmine also cited the prohibition of the Council of Trent (1545–1563), which expressly forbade expounding the Scriptures in a manner contrary to the common interpretation of the holy Fathers, as Galileo had done in his letters to Castelli and the Grand Duchess Christina. "Consider then in your prudence," he concluded, whether the Church can support that the Scriptures should be interpreted in a manner contrary to that of the holy Fathers and of all modern commentators, both Latin and Greek." The Cardinal closed by challenging Galileo on his own ground. If the scientist could truly demonstrate the truth of Copernican theory, then it would have to be admitted that the Scriptures have been misinterpreted by Christendom's greatest thinkers. However, until such time might arrive, "One may not abandon the Holy Scriptures as expounded by the holy Fathers." Bellarmine had thus placed the burden of proof for the Copernican system back on the shoulders of its advocates, where it clearly belonged. And he had done so in a firm but kindly manner. Ever the champion of compromise, he had left Galileo and his supporters with a reasonable choice: either supply proof positive that the new system existed in fact, or treat it as a working hypothesis—nothing more. Had Galileo exercised the prudence of judgment requested of him, he doubtless would have taken the middle ground. But his commitment to Copernicanism went far beyond that of a working hypothesis; indeed, it had become a kind of religion of its own.

Though not yet completely recovered from the lingering illness which had forced him to postpone his planned visit to Rome, Galileo now felt well enough to undertake the journey. The Grand Duke instructed the Tuscan ambassador, Piero Guicciardini, to lodge the scientist in the Florentine embassy and to provide him with "food, a secretary, a servant and a mule." Guiccardini, on hearing of Galileo's coming, vigorously protested. His close contacts with high clerical officials convinced him that Galileo's presence could only weaken his already tenuous case. What Guiccardini sensed was a hardening of attitudes toward the great scientist, whose overweening self-assurance was bound to irritate his detractors all the more. Even the Jesuit ardor for the new science had been significantly dampened by rumors of an impending crack-down on the Copernicans. Galileo would have done well had he heeded this ominous sign; but he was too busy mapping out a viable strategy for the upcoming debate in Rome. When the much needed support of the Jesuits was not forthcoming during this, his greatest hour of need, it was something which, in later years, he never forgot or forgave them for. Still, he should have realized that even in the liberal Jesuit mind orthodoxy must take precedence over matters of science.

Galileo was purring with optimism upon his arrival in the Eternal City. The old fighting spirit had been rekindled, and he employed every opportunity at his disposal to extol the merits of Copernican astronomy. Time after time he made mincemeat of the Aristotelian position; yet in doing so he failed to establish the superiority of his own. Deceived by his belief that reason would triumph over all, he became the victim of a naive optimism which severely impaired his ability to see the situation as it truly was—a conflict between authority and the search for scientific truth, not a conflict between one type of scientific truth and another. Many of his friends, Ambassador Guiccardini included, clearly understood the gravity of the situation and repeatedly tried to warn him of just how dangerous was the course he had chosen to pursue. But their efforts were in vain; Galileo could not be persuaded to desist.

By February, 1616, events were moving toward a dramtic climax. On February 19, the Holy Office sent the following two disputed astronomical propositions to its Consultors, requesting them to render an opinion within a few days:

1. That the sun is the center of the universe and is not moved by local motion.
2. That the earth is not the center of the world, nor movable, but moves as a whole, also with diurnal motion.

Four days later the Consultors handed down their decision. The first proposition was "declared unanimously to be foolish and absurd in philosophy and formally heretical inasmuch as it expressly contradicts the doctrine of Holy Scripture in many passages, both in their literal meaning and according to the general interpretation of the Fathers and Doctors." The second proposition was likewise declared false: "All were agreed that this proposition merits the same censure in philosophy, and that, from a theological standpoint, it is at least as erroneous in faith."

This judgment was duly transmitted to the *Index Librorum Prohibitorum* (Congregation of the Index) which had the responsibility of censoring all printed matter not in compliance with prescribed doctrine. On March 5, 1616, the Congregation issued a formal decree of condemnation. Father Foscarini's Copernican treatise was totally condemned, not to be read by the faithful under any circumstances. However, Copernicus' *De revolutionibus*, still considered a hypothetical work by virtue of Osiander's spurious Preface, was only suspended until certain passages could be corrected. This being done, the book was removed from the Index four years later, in 1620. Thus, the Copernican system, though labeled "false and contrary to Scripture," could still be presented as an hypothesis. Interestingly, none of Galileo's works were placed on the Index, even though his famous letters to Castelli and Madame Christina were obviously more than hypothetical endorsements of the heliocentric system. The reason most commonly given is that both letters had been circulated in manuscript only. But this does not explain why Galileo's pro-Copernican *Letters on Sunspots*, published in 1613, was not included in the censure. The more plausible explanation is that the Holy Office simply did not wish to lay the groundwork for a *cause célèbre* by entering into an open dispute of such a highly sensitive nature with the most famous scientist in all Christendom. Yet Pope Paul v, a man who confessed to a private hatred of "science and polite scholars," had already taken steps to insure Galileo's submission to clerical authority. In February he instructed Cardinal Bellarmine personally to notify Galileo in a closed audience that he must abandon his defense of the censured opinions as scientific truth. Should Galileo refuse, the Commissary-General of the Holy Office was to give him official instructions, before witnesses and a notary, to refrain from dealing with Copernican doctrine in any manner whatsoever. If he still refused, which seemed most unlikely, Bellarmine was empowered to have the scientist imprisoned, a drastic step of last resort which those involved sincerely hoped would not have to be taken.

Galileo appeared before Bellarmine on February 26, 1616. Exactly what transpired at this meeting has become the center of one of the most protracted ecclesiastical and scholarly controversies in modern history. At issue, both then and now, is the question of whether, on the one hand, Galileo was expressly prohibited from dealing with Copernican doctrine in any manner whatsoever, as a disputed account of the meeting contained in the surviving archives of the Vatican clearly states, or whether, on the other, he was only limited to a hypothetical discussion of the subject. Because the surviving evidence is open to widely conflicting interpretation, a detailed analysis of the problem is beyond the scope of this study. A few brief comments are in order, however, for the matter was destined to have a direct and crucial bearing on the outcome of Galileo's trial by the Roman Inquisition in 1633.*

The document in question states that Cardinal Robert Bellarmine, acting in accordance with papal instructions, duly notified Galileo that he must abstain from either teaching or defending the Copernican doctrine. How Galileo responded, or whether in fact he was given the opportunity to respond, the document does not say. For no sooner had Bellarmine communicated the official position of the Church to the stunned scientist than Reverend Michelangelo Lodi, Commissary-General of the Holy Office,

> commanded and enjoined [Galileo], in the name of His Holiness the Pope and the whole Congregation of the Holy Office, to relinquish altogether the said opinion that the Sun is the center of the world and immovable and that the Earth moves; or further to hold, teach, or defend it in any way whatsoever, verbally or in writing; otherwise proceedings would be taken against him by the Holy Office; in which injunction the said Galileo acquiesced in and promised to obey.[30]

The document, if genuine, seems clear enough: Galileo had not only been forbidden to teach or discuss the heliocentric system as truth, he promised not to *defend it in any way whatsoever*, including the realm of

---

* The two most recently published studies among many on the subject are De Santillana's *The Crime of Galileo* and Jerome J. Langford's *Galileo, Science and the Church* rev. ed. (Ann Arbor: 1966). The authors, both respected scholars, employ the same historical sources yet have reached divergent conclusions about the affair, proof that after three and one-half centuries the case is far from settled to the satisfaction of historians and probably never will be.

hypothesis. Yet at his trial seventeen years later he insisted that no such prohibition had been exacted from him. Was Galileo, then, a liar? Perhaps, but there is room for reasonable doubt.

In the first place, why, even before he had the opportunity to reply to Bellarmine, had the Commissary-General been permitted to formally enjoin the scientist from demonstrating any further support of Copernicus? This step, according to papal instructions, was to be taken only in the event Galileo refused to comply with the Bellarmine pronouncement. The Cardinal, a stickler in all matters procedural, was well aware of the proper course of action, yet seemingly violated specific instructions handed down by the Pope himself. Even more important is the fact that though Galileo supposedly appeared "before a notary and witnesses," none of those present, the scientist included, signed the injunction, again a direct and major violation of accepted procedure. As Santillana has pointed out, "we have here only what amounts to an administrative minute, unsigned and casually transcribed." [31] Equally as puzzling are the contents of a letter drafted by Cardinal Bellarmine a few weeks after his meeting with Galileo. Though it had been the Pope's intent to keep the matter as quiet as possible in order to spare Galileo the humiliation of a public scandal, the very secrecy of the proceedings gave birth to a number of ugly rumors, including speculation that he had been forced to abandon his unorthodox opinions or suffer the penalty of torture and perpetual confinement in the prison of the Inquisition. To protect his reputation from further injury at the hands of his more vicious detractors, Galileo requested of Bellarmine an account of the proceedings which the Cardinal graciously provided in the following form:

> We, Robert Cardinal Bellarmine, having heard that it is calumniously reported that Signor Galileo Galilei has in our hand abjured and has also been punished with salutory pennance, and being requested to state the truth as to this, declare that said Signor Galileo has not abjured, either in our hand, or the hand of any other person here in Rome, or anywhere else so far as we know, any opinion or doctrine held by him; neither has any salutary pennance been imposed on him; but that only the declaration made by the Holy Father and published by the Sacred Congregation of the Index has been notified to him, wherein it is set forth that the doctrine attributed to Copernicus, that the Earth moves around the Sun and that the Sun is stationary in the center of the world and does not move from east to west, is contrary to the Holy Scriptures and therefore cannot be defended or held. In witness whereof we have written and subscribed these presents with our hand this twenty-sixth of May, 1616.[32]

No mention is made of a formal injunction; in fact, Bellarmine argues just the opposite. Though the Copernican doctrine cannot be "defended or held," there is no absolute prohibition against its discussion as an hypothesis. Whether in light of this and other evidence the document in the extant files of the Inquisition is a forgery, as Santillana and many other scholars believe, we shall never know for certain. But if one chooses to reject this theory, it is indeed difficult to explain why Bellarmine, a man schooled in the most intimate details of canon law, would have been so incautious as to openly sanction conflicting statements in writing on a matter of such overriding importance. Yet it must also be admitted that the possibility still exists, as set forth by Langford and others, that Galileo had been singled out for special treatment simply because he was Galileo, that it was thought necessary to bind him more tightly to silence than anyone else. Whatever the truth of the matter, it is quite clear from subsequent developments that Galileo never considered himself under an absolute prohibition regarding the discussion of Copernican doctrine. Yet before returning to Florence some three months later, he had become reconciled to the necessity of moving far more cautiously than before. The public campaign on behalf of heliocentric astronomy had been stopped dead in its tracks; no hope remained of its being revived without a significant change in Rome's intellectual climate. The aging Paul v could not live forever; perhaps a more moderate cardinal would one day preside over the Holy See. Galileo promised himself that until such a time he would maintain public silence on controversial issues. "Of all hatreds," he lamented to a friend, "there is none greater than that of ignorance against knowledge."

Philosophic Interlude

Before turning our attention to the final and most dramatic chapter of Galileo's embattled life, it is essential, as in the case of the other major astronomers discussed in this work, that some effort be made to assess the scientist's contribution, both intellectual and philosophical, to modern Western culture. We have already examined in considerable detail his pioneering work with the telescope and to a somewhat lesser extent his brilliant and innovative studies in the field of mechanics, either one of which would have entitled him to a lasting place among the immortals of modern science. Yet, in the final analysis, Galileo, like Aristotle before him and Newton after, must be looked upon as one of those great and rare geniuses whose work is not confined to any special field of investigation. This is because he not only set forth new and fundamental physical facts, he invented and permanently established a revolutionary general method of scientific thought by successfully melding empirical observation with mathematical deduction. It is indeed no exaggeration to state that to Galileo belongs the lion's share of the credit for elucidating the scientific method, the acceptance of which, more than any other single consideration, distinguishes modern man from his ancient and medieval predecessors.

Of primary importance to a basic understanding of Galileo's approach to all questions pertaining to science is his relationship to the two dominant schools of thought of his day—the Aristotelian and the Neo-Platonic. Galileo, along with René Descartes, is the scientist most associated with the overthrow of Aristotelianism. Thus, based upon our earlier discussion of his revolutionary findings in the fields of mechanics and astronomy, the public dissemination of which led to an intensely bitter feud with conservative Peripatetic academicians, one might readily conclude that Galileo lacked even the slightest confidence in the principles of natural philosophy enunciated by their ancient master. However, we must keep in mind the fact that while Galileo rejected the classical physics of Aristotle, with its qualitative approach to nature, in favor of a cosmos reinterpreted in terms of exact mathematics, he never doubted the importance of the Greek philosopher's use of observation and logic, so long as they were integrated with mathematical computation. Galileo was perhaps the first scientist of the modern age to consciously realize that there is nothing absolute in what is termed scientific truth. A system is true only so long as it satisfactorily explains phenomena to those who observe them, and that, he readily acknowledged, is precisely what the geocentric system of Aristotle did in pre-Copernican days. Galileo thus respected Aristotle both for his intellectual prowess and methodological approach, even though he disagreed with many of the latter's fundamental conclusions regarding the nature of the world and universe. What Galileo found most intolerable about Aristotelianism, then, was not so much the teachings of the great philosopher himself, but the manner in which Aristotle's scholastic followers subjected science to the excesses of absolute authority. Ever since his student days, he had recoiled from those who, impervious to reason and common sense, mindlessly wrapped themselves in the protective mantle of the Stagirite. Galileo repeatedly expressed his belief that Aristotle would have been among the first to change his mind about the nature of the universe, if only he had been afforded the opportunity to look through the telescope or to participate in experiments on the free fall of heavy bodies:

> I declare that we do have in our age new events and observations such that if Aristotle were now alive, I have no doubt he would change his opinion. This is easily inferred from his own manner of philosophizing, for when he writes of considering the heavens inalterable, etc., because no new thing is seen to be generated there or any old one dissolved, he seems implicitly to let us understand that if he had seen any such event he would have reversed his opinion, and properly preferred the sensible experience to natural reason. Unless he had taken the senses into account, he would not have argued immutability from sensible mutations not being seen.[33]

In effect, what Galileo is saying is that Aristotle, who relied so heavily on sense experience, would have welcomed the invention of the very instrument which brought his geocentric vision of the universe into scientific disrepute. Whether or not the Stagirite would have been so indulgent toward the man who shattered his crystal spheres with a celestial battering ram fashioned from lead, is, I believe, open to serious question. It is readily apparent from the application of this line of reasoning, however, that Galileo clearly recognized his debt to the ancient master: after all, his own best proof of Copernicanism came from the same sense impressions utilized by Aristotle, albeit magnified many times by the telescopic lens. Thus Galileo, like his Greek predecessor, attached crucial importance to the use of sense experience to unlock nature's carefully guarded secrets. Yet as we are about to see, empiricism —which in addition to logic counted for nearly everything in Aristotle's system of investigation—was only one of several components out of which Galileo constructed a revolutionary methodology for modern science.

The Italian also drew upon Aristotelian physics and cosmology in certain other significant respects. For example, it is important to keep in mind the fact that the Copernican system as advanced by Galileo was the very same one designed by the secretive Canon of Frauenburg nearly a century earlier. Unaware of the Keplerian revolution in celestial physics, Galileo clung to the Aristotelian–Ptolemaic belief in epicycles as an integral part of what he termed "rigorously demonstrated physical reality." Nor was he able to completely free himself from Aristotle's teaching that circular motion is by definition the most perfect form of movement, as is evidenced by his studies on inertia. While Galileo sharply disagreed with Aristotle over the latter's belief that rectilinear motion is an unnatural state and may be thus attributed only to the corrupt terrestrial realm, he nevertheless accepted inertial motion as circular. In other words, Galileo taught that an object set in motion on earth which maintains a constant velocity can be presumed to move around the earth's surface indefinitely. Though he also postulated straight line inertial motion, he ultimately rejected it on the grounds that it is impossible for an object to move indefinitely toward a place at which it can never arrive. In order for this to occur, as Descartes and Newton well knew, one must opt for the infinite version of the universe, something which Galileo did not do despite the fact that the cosmos revealed through his telescope was indeed vast by comparison with that of the naked-eye astronomers.*

* It must be pointed out that the lifelong equivocation expressed by Galileo on the question of infinity was doubtless due to other factors in addition to his lingering attachment to certain aspects of Aristotelian physics. Both his appearance before Cardinal Robert Bellarmine and the condemnation of certain passages of *De revolutionibus* in 1616 caused him to be considerably more cautious in his pro-

We could go on to cite other examples of the direct interrelationship between Galilean and Aristotelian thought. The main point to be kept in mind, however, is that the transformation from medieval to modern science has been too frequently presented as simply a reaction against the teachings of Aristotle. Though far from being Aristotelians, Galileo and the more intellectually alert among his followers recognized their indebtedness to the Greek master by holding on to certain significant parts of his system even as they mounted a major campaign against others. Because there was an entrenched resistance to be combatted, the proponents of the new science felt compelled to produce what was often bitterly anti-Aristotelian polemical literature, while at the same time making the argument that Aristotle himself would have supported their cause had he been living in the modern world. Certainly it was not the fault of the Greek philosopher that those individuals of the sixteenth and seventeenth centuries who claimed membership in the "Aristotelian" party largely chose to ignore the content of his natural philosophy, and concentrate instead on the logical devices of a scholasticism in permanent decline.

Because Galileo's quest for scientific truth became inextricably linked to the belief that only quantitative or mathematical relationships are capable of disclosing the essence of things, several noted scholars, including E. A. Burtt and Alexander Koyré, have presented convincing arguments in support of the position that Galileo, like Kepler, was deeply committed to Neo-Platonic doctrine, whose popularity was steadily advancing at the expense of the moribund Aristotelians. The scientist himself provides strong support for this claim: For example, in his *Dialogue Concerning the Two Chief World Systems*, Galileo expresses a deep admiration and affinity for the mathematical system (namely geometry) employed by Plato and his noted predecessor, Pythagoras.[34] Certainly, Galileo's epistemology (or theory of the nature of knowledge) was deeply influenced by Plato's concept of inborn ideas, which we previously encountered in our discussion of Kepler's intellectual development. According to this theory, the great book of nature is written by the Creator in mathematical characters, which are implanted on the human mind at birth in the form of such geometrical figures as the circle, square, and

---

nouncements about the nature of the universe than had previously been the case. Moreover, even though he never mentions Giordano Bruno in any of his books or personal correspondence, Galileo must have given considerable thought to the adverse reaction of the Church and academic community to the executed monks belief in an unbounded cosmos. Lastly, as Alexandre Koyré and a number of other historians of science have pointed out, Galileo seemed relatively little concerned with questions of cosmology. He generally dealt only with those problems which appeared amenable to practical solutions. Since he knew there was no possibility of satisfactorily resolving the question of infinity one way or the other, he preferred to consign this touchy matter to the realm of the philosophers.

triangle. Since mathematics is conceived of as the bridge between human and Divine thought, it follows that no insurmountable gulf exists between the mind of man and the intellect of his Creator. To be sure, the extent of man's knowledge is inferior to God's, but this is because the individual can reach only part of the truth while God knows all. However, if we accept the fact that all true knowledge derives from quantification, each successive generation can make new discoveries which will build upon those of its predecessors. Man will never attain perfect knowledge, but he can move ever closer to an understanding of the Creator's purpose and function. Thus, for Galileo, man is no longer a passive tool subject to the blind natural forces of ages past; he has achieved a radical new maturity, for he is at last capable of dealing objectively with the world on rational terms. The recognition of science as a cumulative ongoing process is clearly at hand. Indeed the belief in an immanent and indefinite progress has taken its place beside the traditional Christian belief in God's transcendent providence. Though Galileo never dreamed of going so far as to replace God with reason, as did the *philosophes* of the eighteenth century, he took the first decisive step in this direction by placing reason between God and man. By emphasizing man's rational capacity to a far greater extent than did his predecessors, Galileo was unconsciously laying the groundwork for a profound intellectual and cultural revolution. As the first major theologian of the modern doctrine of progress, he wanted man to take into his own hands the direction of human affairs which had heretofore been left to God alone. We are so accustomed to this line of thought, after the passage of three and one-half centuries, that we no longer experience the strangeness and wonder of deciphering nature's secrets with the aid of mathematics. But what is for us a foregone conclusion was a truly novel and disquieting experience for Galileo's contemporaries.

With Galileo's pronouncement that the book of nature is written only in mathematical characters, the move away from Aristotle's closed system of natural philosophy began in earnest. In place of the traditional static image of the world, Galileo had substituted a new and dynamic vision of steady and seemingly limitless scientific development. Swept onward by the necessity of more strongly underpinning his revolutionary philosophy of science, Galileo, like Kepler, was attracted to the Platonic doctrine of primary and secondary qualities, except that with Galileo the doctrine emerges in a much more pronounced and developed form. He believed that the privilege of reading from the book of nature is granted to man on the condition that he spell out the mathematical characters in which it is written. Therefore, it is absolutely essential that the scientist distinguish between that in the world which is objective and immutable, and that which is relative, subjective, and constantly undergoing change. Only primary qualities, meaning those expressible in mathematical terms, truly exist for Galileo: number, size, weight, position, and motion. The scientist's main responsibility is to translate these qualities into the objective language of mathematics in order that they can be experienced in a manner universally acceptable to all who are willing to learn. Those qualities or attributes which cannot be expressed in the language of mathematics are secondary and purely subjective: odor, touch, taste, color, etc. They must be consigned to the realm of fallible sensation where misunderstandings based on wrong conclusions are commonplace.

Having established this dualistic framework, Galileo was faced with the difficult problem of striking some sort of compromise between a Copernican universe, whose truth can be demonstrated mathematically, and God's biblical revelation of the universe, which contains numerous passages that are seemingly incompatible with heliocentric cosmology. A loyal Catholic, Galileo opted for an explanation whereby both the Copernican doctrine and the teachings of the Bible could be presented as truth, albeit truths of widely divergent meaning. He believed the way out of this dilemma required the recognition of two radically different languages: the scientific–mathematical language of primary qualities with its attendant precision and rigor, and the everyday, emotion-laden language of secondary qualities with all its inherent inconsistencies and inexactitude. According to Galileo, when God dictated the Scriptures to man, He knew full well that it would be necessary to employ the language of the common people. The children of Israel, after all, had founded a prescientific culture comprised mainly of shepherds, goatherds, and small farmers, who knew nothing of natural philosophy— Aristotelian or Copernican. Biblical language, by necessity, had to conform to daily experience and hence contains certain unscientific propositions which supported early man's belief that the sun turns round the earth. The perpetuation of such an erroneous impression is not tolerable in the scientific realm, however, where Galileo renounced the language used by God in the Bible, though he has no doubt regarding its truth and moral authority in the realm of spiritual affairs. This position is carefully set forth in his *Letter to the Grand Duchess Christina*:

> I think that in discussions of physical problems we ought to begin not from the authority of scriptural passages. . . . It is necessary for the Bible, in order to be accommodated to the understanding of every man, to speak many things which appear to differ from the absolute truth so far as the bare meaning of the words is concerned. But Nature, on the other hand, is inexorable and immutable; she never transgresses the laws imposed upon her, or cares a whit whether her abstruse reasons and methods of operation are understandable to men. For that reason it appears that nothing physical which sense-experience sets before our eyes, or which necessary demonstrations prove to us, ought to be called in question (much less con-

demned) upon the testimony of biblical passages which may have some different meaning beneath their words. For the Bible is not chained in every expression to conditions as strict as those which govern all physical effects; nor is God any less excellently revealed in Nature's actions than in the sacred testaments of the Bible.[35]

Clearly, as intellectual historian Ernst Cassirer has pointed out, there is for Galileo a double revelation of God—"the one contained in His word, the other contained in His works, the one to be found in the Bible and the other to be found in nature and its general laws." [36] In case of doubt, Galileo believed that the objective, impersonal language of science must take precedence over the imprecision of the common tongue. The penetration of the Divine mind and the contemplation of the wonders of nature are now possible, though the application of mathematical principles, for all human beings who can adapt their mode of thought to the methods and conclusions of correct scientific thinking. Words, as such, are strictly designed for the human purpose of bringing about communication between men. But communication between God and man cannot be based upon so undependable a vehicle as words, which are ambiguous and confusing by their very nature. However, when we turn to the second source of divine inspiration—God's mathematical representation of nature—uncertainty and misunderstanding cease. Galileo thus advocates that man, in his striving to know the nature of the universe and its parts, begin with God's *work* rather than His *word*. No mediator is required for man's cognition of the Divine intelligence so long as he uses the tools supplied by a progressive science. The doctrine of primary and secondary qualities leads to an immense simplification of nature because the physical universe now consists only of those qualities which are reducible to mathematical terms. The religious mysticism previously associated with the natural world has suddenly vanished. The only remaining mysteries are those which the scientist will eventually solve on the basis of experimentation and mathematical deduction—or so it seemed.

Had Galileo given more thought to the theological implications of his bold new doctrine, he might have better understood the hostility with which it was met in ecclesiastical quarters, both Protestant and Catholic. For it implicitly denies one of the major tenets of medieval thought: the belief that an insurmountable gulf separates human from divine truth. While it is true that Galileo maintained that one requires complete freedom from the Bible in scientific matters only, he is implicitly asking for much more. His biographer Ludovico Geymonat poses the question: "What guarantee could he [Galileo] offer that the method of scientific research, once it emerged victorious in the field of physics, would not seek to extend itself into ethics and religion?" The Church perceived this as only the first step in a chain of related developments that could gradually call into the question its very existence. If, as

Galileo believed, where reason speaks, there also speaks God, it seemed inevitable that the new priestly class of the future must be recruited from the ranks of the scientific community rather than from the cloisters and seminaries of Christendom. Because his conscience was never torn by a conflict between faith and science, Galileo was convinced that his Platonism could be easily reconciled with Christianity. As a consequence, he never admitted to himself that there was any major ideological issue at stake—namely, the secularization of what had been previously considered as sacred. While Galileo advocated the separation of science from religion as the best way of securing their coexistence without contradiction or conflict of conscience, the leadership of the Church saw it as yet another in a series of concerted attempts to undermine the authority which it had so jealously guarded for the past thousand years.

The mathematization of nature by Plato and his followers appealed to Galileo for yet another important reason. In light of the extensive astronomical phenomena revealed by the telescope, the majestic hierarchy of the Aristotelian–Thomistic universe, in which a graduation of the cosmological worthiness of the celestial bodies had been carefully set forth, suddenly turned out to be nothing more than the product of fertile imagination. No longer could the earth be considered unique in and of itself; what occurs on the planet inhabited by man also occurs in the heavens, and vice versa. Galileo became absolutely convinced that God, the master geometrician, employed only one set of laws when He created the cosmos and set its various bodies and spheres in motion. As we have observed, it was to the mechanical branch of mathematical studies that Galileo was drawn as a young man, and the Copernican attribution of motion to earth provided him with a sufficiently powerful stimulus to make the study of kinematics his lifework. Thus, he adopted the heliocentric system not merely as an astronomical hypothesis, but as an emancipation from the narrow anthropocentrism that had heretofore seriously impeded the scientific investigation of natural phenomena. Galileo abandoned the old qualitative division of the cosmos into two different worlds, one celestial the other terrestrial, for a mechanistic concept of a universe which operates according to a single set of rigorous mathematical principles. Here the Platonic belief in the mathematical unity and oneness of nature has clearly triumphed.

Unwilling to set limits on the amount of knowledge the human mind might acquire through the pursuit of science, Galileo, though deeply influenced by Plato, also encountered in the mysticism of the Greek philosopher a major barrier to scientific understanding. Nature, Galileo believed, is no mystery but an open book, legible and accessible to everyone. Scorning completely the Pythagorean–Platonic belief in the magical qualities of numbers, he was interested in arithmetic and geometry only for their practical application to the solution of specific problems. No better illustration of this important point can be presented than by comparing Galileo with his brilliant German contemporary, Johannes Kepler. Kepler, an avowed Neo-Platonist, believed that mathematics provides a mystical and transcendent method of communication with the realm of the harmonious and eternal. Despite the revolutionary importance of his three laws of planetary motion, Kepler steadfastly refused to sacrifice his belief in a mystical relationship between the paths of the planets, the five perfect geometric solids, and the music of the spheres. In his mind, mathematical thought and esthetic speculation interpenetrated one another, becoming inseparable. It is primarily for this reason that, throughout his life, Galileo persistently refused to read Kepler's works on pure mathematics and theoretical physics. No abstract theoretician, Galileo's universe, however large, had little if any room for witches, or astrology, or for ghostlike fingers sweeping outward from the sun to move the planets in their orbits, indeed, for any part of Kepler's cosmic flights into the mystical unknown. In 1634, Galileo wrote his friend Fulgenzio Micanzio that when it came to the matter of celestial motion he and Kepler "have sometimes hit on the same ideal . . . but this will not be found to have happened in one per cent of my thoughts." [37] One of the great ironies of the scientific revolution in astronomy is that Kepler, who clung so tenaciously to the mystical element of Platonism, should have formulated the three laws of planetary motion, while Galileo, the champion of applied mathematics, relied almost entirely upon evidence garnered from sense experience, the same method employed by Aristotle, to support his belief in Copernicanism.

On the other hand, Galileo's total abandonment of any reality beyond perceptual or intellectual apprehension paved the way for a modern methodological revolution; one which Kepler, despite certain fundamental differences with Aristotle, could not have brought about, for the reason that he continued to think within the framework of traditional metaphysics. Since nature presented herself to Galileo, more than to any of his predecessors, as an orderly system whose every proceeding is thoroughly regular and totally necessary, the Italian felt compelled to devise a method whereby all natural phenomena could be scrutinized with unfailing accuracy. He employed what we today refer to as the "scientific method," the highly structured process regarded as essential to objective scientific investigation which, in somewhat simplified form, includes the following steps: one, the gathering of data; two, the formulation of an hypothesis; three, observation and/or experimentation; and four, the validation or rejection of the hypothesis. Galileo, as one might expect of a seventeenth-century scientist, had his own favorite terms to designate the steps into which he divided this method: intuition or resolution, demonstration, and experiment. [38] Fully recognizing that the scientist, like any other individual, must face a world whose existence is communicated to him by a multitude of sense impressions, Galileo taught that he must isolate and analyze the specific phenomenon he wishes to study by reducing it to its simplest parts. Once this is done, these parts must be expressed in mathematical terms, in order that no mistake be made regarding their true nature. For Galileo, then, as for Aristotle, sense experience is an important, indeed indispensable, tool of scientific discovery. But unlike Aristotle, Galileo believed that all sense data relating to primary qualities must be expressed in quantitative terms. Once this is done, facts garnered from the often fallible senses are no longer needed because the phenomenon under investigation has been mathematically reduced to its true constituents. If this process is correctly carried out, the quantified data will yield the same results time and again, for although certain qualities, e.g., space, time, motion, etc., may be reduced to number, number can never be changed back into quality.

In Galileo's methodology, a law of nature must be based on mathematical facts, it must contain no element incapable of verification by experimental proof. Following the example set by his ancient hero, Archimedes, he was among the first, if not the first, to use experiment in the study of motion. But even more important, he became the first scientist of the modern era to consider experimentation as the *sine qua non* of checking an hypothesis against facts and discovering truth—though it must be admitted that he did not always translate into the concrete what he believed in the abstract.

Much has been written regarding the question of whether, in fact, Galileo ever reached a full understanding of the true nature of the experimental method. This is primarily due to his frequent use of what he termed "thought experiments"—imagining the consequences of what would happen given a particular set of circumstances, rather than directly observing the result under controlled laboratory conditions. There are numerous documented instances, however, where Galileo actually built the experimental apparatus and carried out carefully controlled demonstrations, both in private and in the company of his students and colleagues. The telescope itself is just one of the many examples whereby Galileo constructed an instrument, carried out his experiments, and published the results. This systematic approach also played a key role in his work with floating bodies, and in his discovery that acceleration is an essential mode of natural motion. Based upon these and other uses of the experimental method as a major tool of investigation, we must credit Galileo's frequent use of this systematic approach with playing a key role in his most important scientific discoveries. Moreover, when he was unable to carry out an experiment because of technical difficulties, he freely confessed it.*

* For a more detailed discussion and analysis of this subject see Thomas B. Settle, "Galileo's Use of Experiment as a Tool of Investigation," in *Galileo: Man of Science*, ed. by Ernan McMullin (New York: 1967), pp. 315–337.

In retrospect, the great merit of Galileo's experimental method, once it was widely applied, is that it opened an important part of the physical world to systematic public examination, while the method itself elevated the results above the possibility of private disagreement. It instantly made obsolete the dialectical method employed by the Peripatetics of his day, which placed a premium on the verbal reasoning, forensic skill, and the personal magnetism of the disputants. Always extremely concerned with the confrontation of reason and experience, theory and practice, the scientist's writings oppose both untested hypotheses and the scholastic abuse of authority. To Galileo belongs the credit for establishing a method that more and more induced open minds to abandon their personal bias and faulty reasoning and, through controlled observation and experiment, to arrive at universally acceptable conclusions.*

Based upon our admittedly abbreviated discussion of Galileo's methodology and philosophy of science, it seems appropriate that this section conclude with a few remarks concerning the great man's relationship to the modern age. To Galileo, more than to any of his contemporaries, the science of nature consciously presented itself as a means of improving life and increasing the prosperity of man. What had been previously called "science" was not directed, consciously or otherwise, to the material enrichment of human life. Neither the doctrines of Aristotle or Plato, nor the scholasticism of medieval times could serve this end. A more exact knowledge of nature based on Galileo's use of experience and experimentation was required. This is made even more apparent when we pause to consider that Greek physics is now of historical interest only; no modern physicist would dream of turning to Aristotle for help in the solution of a perplexing problem. But despite the fact that physics has also come a long way since Galileo's time, certain of his experiments and the laws derived therefrom are still cited in the textbooks assigned college students today. And speaking of textbooks, it is worth noting that Aristotle, brilliant though he was, could not have understood a page from an introductory work on modern physics, but that Galileo, with some brushing up, would encounter no major difficulty in following most classroom discussions of the subject. This is because even though many of the arguments whereby Galileo attempted to prove Copernican theory have long since become obsolete, the truly decisive element in his work was not the defense of a special astronomical doctrine, but the advancement of a new concept, a new systematic ideal of scientific truth.

---

* I am aware that there are those who claim that this honor rightfully belongs not to Galileo but to Francis Bacon. From my point of view, Bacon, though an unqualified advocate of the experimental method, was just that, an *advocate* as opposed to an experimental scientist in the manner of Galileo. This view and an analysis of Bacon's contribution to modern science are discussed in detail in chapter 8.

Abandoning the Aristotelian doctrine of means and end, Galileo depended upon the concepts of space, time, number, and quantity. For him, as for Newton, Einstein, and all modern physicists, the real world is the world of bodies in mathematically reducible motion. Hence, all matter must conform to the same laws because the universe is one.

The roots of modernity penetrated another equally important aspect of Galileo's thought; namely, his answer to the question of what should be done with God. It will be recalled that Aristotle and his scholastic followers had conceived of God as the Final Cause behind the occurrence of all cosmic phenomena. All that exists in the universe—whether animal, vegetable, or mineral—aspires after the Creator's divine perfection, the process by which the universe is kept continuously in motion. Aristotle's doctrine of final causation has no place in Galileo's scientific thought, however. Rather, he believed that for the true scientist causality must be explained in terms of mathematical changes which take place in material elements; a subject we have already discussed in connection with his doctrine of primary and secondary qualities. But where then is God if nature is constructed mechanistically, and its function determined according to quantitative relationships? Even had the thought crossed his mind— which it did not—Galileo was in no position to deny God's existence outright. Yet, as E. A. Burtt observes, the only way to keep God in a mechanistic universe was to invert the Aristotelian belief in Him as the Final Cause and regard God instead as the First Cause. In other words, the God of Galileo's universe ceases to exercise control over the cosmos according to any preconceived plan which may be altered at will. Rather He is a master inventor or mechanic who, once having created the universe and set it in motion, retreats into the background. The great miracle of God is that He originally constructed the universe out of nothing, and fashioned precise mathematical laws by which all cosmic phenomena are controlled. Furthermore, God gave to man the rational capacity to discover nature's immutable principles. Thus, it would make no sense whatsoever for him to violate the very laws He himself set forth, even though Galileo did not deny his power to do so. However commonplace and harmless this doctrine may appear to us today, the theologian of Galileo's time found it impossible to grasp the absolute separation between the world envisioned by the man of science and that contemplated by the Church. How indeed could any Christian clergyman—Catholic or Protestant—remain indifferent to such a revolutionary redefinition of the world; one not merely regarded as the hypothesis of a physicist–mathematician but as existing in reality.

This writer would be among the first to admit that Galileo deserves both our admiration and respect for the revolutionary change of temper he helped bring about as the result of his major contribution to modern science. Yet one wonders whether Galileo, innocent of just how deeply his scientific method would penetrate and influence the course of modern history, might not have had second thoughts had he known that the conceptual world he helped to create would eventually displace traditional values by rejecting all experience and knowledge that did not conform to the pattern of mathematical thought. By dismissing subjective experience as secondary and unreliable, he also discounted an important part of man's humanity and accumulated culture. In his celebrated treatise, *The Assayer*, Galileo comments that without mathematics "one wanders around in a dark labyrinth." From his personal point of view this was doubtless true. But what he and many of his contemporary spiritual heirs have failed to consider is that it is not the objective world of primary qualities but the world of language, symbol, art, politics, religion, and ritual that human beings move about in with some degree of confidence. That this is so may be grasped from the fact that the science and technology to which man has increasingly turned in the centuries since Galileo's death no longer satisfy his questing mind. During the latter half of the twentieth century there has arisen an almost palpable homesickness for the mystical satisfactions of those spiritual, artistic, and moral longings which science does not and cannot fulfill because it is directed to other ends. Doubtless Galileo never dreamed that it would be in this, the realm which he called secondary and sought to separate from scientific inquiry, that the promising new age he envisioned would demonstrate its most disturbing limitatons.

Inquisition

Upon his return to Florence in 1616, an embittered and disconsolate Galileo retired to his villa at Bellosguardo where, in the welcome company of his literary friends, he proceeded to nurse his deep emotional and intellectual wounds. The battle he had been fighting was simply too grandiose for any individual, yet he had come far closer to victory than he or anyone else of his generation realized. All things considered, the Church had treated him with deep respect and a major degree of restraint; before he left Rome, Pope Paul v granted him a private audience, during which he promised Galileo safety from his enemies; from Cardinal Bellarmine he had secured a letter that condemned the malicious gossip circulating about him as false, while several noted Jesuit fathers privately confessed their surprise and displeasure over the injunction handed down against him. Still, this was small consolation for a man of Galileo's sensitive temperament and inexhaustible taste for intellectual combat. Deep down he must have realized that by his submission to authority he had helped the Church avert a major public scandal, but the cost in personal terms had been incalculable. In his mind's eye the scientist could not dismiss the haunting image he had formed of himself as a traitor to the cause to which he had dedicated his entire adult life. Under the circumstances the Pope could afford to be gracious; but he, Galileo, could not. Though he had no desire to gain revenge by embarrassing either the Holy Father or the institution he headed, neither could Galileo accept defeat. Forbidden to present his scientific views openly and frankly, it was not long before he began to cast about for some less direct means of doing so.

Ever since his conversion to the heliocentric system, the scientific problem that had most troubled him was his lack of a decisive physical proof of the Copernican theory. And of all the arguments leveled against Copernicanism none had been more persistently advanced or difficult to counter than the charge that the earth is a stationary body at the center of the universe. Only a few weeks before his celebrated meeting with Cardinal Bellarmine, Galileo believed he had discovered direct physical proof of the earth's movement in the phenomenon of the tides.

Several years earlier, Kepler, in the *Astronomia Nova*, had correctly supposed the tides to be an effect of the moon, a theory which Galileo dismissed on the grounds that action at a distance is impossible in the absence of some physical connection between an object and its mover. Rather, he tried to explain tidal action as a result of inequalities in the velocity at which the various parts of the earth's surface move. The tides, he postulated, are caused by the combined motion of the earth around its axis and around the sun; on the night side of the earth their velocity combines, and on the day side toward the sun they subtract. The difference in momentum resulting from this dualistic movement causes the sea to slosh about much like the water in a tilted tub or basin. The main problem with Galileo's theory is that high and low tides should occur only once each day rather than twice. Though his argument is not very convincing today, it is by no means as silly as many people have been inclined to regard it. It was not until Newton, who was the first to correctly show how the tide-generating force arises through the gravitational attraction of the moon and sun, that Kepler was ultimately proven correct. Yet long before Kepler the astrologers and seamen of antiquity had correctly observed and recorded the direct relationship between the moon's phases and the constantly changing level of the seas. "Galileo flaunted tradition elsewhere and won," wrote a modern scholar, but "in his theory of the tides he lost what was perhaps his greatest gamble." [39]

In January, 1616 Galileo presented a fairly detailed analysis of his attempted physical proof of the earth's double motion to twenty-two-year-old Cardinal Alessandro Orsini. Titled *Discourse on the ebb and flow of the seas*, the scientist hoped that with this document in hand, Orsini, a favorite of the Pope, would be able to win an official endorsement of Copernicanism or, failing that, a promise that heliocentric cosmology could be argued without the threat of clerical intervention. Whatever possessed Galileo to entertain the chimera that Paul v, one of the most antiintellectual pontiffs in all of Christian history, would condescend to read, let alone reflect in depth, on his highly technical scientific treatise is difficult to imagine. Though Orsini was granted the desired audience, he was informed by the Pope at the outset that it would be in everyone's best interests to encourage Signor Galilei to abandon his pro-Copernican views, and the sooner the better. When Orsini then tried to present Paul with a copy of Galileo's little treatise on the tides, the Pope cut him short. The matter, he informed the shaken young cardinal, would be referred to the Holy Office of the Inquisition for further consideration; with that, Orsini was brusquely dismissed.

The papal rebuff, combined with certain distasteful events which occurred shortly thereafter, did nothing to weaken Galileo's faith in his newly developed theory; if anything, he was more determined than ever to make his discovery known. In 1618, he again turned his attention to the paper presented to Orsini some two years earlier. Aware that he must proceed more cautiously than before, he composed a preface which described the work as a "poetical conceit or dream" written during the period when he believed the Copernican doctrine to be true. Now, however, the authorities, who "are guided by a higher insight than any to which my humble mind can of itself attain," may be assured that his previous position has been abandoned. Yet even with this patently false disclaimer attached to the document, Galileo knew better than to undertake its publication on his own. He hit upon the idea of sending it to Archduke Leopold of Austria in the hope that that influential nobleman would bring it to press on his own, thus relieving Galileo of any personal responsibility, something Mark Welser had done for Father Scheiner when he published the latter's findings on sunspots. But the Archduke did not bite, and a disappointed Galileo's "conclusive proof" of the earth's motion continued to go unnoticed.

While he waited for Leopold's response to his trial balloon, the attention of the world was suddenly drawn skyward by the spectacular appearance of three comets in the autumn of 1618. Coming, as they did, at the outset of the Thirty Years' War, apocalyptic specialists interpreted them as a portent of God's terrible wrath and began computing anew the end of human existence. Galileo, the victim of intermittent attacks of rheumatism during his later years, was confined to bed at this time and could not undertake extensive observations of his own, but the many friends who visited his villa discussed the comets in detail.

In 1619 the Jesuit Father Horatio Grassi, a member of the *Collegium Romanum*, delivered a lecture, subsequently published, in which he correctly maintained that the comets are bodies which move in orbits like the planets at a distance from the earth far greater than the moon's.

Grassi supported his study of the phenomenon by citing Tycho's observations of the famous comet of 1577. By means of parallax, the Danish astronomer had been able to establish that the great object was located far beyond the moon, most probably somewhere in the neighborhood of Venus. Five years earlier, in the *Letters on Sunspots*, Galileo endorsed Tycho's findings, because the placing of new stars and comets in the firmament was an excellent way of proving that the heavens were not fixed but mutable.

Much had changed since then, however. The Jesuits, whom Galileo had once counted among his strongest allies, had done little to help him in 1616. After he had been silenced by the Holy Office, they abandoned every concession to Copernicus and threw their full support behind the astronomical system of Tycho. Though this system represented something quite modern in that it was based upon the most exact of observations, it by no means constituted a revolutionary break with the old cosmology, for it still assumed the immobility of the earth. When Galileo read Grassi's newly published treatise his temper literally exploded. He scrawled one expletive after another in the margins of his copy—words which, in Santillana's judgment, would "make a vocabulary of good Tuscan abuse." Typical among his bitter comments are: "piece of asininity," "elephantine," "ungrateful villain," "rude ruffian," etc. Though Galileo was never mentioned by name, he harbored no doubts that Grassi was taking full advantage of his hamstrung condition to further the Jesuit cause by driving yet another nail into the coffin of Copernicanism.

Though he clearly overreacted to the position taken by Grassi, Galileo had cause for deep concern. He was being forced to watch in vain while Tycho's system was pressed into the service of the Vatican as providing the most opportune deviation from the Copernican issue. He knew that somehow this compromise position must be challenged before it gained a stranglehold in the universities; the choice must remain confined to one between the now discredited Aristotelianism and the revolutionary doctrine of Nicolas Copernicus. With this thought in mind, Galileo abruptly reversed his position on the comets by claiming that they are only optical effects born of earthly vapors, like the aurora borealis, mock-suns, and rainbows. He realized, of course, the reactionary appearance of this stance; he was seemingly casting his lot with the ancients, who had never regarded new stars and comets as real celestial bodies. Yet there were serious reasons for this choice of attack. In the first place, the telescope had proven beyond doubt that the Aristotelian concept of celestial immutability was without scientific foundation. Thus, Tycho's theory of the comets had become far less crucial to the veracity of Copernican cosmology than before. Moreover, if comets were real, should they not grow larger as they approach earth and smaller as they recede into the distance? But, as Galileo pointed out, they suddenly appear at their full size and then vanish altogether. Lastly, the astronomer related how Tycho himself had shown the comets to have paths which are totally irregular from a Copernican point of view; most were markedly elliptical, and some even contained major retrogressions. Newton was ultimately able to turn his accurate prediction on the path of a comet into a triumphant proof of the new system, but his hour upon the stage had not yet come. Since Galileo still believed that circular orbits were the only ones physically possible in outer space, the movement of comets could not be reconciled with the perfect paths in which all heavenly bodies must move round the sun.

Despite his strong emotional reaction to Grassi's treatise, Galileo knew better than to directly attack a Jesuit father under his own name. He enlisted the support of his friend and former pupil Mario Guiducci, who generously agreed to sign a written reply directed against his mentor's antagonist. The resulting *Discourse on Comets*—the original manuscript of most of which has survived, and is in Galileo's handwriting—appeared in June, 1619. Though signed by Guiducci, there was no mistaking the distinctive style of its true author. The treatise was greeted with great enthusiasm by Galileo's supporters; their hero had once again overcome great adversity in order to continue his pursuit of scientific truth. No greater praise was forthcoming than that bestowed upon him by his old friend Cardinal Maffeo Barberini, who, in 1616, emerged as a strong opponent of the decree of the Congregation. In 1620, the Cardinal sent Galileo a Latin poem composed in his honor titled *Adulatio Perniciosa* (*Dangerous Adulation*). At the time, neither man realized that in less than three years Barberini would be elected to the papacy, an event which Galileo, much to his later sorrow, prematurely celebrated as the salvation of the new science.

The criticism of Grassi's views, which Galileo offered under the name of Guiducci, so offended the Jesuits they urged their besieged colleague to strike back as quickly as possible. Masquerading as his own pupil, Grassi responded with a new book, *The Astronomical and Philosophical Balance*, written under the transparent anagram of Lothario Sarsi. Ignoring Guiducci altogether, the Jesuit father attacked Galileo and the Copernican system with the vengeance of a zealot. He disputed the latter's claim to priority for a number of basic discoveries and took up Galileo's challenge to the Tychonic system. Since Aristotle has been refuted and Tycho disavowed by the famous natural philosopher, does he then suggest that Grassi and every other faithful Catholic endorse Copernicanism, a system condemned by Holy Mother Church?

Grassi had clearly hit home. Galileo's friends were unanimous in their judgment that he could not let this latest publication go unanswered; it simply contained too many serious and inaccurate accusations. Furthermore, he must step forward in person to protect Guiducci's honor and defend the integrity of the new science. But fearing the consequences of a direct frontal assault upon the powerful Jesuits, they also advised him to reply in an indirect manner. Galileo agreed that the Jesuits must be left out of it; he would limit his remarks to a refutation of the charges leveled by the straw man Sarsi, who had been so conveniently created by his opponent. After all, he had not been enjoined by the Church from showing up a pompous ass, particularly one who in theory was nonexistent.

Galileo's answer to Sarsi (that is, Grassi) took three years to compose. Published at Rome in October, 1623, the slender volume is doubtless the greatest polemic in the history of the physical sciences, and has been justifiably called Galileo's scientific manifesto. He chose to title his work *Il Saggiatore* (*The Assayer*), a continuation of Grassi's pretence of carefully weighing the truth of his opponent's scientific arguments against all available evidence.* Yet judged solely from the viewpoint of astronomical science, we are obliged to admit that Galileo falls into error of fact almost as frequently as does his outmanned opponent. For example, he stubbornly denied the chief scientific merits of Tycho's system by challenging the accuracy of the great astronomer's meticulous observations. In fact, he went so far as to suggest that Tycho had not made many of the observations attributed to him. And not forgetting the controversial celestial phenomena that had originally provoked the four-year-old quarrel, he dismissed the comets as nothing more than "Tycho's monkey-planets." With incisive wit and devastating irony, he then proceeded to dismantle "Sarsi's" scientific arguments root and branch, as is illustrated by the following example. Grassi had attempted to prove that a projectile is heated by friction with the air as it travels from one point to another. This, as we now know, is a perfectly sound scientific concept, but the Jesuit unwisely attempted to prove an excellent case with a bad argument. He borrowed from the writings of Suidas, a Greek lexicographer of the tenth century, a fanciful story of how the ancient Babylonians had cooked their eggs by swiftly whirling them over their heads in slings. Galileo simply could not resist the temptation of making Grassi appear the fool.

> If Sarsi wants me to believe with Suidas that the Babylonians cooked their eggs by whirling them in slings, I shall do so; but I must say that the cause of this effect was very different from what he suggests. To discover the true cause I reason as follows: If we do not achieve an effect which

---

* The implication of Galileo's choice of titles is quite clear. The assayer (meaning himself) measures things on the finest of balances while the common shopkeeper (meaning Grassi) rests content with the crude steelyard of antiquity. Grassi also had his moments, however, and gained a small measure of revenge for this clever allusion in his reply to Galileo's latest work. He substituted *assagiatore*, or wine-taster, for *saggiatore*, an intimation that Galileo had been less than sober when he wrote his book.

others formerly achieved, then it must be that in our operations we lack something that produced their success. And if there is just one single thing we lack, then that alone can be the true cause. Now we do not lack eggs, nor slings, nor sturdy fellows to whirl them; yet our eggs do not cook, but merely cool down faster if they happen to be hot. And since nothing is lacking to us except being Babylonians, then being Babylonians is the cause of the hardening of eggs, and not friction of the air.[40]

Scattered among the lengthy passages of humor and sophistry with which Grassi was pulverized in the public eye are Galileo's answers to the major philosophical issues discussed in the previous section of this chapter. For it is in *The Assayer* that he delineates his culture-shattering views on observation and experimentation, primary and secondary qualities, and the steps of the modern scientific method. Above all, he informs the natural philosophers of his day, and of days to come, that it is their duty to be intensely skeptical about authorities and principles that have so long been taken for granted. The scientist of the present and future must reason in mathematical terms, and by so doing reject that spirit of scholastic compromise which Galileo regarded as the major impediment to the free growth of scientific thought and human progress. In the final analysis, he did much more than expose the many weaknesses and inconsistencies of a Jesuit father's deceptive logic; he had composed a compelling polemic against the intellectually bankrupt dialectic of a dying age.

In 1621, during the battle of the comets, Galileo was deeply saddened by news of the premature death of his good friend and generous patron, Cosimo ii.The strong-minded prince, who had unswervingly supported his controversial court mathematician for over a decade, was succeeded by his young son, Ferdinand ii, who assumed the throne under the regency of his paternal grandmother, Christina of Lorraine. Ferdinand's weakness of character would have important repercussions upon the development of the conflict between Galileo and the Inquisition some twelve years later. Cardinal Bellarmine, who as head of the Society of Jesus had been a restraining influence on Galileo's opponents, died in the same year. In August, 1623, however, these losses were seemingly more than offset by an event that raised high hopes in the minds of Galileo and the more progressive Catholics who supported his scientific studies: Maffeo Barberini was elected Pope. The event occurred just in time for Galileo to dedicate *The Assayer* to his Holiness, the new Urban viii.

After Paul v, an antiintellectual and ill-tempered old man, the brief reign of Gregory xv had brought little if any improvement. But Barberini, a Florentine by birth, had a very acute mind, and was deeply immersed in the cultural developments of his time. The new Pope was a lover of music and the arts; he commissioned the composer Gregorio

Allegri to write a *Misere* for nine voices, which was later reserved for the Vatican. Desirous of making a revitalized St. Peter's the artistic as well as the religious center of Rome, Urban put the brilliant baroque sculptor and architect, Giovanni Lorenzo Bernini, in charge of completing the interior of the great basilica, for which the artist boldly designed a massive and towering canopy over the Papal throne, the Baldacchino. In his younger days this intellectual Pope wrote poems, one of which, *Dangerous Adulation,* was a sonnet of compliments to Galileo on his astronomical discoveries and writings. Barberini also held a coveted membership in the Lincean Academy, and in the past had always defended the new science, even during the crisis of 1616. Considering Barberini's prior conduct and wide-ranging intellectual interests, it is hardly surprising that Galileo looked to the future with renewed optimism and an even deeper sense of purpose.

Yet once in power the new Pope's latent pride and vanity asserted themselves without restraint. His ego was indeed monumental, even for an age which had little use for men of modesty. A lavish nepotist, extravagant, domineering, and grandiose in his schemes, he fortified the Castle St. Angelo at great and needless expense, and had the bronze ceilings of the Pantheon cast into huge cannon, an act which gave rise to the derisive saying: "What the barbarians have not done, Barberini did." He was also the first Pope to allow a monument to be constructed to him during his lifetime; he even had the birds living in the Vatican gardens poisoned because they disturbed his thinking. "I know better than all the cardinals put together," he once declared. "The sentence of a living Pope is worth more than all the decrees of a hundred dead ones." And on hearing of the death of France's powerful Cardinal Richelieu, he remarked, "If there is a God, Richelieu will have much to answer for; if not, he has done very well." One wonders whether it ever occurred to Urban, during the twilight of his papacy, that he might well have said the same of himself.

Having received word from his friends of the Pope's favorable reaction to the dedication of *The Assayer,* Galileo optimistically came to Rome in the spring of 1624. He wanted to sound out the Pontiff's position personally and, if possible, to gain a concrete pledge of a shift in the Church's position on Copernicanism. He was particularly hopeful that the personally humiliating prohibition of 1616 would be withdrawn. The Pope, who seemed genuinely pleased to be in the company of his old friend, granted Galileo no less than six private audiences during the scientist's relatively brief stay in the city. What exactly took place in those meetings will forever remain unknown, but Galileo doubtless soon realized he was no longer addressing the affable Maffeo Barberini, but a domineering and inflexible Urban VIII.

The Pope informed him at the outset that the decree of 1616 would not be revoked under any circumstances. For unlike Galileo, he did not accept the principle that the ultimate test of a scientific theory must be found in nature. Granted, many things seemed to prove that the earth orbits the sun and that man's home is not truly at the center of the universe, but appearances are not the same as absolute truth. There can be no ultimate physical test of God's design because, according to Urban, He is only partially revealed to man by his material creations. The phenomena that might appear to be satisfactorily explained by a given hypothesis may have been produced by some entirely different means, and for a purpose which is not understood by the mind of man. Or in the Pope's words, "It would be an extravagant boldness for anyone to go about to limit and confine the Divine power and wisdom to some one particular conjecture of his own." Thus, there can be no scientific truth as Galileo perceived it, since God can make things appear in any manner he may choose: only calculated conjecture and speculation remain.

Urban's argument, which continues to be the ultimate refuge of those who on religious grounds do not accept as true certain scientifically proven principles such as natural selection and evolution, both surprised and dismayed Galileo. For the first time in their long and cordial relationship he was able to obtain a true measure of the broad intellectual chasm which separated his thought from that of the Pope. He was, of course, already familiar with this type of argument, for it had been employed by his Peripatetic opponents time and again. He also knew from personal experience the difficulties of coping with it. For all his intelligence and education, the Pope was totally unable to grasp the full implications of the new science; Urban still lived in a medieval world of scholastic dialectic, where major differences of opinion were carefully worked out through skillful rhetorical dodges and face-saving compromise. He simply did not and probably could not understand that by consigning inquiry to this realm it would not be getting anywhere, that progress, as Galileo conceived of it, was impossible.

Still they were friends. Urban assured Galileo that if he remained a dutiful son of the Holy Church, as he had hitherto shown himself to be, he could count on being protected from his enemies. Moreover, it was agreed that the forbidden topic of Copernicanism could be discussed hypothetically and impartially, along with the Ptolemaic system. Galileo was further encouraged by the fact that Urban had not raised a single major objection to his latest publication even though there was no escaping its strongly anti-Jesuit flavor. Since he had gone this far, might he not then go still farther? Galileo, unable to face the prospect of yet another major setback, willfully misinterpreted Urban's approval of his previous writings as a sign of the latter's toleration of the heliocentric doctrine. But the Pope's course was set; he wanted to give his friend every liberty short of officially sanctioning the teachings of Copernicus, an act that would almost certainly diminish the authority of the Church. Matters standing thus, Galileo decided to undertake the perilous task of putting the new Pontiff's yet unknown resolve to a direct test.

By June, Galileo, now sixty years of age, was back in Florence, anxious to begin work on a book of "immense design," which had been taking shape in his fertile mind ever since his earliest experimental work with the telescope. Despite his relatively advanced age, the seemingly inexhaustible creative powers with which he had been endowed were never more acute. However, his progress was impeded by a number of nagging obstacles, including gradually failing health and the intermittent surfacing of repressed doubts about the scientific soundness of his pro-Copernican arguments. He recalled Cardinal Bellarmine's statement to the effect that if he could provide conclusive physical proof that the earth orbits the sun, then the Church would have no alternative but to seriously reconsider its position on the matter. Though Urban VIII had made no such promise—in fact had argued just the opposite—Galileo proceeded as though it was Bellarmine who now wore the shoes of the fisherman. The time had come, he decided, to publish his theory of the tides, the "conclusive proof" of the earth's motion which had gone unnoticed ever since its formulation in 1616.

The great Copernican manifesto, which was intended to bear the title *Dialogue on the ebb and flow of the seas,* was finally brought to a conclusion in January 1630, having taken six years to complete. The *Dialogue* is carried on by three characters: Salviati (Galileo himself), who holds the Copernican doctrine; Sagredo, an intelligent man-of-the-world and supposedly impartial listener; and Simplicio, a diehard Aristotelian, always worsted in the argument. (Galileo claimed that the latter's name derived from Simplicius, a sixth-century commentator on Aristotle, yet the double meaning seems obviously intended.)

The conversation takes place over the course of four days. During the first day the reader is encouraged to abandon once and for all the Aristotelian view of the cosmos. Sunspots, the mountains of the moon, and the varying speeds at which celestial bodies move, all combine to show that the traditional distinction between celestial and terrestrial matter is only an illusion. The second and third days are directed to an analysis and refutation of the traditional objections raised by Aristotle, Ptolemy, and Tycho against the earth's motion. Having answered these arguments, Galileo then goes on to show how the hypothesis of the double rotation of the earth (diurnal and annual) formulated by Copernicus offers the most acceptable explanation of all celestial phenomena. Finally, as the capstone to his argument, Galileo brings forth his famous theory of the tides on the fourth and final day. By this time a battered and confused Simplicio is in embarrassed retreat, although he steadfastly refuses to surrender to the quick-witted Salviati. In his final comment the harassed Aristotelian takes refuge in an argument with a very familiar ring:

> As to the discourses we have held, and especially this last one concerning the reasons for the ebbing and flowing of the ocean, I am really not entirely convinced; but from such feeble ideas of the matter as I have formed, I admit that your thoughts seem to me more ingenious than many others I have heard. I do not therefore consider them true and conclusive; indeed, keeping always before my mind's eye a most solid doctrine that I once heard from a most eminent and learned person, and before which one must fall silent, I know that if asked whether God in His infinite power and wisdom could have conferred upon the watery element its observed reciprocating motion using some other means than moving its containing vessels, both of you would reply that He could have, and that he would have known how to do this in many ways which are unthinkable to our minds. From this I forthwith conclude that, this being so, it would be excessive boldness for anyone to limit and restrict the Divine power and wisdom to some particular fancy of his own.[41]

Simplicio, after being shown up as an ass time and again, has trotted out the very argument (even employed some of the very words) used by the Pope during his private conversations with Galileo in 1624. Considered conclusive by Urban, it rings hollow in the mouth of the defeated scholastic philosopher. Though it was clearly Galileo's intent to defer to the Pope's authority by giving him the last word on the subject, it was equally obvious to almost everyone that he had written only a thinly veiled apologia of Copernican astronomy. He would have been better advised, under the circumstances, to have excluded the Pope's comments altogether, rather than to have given them such short shrift. Galileo's enemies attempted, apparently with some success, to make Urban believe that Simplicio stood for the Pontiff himself, something the author had never intended. It was not long before Galileo began to realize just how dire the consequences resulting from the use of this most tactless of stratagems could be.

Galileo went to Rome in May, 1630, for the purpose of obtaining a license to print his recently completed manuscript. He was received in a long audience by Urban VIII, who again endorsed the idea of an astronomical dialogue, provided the subject's treatment were strictly hypothetical. However the Pope objected to the proposed title *Dialogue on the ebb and flow of the sea*, which he thought placed too much emphasis on the question of physical proof. Galileo deferred to His Holiness and a new title was selected, *Dialogue Concerning the Two Chief World Systems*. Since Urban was far too busy to read the book himself, the remaining details were left to be worked out by the licensers.

Father Niccolo Riccardi, Master of the Sacred Palace, was entrusted with the considerable responsibility of either approving or forbidding the publication of all books in Rome. A jovial friar, who because of his generous girth had been nicknamed *Il Parde Monstro*, the Father Monster, by King Philip III of Spain, he was highly sympathetic toward Galileo's cause and assured the scientist through mutual friends that the official *imprimatur* would be quickly forthcoming. But Riccardi was also a conscientious servant of the Church, and felt compelled to live up to the responsibilities of his office. As he read through the lengthy manuscript, the Chief Censor became increasingly aware that something major was definitely amiss; yet he could not establish exactly what it was. He knew, of course, that the Pope himself had encouraged Galileo to write this latest work, but somehow it did not appear to be as hypothetical as Riccardi had been led to believe. At the very least it seemed to contradict the spirit if not the letter of the prohibition of 1616, which the Pontiff had not seen fit to revoke.

What to do? Riccardi finally decided to pass it on to his assistant, Father Raphael Visconti, who was supposedly versed in mathematics. But unfortunately Visconti proved little better equipped for the task than his superior. He suggested a number of minor alterations, designed to make Galileo's pro-Copernican arguments appear somewhat less conclusive, then returned the manuscript with his approval. Poor Riccardi's doubts were anything but quieted. Still, he was in no position to request that the author undertake a major revision of the work when he, as Chief Censor, lacked the specialized knowledge to suggest how it should be written. To make matters worse, the Papal Secretary, Giovanni Ciampoli, and the new Tuscan Ambassador to Rome, Francesco Niccoloni, who was married to the Father Monster's favorite cousin, Caterina, were both pressing him to grant the *imprimatur*. Riccardi continued to temporize, but in the end suggested an unusual compromise. He wanted to personally review the book's contents one last time; however, to expedite this process he would pass each page on to the printer as soon as it was approved. He further insisted that the Preface and conclusions be revised to comply more directly with the Pope's intentions. Prince Cesi, the widely respected President of the Lincean Academy, was to personally assist him, as well as to undertake supervision of the printing. Galileo joyfully accepted these conditions, and Riccardi granted the *imprimatur* at last.

The agreement concluded, Galileo returned to his home in Florence for the summer months. Not long after his departure from Rome, however, he received word of Prince Cesi's unexpected death. The scientist had not only lost a great patron and friend, he cringed at the thought of again having to go through the laborious process of negotiating another deal with the indecisive Riccardi. He had not yet decided what to do when the disastrous plague of 1630 made its appearance in central and northern Italy. The authorities immediately imposed a strict quarantine which made communication with Rome extremely difficult. Travel to the city was completely out of the question. Galileo, who was nothing if not one of the most resourceful men of his age, shrewdly seized upon this otherwise unfortunate course of events and turned it to his personal advantage. Employing Niccolini as his go-between, he informed Father Riccardi that, in view of Cesi's death and his own inability to travel, the conditions under which the *imprimatur* had been granted were no longer applicable. He requested instead that official permission be granted to have the final version of the manuscript published in Florence. This Riccardi flatly refused, but intense pressure was again brought to bear. This time Galileo enlisted the support of the Grand Duke, Ferdinand II, to whom Riccardi, a Florentine by birth, owed allegiance. Niccolini, Ciampoli, and a number of Galileo's other influential friends also got into the act. The hapless Riccardi finally yielded a second time.

Galileo was well aware of the fact that once granted to a book the *imprimatur* of the censor strongly implied that its author had been relieved of any further responsibility. Moreover, it appears that throughout the long and trying prepublication delays he was careful to comply with all explicit instructions given him by Riccardi and his staff. It seems equally certain, however, that the book, as finally published in Florence, would never have been authorized in Rome. Not even Galileo's most ardent admirers can deny that he and his friends resorted to some very questionable tactics to obtain official permission to print the *Dialogue*. In the subsequent proceedings against Galileo it was charged that he had indeed obtained the *imprimatur* by false pretenses. The affable and anxious-to-please Riccardi had obviously been no match for the quick-witted scientist and his conspiratorial friends. When the Pope was finally apprised of the underlying facts of the case, he became understandably enraged. He felt, with considerable justification, that he had been personally hoodwinked by one of the very few men whom he had openly trusted and admired.

Only a few weeks passed from the time of publication until Urban and the Holy Office discovered that they had been outwitted. In August, 1632, an order was issued that all unsold copies of the *Dialogue* be confiscated, and that those already in private hands be bought back. Two months later, under the prodding of the battle-scarred Jesuits, a stunned Galileo was ordered to appear before the Inquisition in Rome. He delayed as long as he could on grounds of ill health, but in February of the following year he was left with no alternative but to comply. If he did not, the Florentine Inquisitor was empowered by the Holy Office "to conduct him to the prisons of this supreme tribunal in chains." As a last resort, Galileo asked the Grand Duke to intervene on his behalf, but this request, though politely received, was denied.

He left Florence on January 20, 1633, during the worst part of the winter season. The miserable journey was interrupted for several days at both Ponte and Centina by the quarantine imposed on all travelers as a precaution against the plague. After nearly a month of arduous, bone-chilling travel, he arrived at the palace of the Tuscan ambassador in Rome. Though technically a prisoner of the Inquisition, he was granted the unprecedented privilege of residing in the embassy until summoned by the Holy Office. There he was visited from time to time by an old acquaintance, Monsignor Ludivico Serristori, who, on the pretext of calming Galileo's fears and suspicions, had been given the task of discovering the type of defense he intended to adopt. Otherwise, the scientist was neither permitted to leave his quarters nor receive friends.

The machinery of the Inquisition was set in motion by an embittered Urban VIII with the appointment of a special commission to investigate the entire affair. Ambassador Niccolini, by making discrete inquiries, was able to learn that a central part of the Commission's findings was Galileo's transgression of Bellarmine's supposed order against teaching the Copernican doctrine in any form. He immediately notified his guest of this but Galileo, to Niccolini's surprise, showed no evidence of alarm. He informed the ambassador that he had in his possession a letter written by the late Cardinal which stated that he had been forbidden from teaching the heliocentric doctrine as truth only. As for the rest, the Pope himself had personally approved of the contents of *The Assayer*, while the *Dialogue* had been safely concluded with Urban's own argument that Copernicanism could never be accepted as absolute truth. There had obviously been a misunderstanding, but Galileo was certain that he would have no difficulty in persuading the Inquisitional authorities that he had fully complied with the letter of the decree of 1616. Unaware of just how serious was his plight, he even expressed the naive hope that he might engage his judges in a friendly discussion on the scientific merits of Copernicanism, and thus save the Church from making the grave mistake of opposing the compelling and irreversible new scientific doctrine. Niccolini was justifiably alarmed, for it was obvious to him that Galileo had not taken the Pope's earlier warnings nearly seriously enough.

On April 12, 1633, Galileo was brought to the Dominican cloister of Santa Maria Sopra Minerva and the hearings began. Father Vincenzio Maculano da Firenzuola, Commissary-General of the Holy Office, was in charge of the proceedings. As was characteristic of inquisitorial procedure, the accused, who was assumed guilty unless proven innocent, was not informed of the charges brought against him; neither was he entitled to a lawyer, thus placing him at serious psychological and legal disadvantages from the outset. When asked whether he knew on what grounds he had been summoned before the Holy Office, Galileo said he believed that it was on account of his latest book. Firenzuola dismissed this subject for the time being, however, and concentrated instead on what had transpired during the audience between Ballarmine and Galileo on February 26, 1616. Galileo replied that the Cardinal had informed him that the Copernican opinion, taken absolutely, must not be defended or held. He further stated that he had "not in any way disobeyed this command; that is, had not by any means held or defended the said opinion." He then presented the letter given him by Bellarmine some seventeen years earlier, when he had asked for a statement in writing with which to protect his injured reputation. Firenzuola examined the document, and then asked Galileo if any other command had been issued to him during the audience in question. He replied that he did not remember any.

At this point the Commissary-General read to the accused the highly disputed minute of 1616 which states that Galileo must "neither hold, defend, nor teach that [Copernican] opinion *in any way whatsoever*." Galileo was stunned, for he had been operating all along on the assumption that it was Bellarmine's only partially restrictive decree that was at issue. Still, he possessed sufficient presence of mind not to brand the unexpected document as false. Instead, he informed the Commissary-General that he could not remember having been notified of any such prohibition by Bellarmine or anyone else present at the fateful audience—which was probably the truth. Firenzuola then went on to question why, during his negotiations for permission to print the *Dialogue*, he had not informed Father Riccardi about the command issued against him. Galileo replied that he did not bring the matter up because "I have neither maintained nor defended in that book the opinion that the Earth moves and that the Sun is stationary but have rather demonstrated the opposite of the Copernican opinion and shown that the arguments of Copernicus are weak and not conclusive."

With this patently false claim the first hearing concluded. Simplicio's peroration in the *Dialogue*, as is pointed out by Santillana, could by no means be taken as proof of Galileo's disenchantment with Copernicanism, no matter which version of the prohibition of 1616 one chooses to believe. Galileo realized immediately how damaging this statement was to his case, for it impaired the value of all his later testimony. Though the Inquisition was called into session twice more, Galileo knew at the end of the first day that his fate rested completely in the hands of his ten judges: all Cardinals and all Dominicans. The public program for which he had fought so courageously and long was clearly in a state of complete collapse.

During the period between the first and second interrogations, Galileo remained a prisoner of the Inquisition, though in the comfortable quarters of one of its officers. On April 17, three Counselors of the Holy

Office, who had been appointed to examine and analyze the text of the *Dialogue*, filed their reports, which all came to the same conclusion: its author had taught and defended the Copernican system. Not only had he disagreed with Aristotle and Ptolemy, he had set forth several new arguments, among them his theories on the tides and sunspots, with the intent of establishing the truth of the forbidden doctrine, though these phenomena had already been satisfactorily explained in other ways. Moreover, there was a "vehement suspicion" that he continued to hold it to this very day. Their reports were liberally documented with lengthy quotations from the book which confirmed their conclusions. Historians since have generally agreed that theirs was a fair appraisal of Galileo's position.

Weeks passed, and tensions mounted as the judges deliberated the case. They had safely gotten around the very shaky question of the disputed injunction because Galileo, according to the Counselor's reports, had knowingly lied about the contents of the *Dialogue*. It was agreed that this constituted a direct violation of Cardinal Bellarmine's explicit instructions to the scientist, no matter what anyone else, the Commissary-General included, might have said during the audience of 1616.

Though a legal case had been made, it was not a particularly strong one. What was probably a false document had been used to cow the accused, a frightened and confused man of seventy, into foreswearing a valid one. Yet everyone directly involved in the case, with the possible exception of the defendant himself, also knew that once the machinery of the Inquisition had been set in motion, any person unfortunate enough to become enmeshed in its powerful gears was completely at the mercy of those who sat at the controls. Fortunately, however, this was not Torquemada's Spain, but Renaissance Italy. Nor was the accused, the most famous natural philosopher in Christendom, just another citizen. It was becoming apparent that at least some of the Inquisitors were finding the proceedings highly distasteful and thus began to drag their feet. Furthermore, it had never been the Pope's intention to destroy his former friend, but rather to make of him a public example for the purpose of illustrating that even the famous must submit to the authority of Holy Mother Church. That the tactics brought into play to accomplish this end were now being questioned, is evidenced by the following turn of events.

Among the ten Dominicans assigned to judge the case against Galileo was the Pope's own nephew, Cardinal Francesco Barberini. Both he and the Commissary-General, Firenzuola, had come to the conclusion that the issues involved were far more personal than doctrinal. Neither man wanted to be the implement for exacting revenge in the Pope's personal vendetta. While they agreed that some type of penance was clearly in order, both were strongly opposed to the prospect of Galileo's abject humiliation. Firenzuola, with Cardinal Barberini's support, decided to seek permission to deal extrajudicially with the accused; in other words, to reach an agreement outside the normal judicial process in much the same manner involved in modern plea bargaining. The Commissary-General reported the results of his efforts in this direction in a secret letter to Barberini dated April 28, 1633:

> I suggested a course that the Holy Congregation should grant me permission to treat extrajudicially with Galileo, in order to render him sensible of his error and bring him, if he recognizes it, to a confession of the same. This proposal appeared at first sight too bold, not much hope being entertained of accomplishing this object by merely adopting the method of argument with him; but, upon my indicating the grounds upon which I had made the suggestion, permission was granted me. That no time might be lost, I entered into discourse with Galileo yesterday afternoon, and after many and many arguments and rejoinders had passed between us, by God's grace, I attained my object, for I brought him to a full sense of his error, so that he clearly recognized that he had erred and had gone too far in his book. And to all this he gave expression in words of much feeling, like one who experienced great consolation in the recognition of his error, and he was also willing to confess it judicially.[42]

Two days later, on April 30, Galileo was called before the Inquisition a second time. When asked whether he had anything to say he admitted, "that in several places the [*Dialogue*] seemed to me set forth in such a form that a reader ignorant of my real purpose might have had reason to suppose that the arguments brought on the false side; and which it was my intention to confute, were so expressed as to be calculated rather to compel conviction by their cogency than to be easy of solution." Two arguments in particular, "the one taken from the solar spots, the other from the ebb and flow of the tide come to the ear of the reader with far greater show of force and power than ought to have been imparted to them by one who regarded them as inconclusive and who intended to refute them." It is evident from these and similar remarks that since Galileo had lied at his first hearing about his true feelings toward the Copernican system, he felt compelled to compound his original indiscretion by continuing to cling to an obvious falsehood. Both he and his judges knew that he was perjuring himself; but there seemed little need to pursue the issue any further, at least for the time being. The Commissary-General had obtained the open admission of error he required, and it had been accomplished by enlisting Galileo's cooperation. The reputation of the Holy Office remained intact, while Galileo finally admitted to himself that the Church hierarchy was far less interested in scientific truth than in envoking strict obedience to authority.

There can be no doubt that Galileo was at least partially responsible for the tragic situation in which he found himself. He had made numerous enemies over the years, often unnecessarily, and had challenged the conventional wisdom of his day with a degree of brashness and self-confidence that served as an open invitation for reprisal. Still, he might have escaped the Inquisition altogether had not certain major social and cultural forces over which he exercised no control been working against him. His was the misfortune to have challenged orthodoxy just at the time when the Church was most convulsed by internal reforms undertaken to meet Protestant criticism. The condemnation of Copernican theory, because of its direct challenge to established doctrine, became one of those reforms. Indeed, Galileo's confession constituted the highwater mark of the Counter Reformation battle against the new science. Ironically, it came at a time when the outcome of the struggle could and should have been foreseen by the clerical authorities. But the trappings of Scholastic logic had become the Catholic Church hierarchy's single uncompromising method of "saving the phenomena" of its monolithic control over the hearts, minds, and souls of the men and women of Western Christendom. That control was, by the first half of the seventeenth century, a fiction, a cherished notion no more viable than the epicycles of Claudius Ptolemy. By 1633, the evidence in support of the heliocentric system was so overwhelming that the verdict was no longer seriously in doubt. Thus, the Church's unwavering commitment to the geocentric universe did great harm to both the pursuit of science in southern Europe and to the prestige of the Church itself, as Galileo repeatedly warned it would. The Church did not sanction the printing of books that treated the earth's motion as a fact until 1822. And by then all except the most conservative Protestant sects had been convinced. Pope Urban viii, whose mentality was essentially medieval and completely authoritarian, sincerely believed that a formal juridical procedure would be sufficient to eradicate doctrine which he conceived of as daring to limit and constrain the power and wisdom of God within the limited scope of human understanding. He never grasped the fact that the Church's vision was in many ways far more limited than Galileo's, who saw science as an open-ended quest for progress and truth. It apparently never dawned on the Pope that the infinite power of God might be best understood by attributing to Him the creation of an infinite universe.

Galileo's greatest miscalculation, on the other hand, was that he assumed the Church would be as rational in its evaluation of Copernicanism as were the scientific principles on which the heliocentric theory rested. He somehow forgot his own concepts of primary and secondary, objective and emotional. Still, Galileo was one of the very few men of his time who was aware of what was happening and what would follow. He knew beyond doubt that the opposition to Copernicanism had been limited to a hopeless rear-guard action; that no matter what the outcome of his trial, or the preventative measures adopted by the spiritual police, the truth could no longer be suppressed.

On May 10, 1633, he was again called before the Holy Office, for the purpose of submitting his written defense. The document offered nothing new. He explained once more how he had complied with Bellarmine's injunction of 1616: "Those faults which are seen scattered throughout my book have not been artfully introduced with any concealed or other than sincere intention, but have only inadvertently fallen from my pen, owing to a vainglorious ambition and complacency in desiring to appear more subtle than the generality of popular writers." He concluded with a touching appeal for mercy, asking his judges to take into consideration his declining health and advanced age when reaching their decision.

> Lastly, it remains for me to beg you to take into consideration my pitiable state of bodily indisposition, to which, at the age of seventy years, I have been reduced by ten months of constant mental anxiety and the fatigue of a long and toilsome journey at the most inclement season—together with the loss of the greater part of the years to which, from my previous condition of health, I had the prospect. I am persuaded and encouraged to do so by the faith I have in the clemency and goodness of the most Eminent Lords, my judges; with the hope that they may be pleased in answer to my prayer, to remit what may appear to their entire justice the rightful addition that is still lacking to such sufferings to make up an adequate punishment for my crimes, out of consideration for my declining age, which, too, humbly commends itself to them.[43]

After this meeting Galileo was permitted to move back to the Tuscan embassy, a privilege never before or since granted in the annals of the Inquisition. Thus, contrary to popular legend, he never spent a day of his life in the dungeons of the Holy Office, or in the cell of any other prison. The psychological strain inflicted upon the aging scientist was an entirely different matter, however. Ambassador Niccolini, who was both surprised and elated to see his friend return to the Villa Medici, wrote the Grand Duke that, "The poor man has come back more dead than alive. It is a frightful thing to have to do with the Inquisition."

On June 20, Galileo was summoned before the Holy Office for a third and final interrogation. At issue was the question of his "intention." In accordance with the deal he had made with Firenzuola, he had previously been permitted to admit that his error was one of conceit rather than of intentional wrongdoing. However, the Pope, with the backing of the majority of the Inquisitors, had since decided that the Holy Office had not taken a hard enough line. Nothing less than a "rigorous examination" of the accused's convictions with regard to Copernicanism would do. Moreover, he was to offer a public abjuration of Copernican theory which, under normal circumstances, would require that he be sentenced to perpetual imprisonment. It was clearly Urban's intent to humble Galileo in the eyes of his ardent admirers. But the Pope also informed Niccolini that once this purpose had been accomplished he would seriously consider lightening Galileo's sentence.

"Rigorous examination," as administered by the Holy Office, could go so far as to include physical torture to persuade the defendant to confess the whole truth. However, in this instance the threat of bodily harm was only a formality since Galileo's age and ill health prohibited its use. He was probably shown the instruments of torture to comply with established procedure, but once this ritual was accomplished both the accused and his judges knew the matter would go no further. Though a rather minor point, it gave rise to the totally unfounded belief, still accepted by many, that Galileo confessed to his accusers only because his resistance had been shattered on the rack.

Galileo again denied any attempt on his part to present the Copernican system as true. The minutes of the meeting, which bear his signature, show that he answered every question put to him in the negative: "I do not hold, and have not held, this opinion of Copernicanism since the command was given to me that I must abandon it; for the rest, I am in your hands, do with me as you please."

If, as Koestler points out, it had been the intention of the Inquisition to break Galileo, this was obviously the point at which to do it. He had both indicted and convicted himself in the *Dialogue*, lengthy excerpts from which lay within arm's reach of his judges. But since physical torture was already ruled out, the interrogation ended at this point. Galileo was not permitted to return to the embassy, but was led back to the apartment previously assigned him to await sentencing. On the following day, clad in the white shirt of penitence, he was escorted to the great hall of Santa Maria Sopra Minerva, where his sentence was read to him in the presence of the Congregation of the Holy Office.* The *Dialogue* was prohibited and its author condemned to formal imprisonment at the pleasure of the Inquisition. For the next three years he was to repeat the seven penentential Psalms at least once a week. Then, forced to kneel before those present, he was presented with the most frequently quoted formula of abjuration in the history of the Inquisition, which he was obliged to read out loud:

> I, Galileo, son of the late Vincenzio Galilei, Florentine, aged seventy years, arraigned personally before this tribunal and kneeling before you . . .

---

* Only seven of Galileo's ten judges signed the document: Cardinals Francesco Barberini, Gasparo Borgia, and Laudivio Zacchia abstained. The Commissary-General Firenzuola, who was responsible for the conduct of the trial, was prohibited by inquisitional procedure from sitting in judgment of the accused.

having before my eyes and touching with my hands the Holy Gospels,
swear that I have always believed, do believe, and by God's help will in
the future believe all that is held, preached, and taught by the Holy Catholic and Apostolic Church. But, whereas—after an injunction had been
judicially intimated to me by this Holy Office to the effect that I must
altogether abandon the false opinion that the Sun is the center of the
world and moves and that I must not hold, defend, or teach in any way
whatsoever, verbally or in writing, the said false doctrine, and after it had
been notified to me that the said doctrine was contrary to Holy Scripture—
I wrote and printed a book in which I discuss this new doctrine already
condemned and adduce arguments of great cogency in its favor without
presenting any solution of these, I have been pronounced by the Holy
Office to be vehemently suspected of heresy, that is to say, of having held
and believed that the Sun is the center of the world and immovable and
that the Earth is not the center and moves:

Therefore, desiring to remove from the minds of your Eminences, and
of all faithful Christians, this vehement suspicion justly conceived against
me, with sincere heart and unfeigned faith I abjure, curse, and detest the
aforesaid errors and heresies and generally every other error, heresy, and
sect whatsoever contrary to the Holy Church, and I swear that in the future
I will never again say or assert, verbally or in writing, anything that might
furnish occasion for a similar suspicion regarding me; but, should I know
any heretic or person suspected of heresy, I will denounce him to this
Holy Office or to the Inquisitor or Ordinary of the place where I may be.
Further, I swear and promise to fulfill and observe in their integrity all
penances that have been, or that shall be, imposed upon me by this Holy
Office. And, in the event of my contravening (which God forbid!) any of
these my promises and oaths, I submit myself to all the pains and penalties
imposed and promulgated in the sacred canons and other constitutions,
general and particular, against such delinquents. So help me God and
these His Holy Gospels, which I touch with my hands.

He then signed the attestation:

I, the said Galileo Galilei, have abjured, sworn, promised and bound
myself as above; and in witness of the truth thereof I have with my own
hand subscribed the present document of my abjuration and recited it
word for word at Rome, in the convent of the Minerva, this twenty-
second day of June, 1633.
I, Galileo Galilei, have abjured as above with my own hand.[44]

Here, at last, was a document whose authenticity was beyond
question.

According to the popular legend, which has no basis in fact, Galileo,
upon rising from his knees, stamped at the earth with his foot and mut-
tered, *"Eppur si Muove!"* (And yet it moves!) Had he truly mocked
his judges in this reckless manner their reaction would have been
both swift and merciless.

Shortly after the trial and abjuration, Urban VIII commuted Galileo's sentence, making good his earlier promise to Ambassador Niccolini. The "formal prison" to which the natural philosopher had been sentenced was reduced to perpetual house arrest, first on his farm at Arcetri and later, following the death of his beloved daughter, Sister Maria Celeste, at his villa in Florence, where he spent the last years of his life. Though Galileo emerged from his prolonged ordeal in a severe state of mental depression, his indomitable spirit was not broken. When word of the Pope's welcome decision reached him, Niccolini observed a marked improvement in his friend's outlook on the future.

Within a few weeks after his return to the country, Galileo began work on a book which, because of his prolonged campaign on behalf of Copernicanism, had been postponed for over twenty-five years—the famous *Discourses Concerning Two New Sciences*. Astronomy, as we noted earlier in this chapter, did not provide the field for his most important scientific achievements, though it is the subject which, for obvious reasons, has most concerned us. His true eminence as a practical scientist rests on his pioneering experiments with moving bodies, from which he founded the science of kinematics and formulated basic laws of motion. Though he made numerous discoveries in mechanics both during the early and middle period of his adult life, Galileo had never collected the results of his work for the purpose of making them readily available to fellow scientists and educated laymen. The publication, in 1638, of the *Two New Sciences* changed that. Among the major discoveries announced in this first great work of modern terrestrial physics was the law of free fall, which he deduced from the supposition that there are equal increments of velocity in equally small intervals of time. In the course of his arguments, he correctly maintained that any time interval is capable of being divided into an infinite number of parts, a breakthrough that foreshadowed the differential method. Unlike Newton, however, he did not think of free fall as being due to the gravitational attraction of the earth.

Even though his new book dealt only with terrestrial phenomena, Galileo harbored no illusions of obtaining the *imprimatur* required for its printing in Italy. He arranged through a friend in Venice to have the manuscript smuggled out of the country to the Protestant city of Leyden, where it was placed in the willing hands of the able Dutch publisher, Louis Elzevir. When the manuscript reached print, Galileo feigned surprise, pretending not to know how it had reached the Netherlands. Though it seems improbable that anyone believed him—least of all the clerical authorities in Rome—no steps were taken against him. A copy of Galileo's other major work, the *Dialogue Concerning the Two Chief World Systems*, had been smuggled to Kepler's old patron and friend, Matthias Bernegger, in Strasburg. Bernegger made arrangements for a Latin translation which was published in 1635, and circulated widely in Europe. In 1636 he also commissioned the printing of Galileo's *Letter to the Grand Duchess Christina*. With the publication of these works outside Italy, whatever remaining hope the Catholic authorities might have had of stemming the flow of Galileo's scientific ideas vanished. At the age of 74, Galileo Galilei was again recognized by all Europe as the master of the new science.

Within a matter of months following the completion of the *Two New Sciences* Galileo's eyes, which had been bothering him for several years, gave out completely. Based on the symptoms as he described them in letters to his friends, medical historians have diagnosed him as a victim of the serious eye disease, glaucoma. First his right eye was blinded by inflammation then, a short time later, his left. He mentioned his condition in a moving yet characteristically prideful passage of a letter to his friend, Elia Diodati:

> Alas your friend and servant Galileo has been for the last month hopelessly blind; so that heaven, this earth, this universe, which I, by marvelous discoveries and clear demonstrations, have enlarged a hundred thousand times beyond the belief of the men of bygone ages, henceforward for me is shrunk into such small space as is filled by my own bodily sensations.[45]

Yet even the crippling combination of advanced age and the loss of his sight could not keep Galileo from his work. He continued to dictate his thoughts on scientific matters to his devoted pupil, Vincenzio Viviani, including the design for a clock complete with pendulum and escapement. He also received a steady stream of distinguished visitors; among them, a young poet from England named John Milton, who was himself destined to become a victim of blindness in later life.

Beginning in November, 1641, Galileo was confined to bed with a fever and palpitations of the heart. His condition gradually worsened, and death came on the night of January 8, 1642. His friends in Florence voted to lay his body to rest in the Church of Santa Croce beside the city's other illustrious sons, Michelangelo and Machiavelli, and to construct a large monument in his honor. But Urban VIII, whose enmity toward Galileo had not subsided, remained implacable. He had consistently refused the scientist's petitions for greater freedom during his years of house arrest, and now denied him this last honor. When the Tuscan Ambassador attempted to persuade the Pope to change his mind, he was told that His Holiness would not be a party to such a bad example, for Galileo "had altogether given rise to the greatest scandal in Christendom." Thus the scientist's remains lay in the basement of Santa Croce for nearly a century. Finally, however, the Church, its attitude softened by time, relented, and Galileo's body was properly entombed in a marble sarcophagus. Above it stands a large monument constructed in accordance with the last will and testament of Viviani.

Though the spiritual authorities were temporarily successful in their efforts to silence one of the truly great scientific intellects of all time, no episode in the history of the Catholic Church has so frequently been cited against it as the abjuration forced upon Galileo in 1633. It did not take long for posterity to choose which party in this great controversy it would support: Galileo's epitaph was written for him by countless admirers of later generations—*Eppur si muove!*

8. Sir Isaac Newton: The Grand Synthesizer

A Considerable Gift

WHEN GALILEO WAS BORN IN 1564 THE ITALIAN RENAISSANCE was clearly on the wane; by the time of his death in 1642 it had completely run its grand course. This rather precipitous decline in Italy's economic, political, and cultural fortunes was a matter of more than passing concern to her greatest scientist, for he knew that any unfavorable changes in these areas—especially when coupled with the religious repression brought on by the Counter Reformation—would doubtless adversely affect the further development of Italian science. Twice in the opening pages of his *Dialogue Concerning the Two Chief World Systems* Galileo expressed his fear that Italy was in danger of being eclipsed by powerful rivals to the north. The die had been cast far in advance of Galileo's expression of concern, however. Although its effects were slow to materialize, the first major blow to Italian hegemony in Europe was struck over half a century before his birth, when the Portuguese explorer Bartolomeu Dias, sailing under the flag of his native land, discovered the South African Cape of Good Hope. Nearly every year after 1500 saw armed fleets sail for the Far East from ports on the Atlantic coast of western Europe to collect prize cargoes of spices, silks, and other exotic commodities. Up until this time such goods were shipped first by water across the Indian and Arabian oceans by Arabs, then overland through the Middle East by caravan to the famous ports of the eastern Mediteranean. From there, in a monopolistic enterprise that brought great wealth, prestige and political power both to themselves and to their homeland, the merchants of Italy carried the goods westward to an economically revitalized Europe eager to do business at almost any price. But no more.

During this same period much of western Europe underwent a revolutionary political transformation, which saw the rise of powerful national states and the death of feudalism as the dominant political and economic system. Spain and Portugal took the lead and were quickly followed by England, Holland, and France. Meanwhile, the Italians were unable to resolve their many long-standing local and regional differences and thus remained divided into five independent states, four of them ruled by secular leaders, and one by the Pope. These divisions had not proven a serious disability to Italian economic and cultural growth during the early Renaissance; but as the northern European states grew ever stronger politically and militarily, Italy gradually faded. The political divisions which so bedeviled the Italians also kept them from reaping the benefits of major exploration and discovery. It was the powerful countries of northern Europe, not Italy, that carved out huge territories and great wealth for themselves in the New World. Ironically, it was an Italian, Christopher Columbus, sailing under the flag of Spain (his homeland had no single banner to give him) who paved the way for this discovery. Italy lacked neither the men nor the knowledge to participate in this momentous venture, but it did lack a sense of national purpose and destiny. Thus, the leadership once so proudly exercised by the Italian city-states passed northwest. No longer was the Mediterranean, as its name implies, the middle of the world.

Of all the nations eager to fill the political and cultural vacuum left by Italy's decline at the end of the sixteenth century, and reaching far into the seventeenth, none was in a better position to do so than England which, by then, was already on its way to becoming a Republican and Puritan state. During the reign of the Tudor monarch Henry VIII (1509–47) England had broken with the Church of Rome, and under the rule of Henry's daughter Elizabeth I (1558–1603) the country had established its own peculiar brand of Protestant worship called Anglicanism, an interesting admixture of the religious thought of Martin Luther and John Calvin, coupled with certain carefully selected elements of Roman Catholic church government. In the political realm, royal absolutism after Elizabeth's death was gradually pared away by the growth of Parliamentary government, especially in the area of economic affairs. Within less than a century both the religious and political questions would be resolved to the satisfaction of a majority of Englishmen, although it took a bloody civil war (1642–49), during which royal tyranny was temporarily replaced by that of Cromwell and his Ironsides, and a bloodless revolution (1688), which saw the establishment of a constitutional monarchy and religious freedom under William and Mary, to accomplish it.

Thus England during the seventeenth century was truly a society in ferment: politically, religiously, and culturally. History has shown that it is usually in such a climate of open discussion, cultural transformation, and dissent that scientific activity takes root and flourishes. At the very time the humiliation of Galileo saw the climax of the scientific revolution in astronomy in Italy, and the hatred of Luther and his successors in Germany for the Copernican system forced Kepler to take refuge outside his hostile homeland, England was giving birth to a cluster of scientific geniuses without peer in any similarly limited time period or tightly circumscribed geographic region: Isaac Barrow, Robert Hooke, Edmond Halley, Robert Boyle, John Flamsteed, Christopher Wren, John Wallis and, towering above them all, Isaac Newton. Theirs was the illustrious age of intellectual accomplishment which Whitehead so rightly described as "the century of genius," the greater share of which, between 1650 and 1700, was concentrated on an offshore island of the European continent.

For those who believe that the course of human events is somehow controlled by powerful underlying forces whose presence and purpose perpetually elude the historian's flawed grasp, there is much to contemplate in the unusual circumstances surrounding the birth of Isaac Newton. To begin with, the child was born on Christmas Day, 1642, the year of Galileo's death. Thus one genius departed the scene upon the arrival of another, as though nature itself had taken a special interest in seeing the unfinished intellectual revolution initiated by Copernicus carried through to its dazzling conclusion. Yet from the surviving accounts of his birth, no one, with the possible exception of his mother, Hannah, could have entertained serious hopes that this frail infant would live to accomplish anything—let alone attain a position of greatness in the field of science. Born premature by several weeks, Isaac was so weak and tiny that two women who were sent to a Lady Packenham's at the nearby village of North Witham to obtain some medicine did not expect to find him alive on their return. Years later Newton was fond of repeating his mother's account of how he would have quite easily fitted into a quart mug. Too weak to hold his head upright for proper feeding and breathing, he wore a small supportive collar round his neck for several months, a circumstance which makes one shudder to think how easily the infant's magnificent brain could have been damaged, forever altering the course of modern thought and history.

In addition to the unexpected but deeply meaningful date of his birth and the exceptional physical conditions that threatened his very survival, the infant's father, Isaac Newton, a yeoman farmer of thirty-seven, had died some three months before the birth of his gifted son. Frank Manuel, the author of a fascinating psychoanalytic biography of Newton, suggests that the image of the father Isaac had never known and whom he was taught from childhood to believe was in heaven "soon fused with the vision of God the Father, and a special relationship was early established" between them; that in attempting to please his Creator, Newton subconsciously sought the favor of his deceased father.

Manuel also alludes to the quite common belief among many societies, including seventeenth-century England,‖ that a posthumous child "is endowed with supernatural healing powers" [1] and "destined to outstanding good fortune." [2] In Manuel's opinion, Newton felt a "deep sense of being one of the elect," which was reinforced because of his birth on Christmas Day. Indeed Dr. William Stukeley, a friend of Newton's and his first biographer, mentions the popular folk belief that the Christmas birth is an omen of future success.

Were we able to confirm Manuel's intriguing analysis with some form of concrete historical documentation, it would go very far in explaining the driving force behind Newton's unparalleled intellectual accomplishments; namely, that he believed himself specially chosen by God to reveal the secrets behind hitherto unsolved mysteries surrounding the operation of the universe. But in this and virtually all other aspects of Newton's private life of the heart and mind, we know very little that is proveable. In terms of his dreams, his urges, and his inner visions, the great thinker was mute as a stone. Like Thomas Jefferson, he left a record of only those things that he wanted posterity to scrutinize, the rest he consigned to the flames before his death.*

Newton, a country lad, was born in the small Manor House of Woolsthorpe, near the hamlet of Closterworth in Lincolnshire. The nearest population center of any importance was Grantham, a town of about two or three thousand residents located some six miles from the Newton farm. Despite his premature birth and delicate physical condition, the Closterworth parish record shows that on January 1, 1643, exactly one week after he was born, the infant was carried to the local church by his mother and there baptized "Isaac soone of Isaac and Hanna Newton." Of Isaac Newton, the father, very little is known except that he was a modestly successful farmer who could neither read nor write. According to the Reverend Barnabas Smith, Hannah Newton's second husband and the stepfather of young Isaac, Newton's real father "was a wild, extravagant, and weak man;" [3] but there is no surviving evidence with which to corroborate this harsh judgment. We also know that the elder Newton had only gained proprietorship of Woolsthorpe on the death of his father, Robert Newton, in September of 1641. In a little more than a year Isaac too was laid to rest in the family plot in Closterworth churchyard, never to look upon the face or marvel at the astounding achievements of his illustrious progeny.

---

* There is, of course, no way of knowing beyond scholarly speculation the content of the papers Newton chose to destroy. They doubtless contained some material which could have shed additional light on his private thoughts. However, judging from the general aloofness of his surviving correspondence and the accounts of his few close friends, it seems improbable that he ever put his deepest feelings on paper, except in certain widely scattered instances of his student notebooks.

As an only child whose father and grandfather were both deceased before his birth, Newton's mother became the central figure in his life, a position she occupied until her death many years later. Most of what we know about her is based upon a character-portrait drawn by John Conduitt, who, in 1717, married Catherine Barton, Newton's niece, and lived with the great man for most of the final decade of his life. Hannah Newton was the daughter of James Ayscough of Market Overland in the county of Rutland. Her early ancestors had apparently controlled considerable wealth, but the fortunes of more recent generations of Ayscoughs had gradually declined, as is evidenced by Hannah's marriage to a yeoman rather than a man of gentle birth. "She was a woman of so extraordinary an understanding, virtue and goodness," wrote Conduitt, "that those who think that a soul like Sir Isaac Newton's could be formed by anything less than the immediate operation of a Divine Creator might be apt to ascribe it to her." [4] Though Conduitt's deep admiration for the accomplishments of his wife's uncle undoubtedly contributed to his overdrawn appraisal of Hannah Newton's character, she nevertheless appears to have been a woman of considerable virtue and intelligence. Manuel describes young Isaac's fixation upon her as "absolute" and it appears that she was also deeply devoted to her son. Even before her second marriage to Barnabas Smith, Rector of North Witham, Hannah arranged for the Reverend to have the manor house remodeled for the benefit of Isaac; she also persuaded Smith to purchase additional farmland to increase the child's inheritance. The gray, stone two-story house, as refurbished by Smith, still stands for the visitor to see.

Hannah Newton's second marriage took place on January 27, 1646, a month after Isaac's third birthday. Though the date of her birth is unknown, she was probably thirty, whereas the groom, a well-to-do country clergyman of sixty-three, was over twice her age. The marriage, which came as a painful shock to her sensitive son, soon grew into a perpetual nightmare: the object of Isaac's deepest affection had been stolen away by an unwelcome stranger. Stolen, in this instance, is not too strong a word, because after the wedding Hannah, for reasons never clearly explained, but which I suspect had far more to do with Smith's wishes than her own, left Isaac under the care of her mother, Mrs. Ayscough, and moved to North Witham, about a mile south of Woolsthorpe. Manuel observes that, "The loss of his mother to another man was a traumatic event in Newton's life from which he never recovered." [5] As late as 1662, at the age of twenty, Newton, after the fashion of the day, carefully compiled a list of his sins which includes the following brief passage: "Threatening my father and mother Smith to burne them and the house over them." If truly threaten them he did, it was a transgression of considerably long standing, for by this time the Reverend Smith had been in his grave for some nine years and Newton's mother had long since moved back to Woolsthorpe, bringing with her the three children of her second marriage—Benjamin, Mary, and Hannah. Yet the very fact that Isaac still harbored bitter memories of this disturbing childhood desire says a great deal about the pain and frustration he endured on being "abandoned" by his beloved mother.

From the time of Hannah Newton's second marriage until Isaac's matriculation at Cambridge University in 1661, at the age of eighteen, the information available to historians about his life is almost as much anecdotal as factual. Still, an attempt must be made to piece together some of the more salient experiences of these extremely crucial years. William Wordsworth wrote that the child is father of the man, and in Newton's case, at least, the poet's words ring very true indeed. Though the once warm relationship between mother and son was quickly rekindled after the Reverend Smith's death, the loneliness, trauma, and frustration of Isaac's formative years left an indelible mark on the personality and temperament of the young boy. This is at least partly owing to the fact that though the relationship between himself and his maternal grandmother appears to have been a mutually respectful one during the period of his mother's absence, it also seems to have been entirely devoid of deep personal affection. As a child, Isaac Newton communed chiefly with himself, and this youthful habit of withdrawing into his shell carried over into and completely dominated his career both as a scientist and major religious thinker. All his adult life, he leaves the impression of an unloved and unloving man.

While the major influences on Newton's early life were predominantly familial, the year of his birth coincided with the outbreak of the bloody English Civil War followed by the establishment of the Commonwealth and Protectorate under Cromwell. Newton's native district of Lincolnshire was a major theatre of political and military operations for both the Parliamentary forces and the Royalists. On May 13, 1643, when Newton was an infant of six months, Cromwell won an important victory over a Cavalier army on the fields of Lincolnshire only ten miles from Woolsthorpe. During the turbulent years that followed, local property-holders stood by helplessly while their farms and homes were ravaged by the invading armies of both factions as they foraged for food and other provisions. Cromwell eventually consolidated his position with a number of crushing military victories after which Lincolnshire became Roundhead country. Isaac had just turned six in January of 1649 when the shocking news arrived that the Stuart king, Charles i, had been beheaded in front of the royal palace at Whitehall. To exactly what degree the civil disorder and its attendant violence influenced young Isaac's political and religious outlook, one cannot say for certain. His stepfather, who was a wealthy man and a Royalist supporter (but not a papist), was able to retain his position as Rector of

North Witham during the Commonwealth, though a number of other similarly minded clerics in the area were not so fortunate. The score was evened during the Restoration, however, for many Dissenters in and around Grantham lost their positions after the coronation of Charles II. Though Newton remained a member of the Anglican Church all his life, the world in which he grew up, with its strong emphasis on self-discipline, Bible study, and devotion to duty, bore marked similarities to the Puritanism of his day. There can be no doubt that he was deeply influenced by the Puritan religious tradition, as were a majority of his brilliant contemporaries including Boyle, Hooke, and Flamstead, a subject we shall discuss in greater detail in the next section of this chapter.

Newton's introduction to formal education came by attending two little day-schools at Skillington and Stokes, hamlets near enough to Woolsthorpe so that he could walk back and forth each day. When he reached his twelfth year he was enrolled in Old King's School at Grantham, where he remained a student until the age of sixteen. There he came in contact with his first teacher of note, Henry Stokes, the Master of Grantham School. It was under Stokes' perceptive guidance that Newton's intellectual interests were awakened, as is evidenced by his newly acquired fondness for books and the construction of a number of ingenius mechanical toys and models. Since Woolsthorpe was too far from Grantham to travel every day, Isaac was boarded at the house of a Mr. Clark, the town apothecary, whose wife and Newton's mother were intimate friends. Newton, as we shall see, became deeply immersed in both the study and practice of alchemy while a student and professor at Cambridge, an interest some believe dates from the many pleasant hours he spent watching his affable landlord concoct remedies for the many illnesses, both real and imagined, of the local populace. Though Newton enjoyed remarkably good health throughout his eighty-five years, a venerable age for a man or woman of the seventeenth century (or, for that matter, our own), like many other highly original thinkers he was prone to hypochondria. He frequently mixed his own medicines to alleviate the various symptoms of his imaginary maladies, leading one to suspect that this behavior also had roots reaching all the way back to his youthful experiences in Clark's apothecary shop.

Conduitt tells us that when Newton first entered school at Grantham he seemed very indifferent toward his studies, and because of this was next to the lowest student in his class. However, one day, as he was walking to school, the boy who ranked immediately above him administered a painful kick to Newton's stomach which stirred the usually passive youth's desire for revenge. That same afternoon, when classes were dismissed, Newton challenged the aggressor to a fight, and they went into the nearby churchyard. Newton whipped his rival and then, egged on by Master Stokes' son, pulled his embarrassed tormentor along by the ears to the church where he rubbed his nose and cheeks against the wall. Still not satisfied, Newton became determined to beat his rival in their studies, a goal he accomplished in style by rising to become first student in the school. Conduitt's story, whether completely factual or not, has a prophetic ring of truth about it. For in the words of Newton's biographer Louis T. More: "All during his life he required an external stimulus to arouse his latent power, and to exert himself to complete his work or to make public the fruits of his meditation." [6] One must also add that Newton was a morbidly suspicious man, whose temper occasionally flared up like the furnaces he built in his alchemical laboratory at Cambridge, especially when he felt his work or privacy threatened by the probings or criticism of a fellow scientist. He always feared others would steal from him, and he frequently sought revenge for even the slightest of transgressions, more than a few of which were figments of his highly creative imagination. Newton's otherwise brilliant career was marred by a series of unworthy and protracted quarrels with such luminaries as Hooke, Flamstead, and Leibniz. In his defense, one must point to the deep emotional scars of his unhappy youth, for he doubtless feared that others sought to claim for themselves scientific discoveries that were rightfully his own, just as the hated interloper Smith had carried off his mother.

Newton's scholarly accomplishments at Grantham school notwithstanding, the magnificent intellectual achievements of his later life were most strongly prefigured by the countless private hours he spent in the design and construction of kites, model windmills, water clocks, and sundials. Newton, like his archrival John Hooke and his American admirer Benjamin Franklin, possessed a natural mechanical aptitude which enabled him to create with his hands as well as his mind—a skill all the more important at a time when most natural philosophers were called upon to construct their own experimental apparatus. In Newton, these talents were securely underpinned by a driving curiosity to know not only how but more importantly *why* things function as they do. For example, one of his favorite pastimes involved frequent visits to the construction site of a windmill near his school at Grantham. Newton carefully imitated the operations of the workmen by building a functional model of the finished structure, which he sometimes placed atop the house where he boarded, so that its sailcloth blades could better catch the wind. "But what was most extraordinary in its composition," wrote Dr. Stukeley, "was that he put a mouse into it, which he called the miller, and that the mouse made the mill turn round when he pleased; and he would joke too upon the miller eating the corn that was put in." Stukeley also described another of the ingenious devices constructed by Isaac while at Grantham, his famous water clock:

Sir Isaac's water clock is much talked of. This he made out of a box he begged of Mr. Clark's wife's brother. As described to me, it resembled pretty much our common clocks, and clock-cases, but less; for it was not above four feet in height, and of proportionable breadth. There was a dial plate at the top with figures of the hours. The index was turned by a piece of wood, which either fell or rose by water dropping. This stood in the room where he lay, and he took care every morning to supply it with its proper quantity of water; and the family upon occasion would go to see what was the hour by it. It was left in the house long after he went away to the University.[7]

Newton's youthful fascination with time and motion, the chief pre-occupations of the modern physicist, had other equally significant applications, especially when combined with his extraordinarily acute powers of observation. Isaac kept careful records of the shadows cast by the sun against the wall and roof of his home. He then drove in pegs to mark the hours and half hours "which by degrees from some years observations he had made very exact, and any body knew what o'clock it was by Isaac's dial, as they ordinarily called it." He also experimented with kites to determine the shapes most suitable for sustained flight. Sometimes, at night, he placed a burning candle inside a paper lantern which he then tied to the kite's tail, a practice which according to Stukeley, "affrighted the country people exceedingly [into] thinking they were comets." Newton's childhood experiments in the field now known as aerodynamics cannot help but remind one of his conversion, while a Cambridge student, to the Cartesian belief that all celestial bodies move through space carried along by vast whirling vortices of compacted matter. When he read Descartes, did he not perhaps think back upon the kites he had constructed when a boy as miniature celestial objects sustained in their flight by continuously swirling clouds of invisible matter?

If Stukeley's memoirs of Newton's childhood are to be believed, Isaac's interest in the effects of moving air on variously shaped objects did not end with his experiments on windmills and kites. On the day of Oliver Cromwell's death in 1658 a major storm swept across England, causing considerable property damage and giving rise to predictions of other manifestations of divine retribution to follow. Newton, as the story goes, had entered into a rare competition with his more physically adept classmates to see who could jump the greatest distance. "Observing the gusts of wind, [he] took so proper an advantage of them as surprisingly to outleap the rest of the boys." Years later Newton referred to this as one of his first experiments. But to Manuel, his biographer, Newton's victory was a foreshadowing of his many contests with other famous scientists: "Arrogant, he disdains to join with them in their tomfoolery;

timid, he fears them. But when he knows the way the wind is blowing, he can jump and outleap them." [8]

From hindsight, it is obvious that as a child Isaac Newton was endowed with special creative powers and mechanical skills. "Every one that knew Sir Isaac, or have heard of him," wrote Stukeley, "recounts the pregnancy of his parts when a boy, his strange inventions and extraordinary inclination for mechanics." However, we must remember that Stukeley prepared his account of the great man's life long after Newton's discoveries had brought him enduring fame. It would be a mistake to conclude that Newton was looked upon as a genius by those who were best acquainted with him during his youth. Not even an admiring Stokes, to whom Stukeley gives credit for the discovery of Newton's brilliance, could possibly have foreseen the magnitude of his pupil's latent gifts. Though Isaac Newton was a precocious child, he gave only scattered hints of the staggering intellectual powers that were to burst forth in fullest maturity within a few years after he entered Cambridge.

As the first son of Hannah Newton's two marriages, it seemed only natural that Isaac would return one day to Woolsthorpe to manage the farm left to his mother by the father he had never known. He was sixteen when Hannah decided that her son possessed sufficient formal education for the task at hand. (His father, after all, had been unable to read or write.) The youth, from all accounts, acceded to his mother's wishes without protest; yet it was a plan destined to fail from the outset. For once Isaac's intellectual interests had been aroused, there was no possibility that they could be long suppressed.

The sheep and cattle placed in his care strayed into the unharvested fields of wheat owned by the Newton's and their neighbors, while Isaac, oblivious to such mundane concerns, passed the time by reading books, or building models of water wheels which he tested in a nearby stream. He was sent to Grantham on Saturdays in the company of an elderly servant to purchase the necessary goods for the farm, and to master the ancient art of barter and trade. But the two would no sooner reach town when Isaac, content to leave the affairs of business to his older companion, would head for the Clark house and his old room above the apothecary shop, where he passed the day reading books stored there after his departure from school. At other times he found a comfortable place under a hedge by the roadside, where he read until the servant picked him up on the way home. The first of a number of tales dealing with Newton's absent-mindedness date from this period. Just outside Grantham is a steep natural incline known in Newton's day as Spittlegate Hill. It was the common practice for riders to dismount at the bottom, lead their horses up the sharp grade, and remount at the summit. We are told that on at least one occasion Newton forgot to get back on his horse after trudging to the top, and instead led the animal the remaining six miles home. In another instance, his mount, while being led, supposedly slipped its bridle and found its own way back to Woolsthorpe, while Newton walked the remaining distance without noticing that the empty bridle was still in his hand.

Whether exaggerated or not, the abstracted behavior illustrated in these and similar anecdotes reveals the graphic case of a young man alone in a private world, largely impervious to the demands of everyday reality. By this time, Newton must have realized that he was different from other boys, but he seemed little bothered by it. On the contrary, he gladly took refuge in the fantasy world created by his vivid imagination and phenomenal inventiveness. This pattern of withdrawal continued throughout his life as a scientist, for he only rarely discussed his work in depth with anybody, nor did he ever take anyone into his deepest confidence. But most important of all, these long periods of boyhood abstraction are the first significant intimations of the coming eruption of an unparalleled reservoir of singularly creative scientific genius. We observe in the youthful preoccupation of Newton—as in the often erratic behavior of Kepler and Galileo—incipient qualities of unsocialized behavior which helped open the door to creativity in the sciences of the highest order.

Newton's lackluster performance as a would-be yeoman farmer lasted for about a year. In the meantime, Henry Stokes, who realized that his former pupil would make a far better scholar than man of the soil, made repeated attempts to convince his mother that she should return her gifted son to Grantham School to prepare him for a university education. Stokes was so anxious that this be done he even offered to remit his instructional fee of forty shillings a year in order to ease Hannah's financial burden. It seems that Newton's maternal uncle William Ayscough, Rector of Burton Coggles, also lent support to Stokes' plan. One day Ayscough found Isaac totally engrossed in a mathematical problem while the livestock he was supposed to be tending had strayed in all directions. Enough was enough! Hannah was finally persuaded to give her son up to a higher calling, and Isaac happily returned to Grantham where he resumed his education. He remained there for about a year, until 1661, when at the age of eighteen he was accepted at the famous and influential Trinity College at Cambridge. Though he had made few if any close friends among his Grantham classmates, Stukeley informs us that the occasion of his departure was a deeply emotional one. Stokes "set him in a conspicuous place in the school, tho' not agreable [sic] to the lad's modesty, and made a speech to the boys in praise of him, so moving as to set them acrying, nor did he himself refrain his tears." There was others, however, who were more than happy to see him go. The Woolsthorpe servants "rejoic'ed at parting with him, declaring, he was fit for nothing but the 'Versity." As with many others of far keener insight who were to misjudge his capabilities, Newton lived to prove them wrong: after leaving his Cambridge professorship in 1701 he first became Master of the Mint and later President of the prestigious Royal Society.

Aside from his mother and teacher, there was probably only one other person whose heart was torn by Newton's departure for the University, a young woman who perhaps knew him more intimately than anyone else during his student days. We know her first as Miss Storey and later, after her marriage, as Mrs. Vincent. Her mother, the second wife of the apothecary Clark, was a close friend of Hannah Newton's. As with so much of our scant knowledge of Newton's personal life, Stukeley is the sole source of information on Newton's relations with the young woman, and his account is based on a belated conversation he had with her when she was in her late seventies or early eighties. Mrs. Vincent remembered Newton as "always a sober, silent, thinking lad, never known scarce to play with the boys abroad," but "would rather choose to be at home, even among the girls, and would frequently make little tables, cupboards, and other utensils for her and her playfellows." But what is most fascinating about their relationship is that the old woman revealed a romantic attachment between herself and the future scientist.

> Sir Isaac and she being thus brought up together, 'tis said that he entertained a love for her; nor does she deny it: but her portion not being considerable, and he being a fellow of a college, it was incompatible with his fortunes to marry; perhaps his studies too. 'Tis certain he always had a kindness for her, visited her whenever in the country, in both her husbands' days, and gave her forty shillings, upon a time, whenever it was of service to her. She is a little woman, but we may with ease discern that she has been very handsome.[9]

Whether, like so many of Newton's acquaintances, she read more into their youthful relationships than had actually occurred we will probably never know. It appears certain, however, that this was as close as Newton ever came in his life to matrimony. Dame Science was destined to become his all-consuming mistress, at least for the next three decades. With her as his inseparable companion he lived the life of a semirecluse who could perform almost any task with great efficiency, so long as no emotional element was involved. It was Newton's nature to command reverence, respect, and even awe on the part of others, but never love. Even when in the company of his few friends, he always seemed to be detached and alone.

The World of the Virtuosi

Isaac Newton's impact on modern physics has been so profound and long lasting that much of what he learned from others is remembered as being completely his own. This is because so many of the works which greatly influenced the young scientist became almost totally obsolete with the publication of the *Principia* and *Optics*. From a careful reading of their fertile pages it would be easy to form the impression that the "law of universal gravitation," the atomic theory of matter, the calculus, and the theory of the composition of light sprang full blown from the mind of this extraordinary human being.

But the great acts of scientific innovation and synthesis are never accomplished in a vacuum. The political and religious influences, the major personalities, the books surrounding a young genius are of utmost importance to an understanding of the manner in which his or her intellectual gifts come to be shaped. Newton was heir to a rich and dynamic mathematical–experimental tradition stemming from the work of Copernicus, Kepler, Galileo, Bacon, Gilbert, and Boyle. And from across the Channel, the mechanistic speculations of René Descartes were well known to him. Moreover, he had fallen under the spell of the great seventeenth-century virtuosi of his own country; men who, despite their many individual differences, emphasized a rational approach to knowledge as the best means of more fully exploring the works of the Creator. It is this intellectual world which Newton entered as a country boy of 18, a world we must pause to ponder in some detail if we would truly understand the major currents which shaped the scientific and religious outlook of the towering genius who, in his turn, largely determined the course of modern Western thought.

We will begin by briefly examining the work of the two men who most profoundly influenced the thinking of the English intelligentsia during the early decades of the seventeenth century—Francis Bacon and René Descartes. Francis Bacon, Lord High Chancellor of England, was born in London on January 22, 1561, the second of two sons of the Lord Keeper, Sir Nicholas Bacon. A lawyer by training and profession, Bacon's keen mind and restless ambition enabled him to become a courtier, statesman, philosopher, and master of the English tongue. He is today remembered primarily for his eloquence and persuasive power as a speaker in Parliament, the incisive wisdom of his beautifully written essays, and as a man who claimed all knowledge as his province. What is frequently forgotten is that Sir Francis was also extremely interested in establishing a new method by which men might assert their command over nature, both for the greater glory of God and the general improvement of the human condition. It is upon this latter aspect of Bacon's thought that we shall focus our attention.

It must be stated at the outset that Francis Bacon's place in the history of science remains today—some three and one-half centuries after his death—a subject of lively scholarly debate. He has been both zealously over-praised as a natural philosopher of the first rank, and derided as an amateur whose claim to membership in the exclusive scientific fraternity of the seventeenth century is wholly without merit. The truth, I believe, lies somewhere in between. The crux of the problem is that during his entire lifetime Bacon made not a single scientific discovery, nor was he strictly speaking a scientific man. "If Bacon with all his writings were to be removed from history," observed E. J. Dijksterhuis, "not a single scientific concept, not a single scientific result would be lost." [10] Moreover, Bacon seems to have cared very little about the major scientific developments of his age. He rejected the Copernican system, naively criticized William Gilbert's brilliant treatise *De Magnete* (*On Magnetism*), and apparently read little if any of the work of his famous Italian contemporary, Galileo. Wherein, then, lies Bacon's claim to a position as a major figure of early seventeenth-century science?

For an answer to this question, we must look to Bacon's early educational background. In 1573, as a boy of twelve, he was enrolled at Trinity College, Cambridge, the same institution of higher learning attended by Newton nearly a century later. Bacon informs us that he became thoroughly disgusted with the scholastic Aristotelianism then dominant within the University, because it had led only to meaningless disputations, and had for several centuries blocked the discovery of new and more valuable knowledge, especially in the area of natural philosophy. He decided to undertake the more practical study of law and left Trinity two years after his arrival. Gradually, over the course of the next several years, Bacon, as Galileo, came to the conclusion that mankind's greatest need was a new method for revealing the knowledge stifled by the arcane instructional techniques and strict regimentation of the archconservative Schoolmen. He eventually proposed an alternative manner of investigating natural phenomena, and by so doing became the first major advocate of what we now refer to as the inductive method of science: the collection of data by observation, the drawing of general conclusions from it, and the testing of these generalizations by further observation. It was Galileo, however, not Bacon, who first put this method, or I should say a modified form of it, to practical use in a systematic way. Bacon forever remained the theoretician. Nevertheless, the significance of his advocacy of the inductive method is this: he provided English science with a strong utilitarian motive which directed attention away from the ultimate metaphysical explanations pursued by the medieval Aristotelians and toward a descriptive knowledge of natural phenomena. Science, when developed along Baconian lines, would give man the capacity to control nature for his own benefit, a restoration of the power he had lost with the fall of Adam.

As Christians, Bacon and his followers among the virtuosi of the next generation were quick to point to what they believed to be the deeply religious significance of the new method: God had indeed created an orderly universe, and He had given to man alone the intellectual capacity needed to reveal its most intimate secrets. Working under the influence of Bacon's methodological approach, the English natural philosophers of the seventeenth century abandoned the ultimate question of "why" things are as they are and began to concentrate instead on the more practical consideration of "how" nature operates. As a result, practical secular learning in England was largely freed of the theological trammels suffered elsewhere in Europe. When, on July 15, 1662, the Royal Society received its formal charter from Charles II, its coat of arms bore the motto *Nullius in Verba** to signify that the Society's purpose was to discover scientific facts, rather than to engage in the exchange of personal opinion and unfounded conjecture. Francis Bacon became the idol of the original Fellows: it has been calculated by historian Robert Merton that upwards of 60 percent of the problems handled by the Royal Society during its first three decades were prompted by practical considerations, whereas only 40 percent were problems of pure science. The membership was intrigued with any and all types of gadgetry and mechanical devices, no matter how exotic. Robert Hooke, who at twenty-seven was appointed the first Curator of Experiments, sometimes presented as many as six weekly demonstrations of new concepts to members, an extremely demanding undertaking even for a man of Hooke's considerable ability. Newton himself gained admission to the Royal Society by virtue of his construction of the first working model of a reflecting telescope, which is still on display at the Society's headquarters in London.

Thus, while it is indisputable that Bacon's standing as a scientist, at least in the modern sense of the word, is quite low, I consider it unfair to judge him by contemporary standards alone. Rather, we must think of a field still in its infancy, a delicate state during which the advisability of applying experimental methods to the solution of practical problems remained very much in doubt. Bacon, whatever his shortcomings, must be credited with the formulation of the inductive method which Newton and many of the most gifted English intellectuals of his generation wholeheartedly embraced. In Bacon, the practical bent of the English scientific mind stands in stark contrast to the theoretical mind of its Continental counterpart. Newton, Boyle, and Hooke, unlike Descartes and Leibniz, asked not for knowledge of the deepest kind—the knowledge

---

* Literally, "at the dictation of no one."

of ultimate causation—but instead sought practical results, which were boldly applied to the solution of new problems in a repetitively inexorable pattern, the very basis of modern scientific thought and progress. They were truly Bacon's heirs, and they employed a simplified version of his complex scientific method to the full.

One might even argue that Bacon died for the cause of modern experimental science, though certainly not as a result of being condemned by any tribunal. On a cold March day in 1626, while driving through the Highgate district north of London, he decided on an impulse to test whether snow would retard the process of decay in fresh meat. He ordered his driver to stop the carriage, purchased a recently butchered chicken, and stuffed it with snow. Shortly thereafter Bacon was seized with a violent chill which brought on the bronchitis from which he died on April 9, 1626.

Francis Bacon was not the only prominent scientific thinker to fill in the gap between the death of Galileo and the maturity of Isaac Newton. Rather, it was a European from the Continent who dominated the intellectual history of the early seventeenth century. René Descartes, French mathematician and philosopher, was born on March 31, 1596, in the village of La Haye (now La Haye-Descartes) in the province of Touraine. René's mother died in childbirth a year after he was born; his father Joachim, a lawyer, was a quite wealthy member of the lower nobility. As in the case of Newton, Descartes was left in the care of his maternal grandmother, while Joachim, who remarried, spent most of his time in the province of Brittany, where he served as counsellor and judge at the local parliament of Rennes. Based primarily on the actions of the mature man, Descartes' biographers have created a picture of a solitary but happy and extraordinarily gifted child.

In 1604, at the age of eight, Joachim Descartes enrolled his son in the recently opened Jesuit Royal College located in the town of La Fleche. The youth was placed under the direct supervision of the institution's rector, Father Charlet, a distant relative and distinguished scholar to whom Descartes frequently referred as "my second father." He spent the next ten years of his life studying at La Fleche which, in Descartes' estimation, was "one of the most celebrated schools of Europe." Yet despite the fact that his admiration for the institution and its faculty was both genuine and deep, he left the Royal College in a state of considerable frustration. The curriculum of La Fleche, as elsewhere in Europe, was centered on the teachings of the medieval Scholastics, and Descartes, no less than Bacon and Galileo, harbored an abiding disdain for their rapidly degenerating system. Unlike his teachers, who looked upon their brilliant pupil as a highly learned man, Descartes thought of himself as ignorant because he had acquired no certainty about anything. He was sustained only by his deep religious faith as a Catholic, which he retained to his dying day.

Endowed as he was with a fine mind and more than adequate economic resources, life came easily to the young nobleman. Perhaps this is why Descartes never took it very seriously. He preferred the role of spectator to the part of an actor on the stage of the world. Indeed, there were many days when he scarcely left the security of his bed, where he preferred to do his meditating with paper on knee and pen in hand. After leaving La Fleche he entered the University of Poitiers, where he took a degree in law in 1616. Then, restless to read from "the book of the world," he enlisted as a gentleman in the Dutch army commanded by Prince Maurice of Nassau, who was opposing the efforts of Spain to recover what had been her richest European territory.

It so happened that Holland was in a period of intellectual and scientific development which exerted a strong influence over Descartes' naturally skeptical yet highly inquisitive mind. He became a close friend of the Dutch doctor and physicist Isaac Beeckman who, we are told, was struck by the ability of the young French officer to solve an elaborate and hitherto baffling mathematical puzzle within a short time. Beeckman so encouraged Descartes' interest in mathematics and physics that the young thinker began to seriously consider the possibility of formulating a single method of thought with which to reveal all the facts and secrets of the universe. This concept was foremost in his mind when, in 1619, he returned to France where he underwent his famous mystical experience, the memory of which came to dominate his intellectual life.

Though the facts remain clouded, Descartes, then aged twenty-three, tells us that on the night of November 10, 1619, he experienced what can only be described as an illuminating flash of mystical insight. During a few brief but intensely overpowering moments, he became convinced of the unlimited potential of the mathematical method by which all possible human knowledge could be linked together into an all-encompassing wisdom. Here, at last, was the necessary proof for his gradually deepening conviction that mathematics in general, and geometry in particular, holds the key needed to unlock the vault wherein nature's most elusive secrets lay stored. So profound was his sense of revelation that he undertook a pilgrimage to the Lady of Loreto in Italy to offer his personal thanks to God, who alone could have provided him with such insight.

The author of a number of important books and essays, Descartes' new method was most simply and elegantly expounded in his first work on philosophy, the revolutionary *Discourse on Method*, a brilliantly articulated introduction to modern thought. In the *Discourse* he takes the position that the first step in the exercise of reason requires that the individual clear his mind of all the ideas, facts, opinions, and prejudices acquired over a lifetime of interaction with other human beings. In other words, Descartes advocates the adoption of a skeptical attitude toward knowledge as the only acceptable basis for fruitful inquiry. One must begin, he writes, with the most fundamental of all questions: How do I know I exist? Is it, asks Descartes, because I am essential to the day to day operations of the world and universe? No, because the machinery of the cosmos would continue to function were I and all other living things to suddenly disappear from the face of the earth. Is it, then, because I can see, and hear, and touch, and smell? His answer is again in the negative because, as Galileo demonstrated on several occasions, the senses can and do lie. Besides, a person who cannot see or hear still believes that he exists. The principle that ultimately convinces Descartes of his own existence is his ability to think. He is the author of the famous aphorism, "*Cogito, ergo sum,*" or "I think, therefore I am." Thus having arrived at a clear and definite conception of his own existence, it becomes the first principle from which the mathematician–philosopher will deduce all that follows.

Descartes' deeply ingrained skepticism next caused him to ponder a question of equal importance to a seventeenth-century Christian thinker: Does God exist? Descartes concludes that he can no more reason out God's existence by using his senses than he could employ them to prove his own. God cannot be seen, tasted, touched, or smelled and, even if He could, such evidence is totally without objective foundation from the Cartesian point of view. But, Descartes observed, I am a doubting creature, and having doubts makes me imperfect. Since I am imperfect, I am not capable of fostering the idea of perfection. Only a perfect being could have instilled this idea in me. Therefore, since I know of perfection, God, the perfect being, must truly exist. By thus reasoning out his own and God's existence, without taking into consideration anything about the external world, Descartes reached the conclusion that the mind is separate from the material body which imprisons it. Moreover, the mind is superior to the things of the corporeal world because it does not need material objects to testify to its own existence. It stands alone, isolated from the world of tangible forms.

Having rejected the use of the senses as a very reliable method of understanding nature, Descartes wholeheartedly embraced the abstract and objective world of mathematics. The first intensive studies in which he submerged himself after the mystical experience of 1619 were in the field of geometry. His efforts were rewarded within a very few months by the invention of a crucial mathematical tool—analytical geometry— which proved far easier to wield than the ancient geometry of Euclid. Without it Newton could not have discovered the law of universal gravitation or written the *Principia*. Descartes knew, of course, that mathematics is no more than a complicated system of symbols, but he also knew that it is the only symbolism which totally resists the attempts of the mind and the senses to transform or shift its meaning. It is exact, and by being exact it is totally self-correcting. It does not lie; it does not confuse. To the dedicated Cartesian, the only truly scientific knowledge is always mathematical knowledge.

In pursuance of this design, Descartes took upon himself the task of describing the happenings of the physical world in a strictly mathematical language. He believed that all bodies have moved in perfect order since the creation of the universe. There is no spontaneous motion; everything continues to function in strict accordance with fixed principles established by God. Unlike many of his contemporaries, who held that all matter is composed of tiny particles or atoms, Descartes regarded space as an infinite plenum filled with an all-pervading fluid, or ether, which is continuous and has unlimited divisibility. Within this subtle and infinite matter there is to be found a denser matter of globular clusters which make up the material bodies with which we are familiar, including the sun, planets, and stars. At the creation, God formed the all-pervading "first matter" into innumerable whirlpools or vortices, providing each planet, star, and comet with its own eddy in which it is spun round and round, much as a piece of wood caught in the current of a swift-flowing mountain stream. The size of the vortex depends upon the size of the body it must move: the rotating sun, for example, is surrounded by a much larger vortex than are any of its companion planets. The only forces exerted between the individual bodies are those produced by the spinning of the vortices of matter in which they ride. When an object, no matter the size, is dropped to the ground it falls because the small vortex surrounding it is drawn toward the far larger and more powerful vortex encompassing the earth.

When viewed from the Cartesian perspective, the world and universe become amenable to rigorous geometrical analysis. All bodies, no matter what their size, shape, or position, yield certain objective data. Of these quantitative attributes, Descartes is most concerned with the spatial dimensions, or what he called the "extension" of bodies. By extension he means the amount of space occupied by a given object based on the calculation of its volume. This is done by reducing it to a mathematical formula. In addition to being extended, or taking up a calculable amount of space, all bodies in the Cartesian universe are in motion. And motion, as Kepler and Galileo so clearly demonstrated, is also subject to quantification. Time itself is nothing more than the measurement of objects in motion. Descartes truly believed that only through the mathematization of the universe can man understand it in any objective sense.

The capstone of Descartes' geometrical analysis of the physical world is contained in his famous philosophical doctrine of duality: the *res extensa* and the *res cognitans*. The *res extensa* is the geometrical world of bodies in motion—each a part of space and each with a limited spatial magnitude—knowable only in terms of pure mathematics. What is more, it is totally mechanistic in character. To Descartes, nature is a vast machine whose movements are predictable and caused of themselves by the mutual attractions and repulsions of its whirling vortices. There is no spontaneity at any point; all continue to operate in fixed accordance with the principle of extension and motion. This means that the external world employs no explanatory principles other than those of mechanics; geometric concepts such as size, shape, and quantity. Once God created the innumerable parts of the universe, He set those different parts in motion relative to each other, and for all time. Because God is perfect, His creation is also perfect, without need of further refinement, or even occasional repair. The fundamental laws by which His magnificent creation operates only await discovery by the rational creature—man. Descartes' was a description of the universe which, as we will soon see, proved far more mechanistic than Isaac Newton's.

In likening the universe to a great machine, nothing resisted Descartes' mechanical–mathematical explanations; nothing, that is, except the thought of man himself. Since the human mind could not be brought into any meaningful relation to Descartes' mathematical analysis of the physical world, he found it necessary to create a special subjective domain for it—the *res cognitans*. All those processes which take place in the mind, and are not expressible in mathematical terms, are assigned to this category. These include qualities such as color, texture, warmth, cold, and, of course, human emotions. Man is therefore distinguished from extended matter but, except for his mind, like other animals is a machine.

We note certain similarities between Cartesian dualism and Galileo's division of phenomena into the categores of primary and secondary. But there is a fundamental difference between their positions. Galileo, while distrustful of sense experience when unsupported by quantitative data, never thought of renouncing empiricism, for to him it was an indispensable tool of scientific investigation. Descartes, on the other hand, wholly renounced the empirical method in his attempt to answer all scientific questions in the language of mathematics. The senses and the information derived therefrom are given no status in Cartesian methodology; all "secondary" qualities are written off as the fruitless product of subjective speculation. Had Galileo been this rigid in his thinking, he would never have constructed his telescope and turned it toward the distant stars.

The rigidity of Cartesian methodology notwithstanding, it would be difficult to overestimate the influence it exerted on the history of modern Western thought. The *Discourse on Method* left an indelible imprint on the work of the most important physical scientists of Descartes' and later generations, especially that of Newton. For Descartes' was the first comprehensive attempt to formulate a picture of the whole external world in a manner fundamentally different from the Aristotelian–Christian view. The new and infinite cosmos he envisioned was no longer spun round by the right hand of God, who in Descartes' system is relegated to the position of creator alone. The happenings of the physical world are self-perpetuating and eternal, nature is a vast machine, whose every movement is regular and predictable. If ever the cosmology of a single individual was deserving of the title "world-shaping," Descartes' most certainly is. As a result of his radical break with the static view of the universe accepted by the ancient and medieval worlds, the scientist of modern times has been elevated to the position once held by the Creator, that of absolute law-giver. "By turning man into a machine made by the hands of God," wrote Lewis Mumford, Descartes "tacitly turned into gods the men who were capable of designing and making machines." One must add that the mechanization of the world picture, with its idealization of mechanical progress, also gave birth to an unjustifiable sense of moral superiority in the West, an attitude that still prevails despite mankind's terrible misuse of technology in the twentieth century.

Ironically Descartes' demise, like Francis Bacon's, was indirectly the result of his commitment to the new scientific spirit of the age. Queen Christina of Sweden, a patron of the arts and collector of learned men for her court, had read some of the now famous philosopher's works and initiated a correspondence with him. She extended several invitations to Descartes to visit her country, but he put her off time and again. Finally, he could respectfully decline no longer, and set out for Stockholm in the autumn of 1649. His chief obligation at court was to instruct Christina in philosophy, a simple enough undertaking, except that tutorial time was set at five o'clock in the morning. Descartes had always been used to rising late of a morning, thus the early hours and rigorous climate soon took their toll. His health steadily declined, and in February 1650 he caught a cold that turned into pneumonia. He died a short time later, a month before his fifty-fourth birthday.

The main contribution of Cartesianism to the new world view lies in the fact that its founder became the first to develop a comprehensive scheme of a universe governed by immutable laws which could be reduced to mathematical formulae. But this method only became fruitful when combined with the spirit of empiricism fostered a generation earlier by Galileo and Bacon. The crucial synthesis of these respective methods was first accomplished in England during the early decades of the seventeenth century by a group of thinkers commonly called the virtuosi.

To us the "virtuoso" is a musician whose masterly ability or exquisite technique makes him a brilliant individual performer. But the use of the word in its present context is a development of the eighteenth century. During the period that most concerns us, the mid-seventeenth century, the term "virtuoso" was most commonly applied to the practicing scientist—or, in the words of Robert Boyle, to "those that understand and cultivate experimental philosophy." Yet, as Newton scholar R. S. Westfall has shown, the English virtuosi were a far less homogeneous group than are the scientists of today: "a virtuoso," writes Westfall, "might answer to other names as well." [11] Many, like the Astronomer Royal John Flamsteed, were practicing clergymen, while others were doctors, government employees, or private entrepreneurs. Because science had not yet become the institutionalized and professional activity with which we are currently familiar, only a very few individuals, meaning those of independent means, were able to devote their full time to its pursuit.

Nor, according to Westfall, was a single version of the world accepted by all of these men. "There were as many mechanical conceptions of nature, as there were virtuosi, and no two presented exactly the same problems." Indeed it could hardly have been otherwise given the fact that mechanistic philosophy was still very much in its formative state. There was, however, fairly general agreement on a common goal. "With varying success according to their different intellectual powers the virtuosi tried to construct a theory picturing nature as a machine running by itself without external aids, a machine which human science could study and comprehend." [12]

The adoption of Descartes' mechanical ideal of nature necessitated, among other things, a major revision of the religious tradition inherited from the late classical and early medieval periods. In order to better understand this development, it is important to keep in mind the fact that in the history of Christianity the rational basis for a belief in God has taken two divergent forms. On the one hand, many early Christian thinkers were most inspired by the apparent unpredictability of a world they did not and could not understand in the modern physical sense. Such seemingly spontaneous occurrences as plague, famine, and flood could only be satisfactorily explained, as in the Bible, by postulating a God who caused "unnatural" things to happen, and for reasons beyond the power of human comprehension. On the other hand, Christians of the modern period have generally based their belief in God on the natural order and harmony displayed in the physical world, as demonstrated by scientific inquiry. Implicit in the concept of an orderly universe is the idea of an intelligent and purposeful Creator, one who only rarely defies nature's laws by intervening in the historical process. The virtuosi were among the very first to exchange the belief in a persistently interventionist God for that of a Deity who only on rare occasions acts directly upon nature and man.

The fact that this shift in thinking occurred in England was by no means an accident. Henry VIII broke with the Church of Rome during the sixteenth century and, by 1650, Protestantism had become the dominant religion of the country. Seventeenth-century English Protestant thought was most profoundly influenced by the religious teachings of the Swiss theologian, John Calvin (1509–64). Calvin and his Puritan followers in England seriously questioned the traditional Roman Catholic belief that God and His celestial agents continually intervene in the historical process by performing miracles or directing the day to day events in the lives of individual men. Only the miracles of Biblical times were accepted as part of Divine revelation. All others were generally regarded as the capricious illusions of zealots or the fabrications of corrupt and willful clergymen. This meant, in effect, that the Calvinist God was more remote than the active and frequently wrathful Creator of the Old Testament. In this sense, at least, He was like the watchmaker God of Descartes' mechanistic universe, a master creator who designed and built the world of nature and then quietly receded into the background. It was this vision of God that most strongly appealed to the virtuosi, as is evidenced in the religious thought of clergyman–mathematician John Wallis, whom Westfall calls "the most rigid Calvanist among the English scientists." In his sermons Wallis presented "the image of God as arbitrary will laying down eternal and unchangeable laws for the creation," a Deity who "seems ready to merge with Robert Boyle's portrait of the Divine Mechanic constructing His inexorable machine." [13]

Calvinist religion was also known for its deeply ingrained sense of skepticism, not to mention suspicion. Faith alone did not sustain the Puritan mind, which demanded that religious doctrine of a fundamental nature be underwritten by reasonable evidence attesting to its authenticity and authority. This view, too, had a strong appeal to the naturally doubting mind of the English virtuoso, who was in quest of rational answers to the fundamental questions he daily addressed to nature. Neither should we overlook the hubristic side of Puritan thought. The Calvinists believed that God, even before He created man, had chosen those few upon whom He would bestow the ultimate blessing of eternal salvation. It is reasonable to assume, I think, that intellectuals of the caliber of a Boyle, a Hooke, and a Newton believed, at least in private, that they, too, had been specially selected by the Almighty to undertake the exalted function of revealing the fundamental laws of the universe to their fellow men. Each of them had a specific mission to perform, and they undertook their divinely ordained duties with unflagging zeal and devotion.

There are a number of other points of conjunction between Calvinist thought and the philosophical outlook of the virtuosi, but they need not be pursued here. Suffice it to say that there can be little if any doubt that the new spirit of Puritanism—vague and poorly defined though it was—exercised a crucial, if not deciding, influence on the pursuit of modern science in England and its reconciliation with Christian thought.

Yet in at least one important respect it would have been quite natural had the virtuosi turned their backs on the Christianity of their medieval forbears. For as the telescope of Galileo and other astronomers probed ever deeper into space, it became evident that the solar system inhabited by man is but a minute speck when compared to the Milky Way, which in turn appeared to be part of a far vaster system of apparently endless stellar clouds. By virtue of Descartes' infinite extension of space and time, man, with his pitifully limited span of existence, seemed frightfully insignificant. Indeed, did not this colossal magnification of the cosmos abolish man's claim to being of central importance? Was it not possible that God no longer cared enough about man to watch over and protect him? The virtuosi, unlike the Deists of a century later, answered these and similar questions in the negative, because from their point of view the physical universe, no matter how grandiose, is unable to behold itself, except through the eyes of man; unable to speak for itself, except through the human voice. And most important of all, it is unable to know itself, except through the intelligence of the human mind. God may not have placed man at the physical center of His creation, as St. Thomas and Urban VIII both argued, but intellectually man remained absolutely central to God's plan—the only known sentient being capable of shedding light on the benign darkness and dumbness of infinite space. It was this reevaluation by the virtuosi of man's relationship to God and the physical world of His creation that provided the sanction for the modification of Christian principles in order to bring them into harmony with the new science.

Exactly what form, we must ask, did this modification take? To begin with, the virtuosi based their belief in God upon the apparent orderliness of nature, rather than upon any supposed divine intervention of a miraculous kind. They were convinced that the Almighty had planned and executed the creation of all that exists. Once set into motion, nature, as Descartes had concluded, is governed by the unfailing rule of natural and eternally immutable laws which, in and of themselves, provide over-whelming evidence of God's rational purpose and design. The supreme challenge facing man is that he must use his reason to determine exactly how the vast machine called the universe operates. And once having understood its operation, he must employ that knowledge to control nature for the benefit of his own kind. Here the Baconian legacy truly begins to assert itself by turning the natural philosopher's attention away from the scholastic concern with questions of ultimate causation and toward a utilitarian knowledge of natural phenomena.

Indeed, the more the virtuosi examined the world around them, the more they were awe-struck by nature's incomparable wonders. Tangible evidence of God's presence abounded; He was more and more worshipped through the glories of His creation than in the words of the ancient

Scriptures. Thus, a natural religion—based on the empirical examination of physical phenomena, or what Galileo referred to as the "Book of Nature"—arose to challenge the institutionalized theology of the past. "The search for design in nature," comments anthropologist Loren Eiseley, "soon became a mania and everything was made to appear as though created specifically to serve man. There were Bronto (thunder) theologies, Insecto-theologies, Astro-theologies, Phyto-theologies, Ichthyo-theologies, Physico-theologies." Intellectually man stood at the center of all things—"the entire universe had been created for his edification and instruction: hills had been placed for his pleasure, animals ran on four feet because it made them better beasts of burden, and flowers grew for his enjoyment." [14] In brief, the religion of the virtuosi had become externalized, focusing itself ever more sharply on the manifestation of divine purpose in the physical world, while neglecting to an ever greater degree the internal world of the spirit. This was possible because in Protestant England science had begun to generate an irresistible momentum of its own, a force destined to become so powerful that it no longer needed to accommodate itself to institutionalized religion—rather institutionalized religion gradually became subservient to science. Watching this historic process unfold once again reminds one of the poet's penetrating insight: "Old gods wither on the stem, but new gods will arise from them."

To better understand this crucial development, we will pause briefly to explore the thought of chemist, physicist, and theologian Robert Boyle (1627–1701), the scientist who is most commonly remembered for the law that bears his name: the principle that at a fixed temperature the pressure of confined gas varies inversely with its volume. Of all the Christian virtuosi, it was Boyle who most strongly influenced the scientific and religious outlook of Newton and his generation. Boyle was both an early convert to the Baconian philosophy of experimentation and a supporter of Descartes' mechanical cosmology. He firmly believed that man inhabits a universe harmoniously governed by natural laws which can be discovered by intensive investigation. In fact, it was Boyle who popularized the belief, dating at least as far back as the writings of Cicero, that nature can be best understood by comparing it to a finely constructed clock. Time and again he referred to the intricate time-piece located on the face of Strasbourg Cathedral to illustrate this point. The huge machine, composed in turn of many lesser machines, seemed to function as though it somehow understood the purpose of its movements; yet, as Boyle pointed out, neither the Strasbourg clock nor the clocklike universe could comprehend what they were about. It was man's place and man's alone, to explain both.

On the other hand, Boyle stopped short of accepting Descartes' total mechanization of the natural world. He was especially loath to limit

God's role to that of Creator alone. Such a view of the Almighty seemed highly dangerous to Boyle, because once the creative act had been accomplished God became nothing more than an absentee deity. The laws according to which the universe operates, by virtue of their immutability, would, in effect, be more powerful than the supreme intelligence who formulated them, something Boyle could not accept; nor, one must add, could Newton. Though the latter's name has become synonymous with the idea of a completely mechanized universe, this is an inaccurate analysis (some would say a perversion) of Newtonian thought: one for which we can thank the Deists of the eighteenth century.

In Boyle's conception, spiritual nature is nobler than the physical, and moral character more valuable than mechanical power. Hence, science is necessarily of a lower degree than religion. It is, of course, entirely permissible for a Christian to pursue the study of nature; in fact, it is his moral obligation to do so. God has revealed Himself to man in nature by the obvious efficiency of the creation and the inconceivability of its being the result of fortuitous accident or blind chance. But God has also given to man an even more authoritative revelation of His purpose of creation in the Bible. To implement that revelation He has not hesitated to temporarily change or altogether stay the operation of physical law, whether by fiat or because of the legitimate petitions of pious men. Admittedly, such supernatural acts were most commonly performed in the early stages of human development—the period during which the Bible was written—but this is no guarantee that God will not set aside natural law in the future. Certainly, He has never promised man that He would not do so; thus it would be foolish of man to deny God's potential to act whenever and however He might choose.

Boyle did not advocate a return to the medieval position wherein man was far more impressed by the "freakishness" of nature than by its harmonious regularity. Yet he clearly recognized what many complacent apostles of modern science and technology have apparently long forgotten: that man is finite in his apprehension. Though he may attempt to grasp the entire scheme of things, as did the Cartesians, an ultimate knowledge of all phenomena and their underlying causes is beyond him. Boyle accepted the mechanization of nature for what it was—a challenging and enlightening method of looking at the world—but to him it was only a process within a much greater and divinely ordained master process, one which man did not and never would comprehend. In his attempt to establish the basis for a new scientific methodology with which to gain command over nature, man must be continually reminded of his limitations; he must not take upon himself the aura and trappings of the gods.

To support his point of view, Boyle maintained that one should not be deceived into thinking that simply by applying the word "law" to inanimate processes that they will inevitably behave in accordance with certain precepts. Bodies such as planets, for example, need Providential care merely to persist in their movements. God is the cause of motion as He is the cause of all things. Granted, the Almighty speaks to man in mechanistic and mathematical terms, but this does not mean that the clock of nature has no need of constant observation or occasional readjustment and repair. Moreover, as Newton would so well demonstrate, the discovery of a scientific law tells us "how" a specific phenomenon like gravity works but not "why" it works. Nature indeed functions like a great machine, but it must never be forgotten that God governs nature —and in ways frequently mysterious and totally incomprehensible to men.

To Boyle, then, man is obliged to gain a better knowledge of the world in order to be better able to glorify God. He accepted Descartes' mechanical conception of the physical world, but rejected the French philosopher's belief that matter can be made totally independent of the Creator. Boyle also refused to read man out of nature, as did Descartes, or belittle his importance, as had Galileo. Man, in Boyle's view, is God's child and this gives him a unique dignity not grasped by the arch-mechanists. Descartes made the mistake of attempting to design a single, all-encompassing conception of the universe, but this is not how science and technology are advanced. Boyle, like Newton, preferred the empirical method of Bacon with its marked utilitarian strain. To them it was far better to have some knowledge based on experiment—no matter how incomplete and fragmentary—than to indulge in the formulation of grandiose theories of the Cartesian kind.

Still, Boyle's attempt to frame a view of the universe that would accommodate both an active God and the new science enjoyed only limited success, even though he was the most respected and widely-read virtuoso of his day. The irresistible process of secularization, unknowingly set in motion by Copernicus, was already well advanced by the middle of the seventeenth century. More and more Christian beliefs were adjusted to conform to the discoveries of the natural philosophers: each time the cause of science advanced, that of religion retreated. After Boyle's death, fewer and fewer of the virtuosi were willing to give more than lip service to the assertion that God can overrule the inexorable mathematical laws of His creation. And even if God did possess this power, it was increasingly felt that He would be irrationally contradicting Himself were He to exercise it. It would be erroneous to assert that the English virtuosi had arrived at the point of looking upon the creation as God's first and *only* intervention in the operation of the natural world. This they left to the Deists of the next century. But they shared a growing confidence in the power of human reason, a confidence that inclined them more and more toward worshipping God through His creation. As a consequence, natural religion became primarily a response of the intellect to the physical world, rather than a concern for the welfare of the inner spirit. Bacon and Descartes had left deep marks on the thought of seventeenth-century scientists: on the theologians they left ineradicable scars. It was Isaac Newton's destiny to reach his intellectual maturity in the midst of this most challenging yet highly unsettled age.

Up to Cambridge

Newton gained admittance to Trinity College on June 5, 1661, and matriculated on July 8 at the age of eighteen. Despite a modest inheritance, he entered Cambridge as a Sizar, a title that probably derived from the "size," meaning the amount of bread and drink allotted to its holder. In Newton's day, it was applied to those students who paid for their tuition and board by performing such commonplace tasks as waiting on tables, doing odd jobs, and running errands for their tutors. His lowly status in a highly class-conscious institution immediately set him apart from the majority of his fellow students, who had either been educated at the great public schools or privately tutored in the homes of their wealthy families. Though Newton always remained a man of very modest tastes, he secretly envied men of position and wealth. When, in his mid-fifties, he decided to resign his professorship at Cambridge, it was to accept the more prestigious government appointment as Warden of the Mint. It would appear that one of the primary reasons for this action was his desire to overcome the sense of social inferiority that had haunted him since his earliest days at Cambridge. This also explains, at least in part, why he so thoroughly enjoyed his role as the autocrat of English science during the last quarter century of his life. Neither must we forget that Newton was far more studious and introspective than most young men of his age. The rural environment from which he came did not place a very high value on the social graces cultivated by the rich, least of all the art of polite conversation and repartee. Rather it emphasized a reverence for God, a respect for hard work, and a deep sense of moral duty. By contrast, the morals of the Restoration period were, as More observes, notoriously loose and clashed with Newton's "strict religious training and temperamental disgust for all forms of moral laxity." This, too, contributed significantly to the young man's estrangement from his peers.

The accounts of all facets of Newton's undergraduate days are far less complete than the historian would wish, but we know considerably more about his academic pursuits and intellectual development than we do of his personal life. As was the custom throughout Europe, the focus of his preparation for admittance to the University centered mainly on the classics. Latin was of special importance because it was the basic language of scholarly books, conversation, and correspondence. Newton also learned some Greek, but he did not master the subject while at Grantham. His introduction to French was even more cursory, and he always read the language with a dictionary in hand. He seems not to have known German at all. Hebrew was in fact considered the most important language after Latin, and one of Newton's early notebooks illustrates that he had an acquaintance with the Hebrew characters. His eventual mastery of the language later proved invaluable in furthering the Biblical studies which, next to science, occupied the majority of his time. He also received lessons in ancient and Biblical history, grammar, and Christian exegesis. However, his preparation in mathematics, the field in which he made some of his greatest contributions as a scientist, was slight indeed. He learned arithmetic and perhaps a little geometry but had not yet been introduced to Euclid. The mathematical problems of his childhood, which biographers have frequently pointed to as a foreshadowing of his latent genius, were most probably nothing more than simple arithmetical calculations incident to his mechanical inventions. Henry Stokes' interest in his intellectual development notwithstanding, Newton was not very adequately prepared when he entered college, where he was placed in direct competition with students whose knowledge was considerably in advance of his own.

On the other hand, Stokes had not erred in his judgment that Newton's intellectual gifts were considerably beyond the ordinary. Indeed, one of the significant features of his early intellectual development is that he attracted and held the attention of highly distinguished older men, scholars who seem to have sensed the promise of something inexplicable within him. Of Newton's admirers among the Cambridge intelligentsia, two men in particular exerted a great influence on his thinking, Platonist Henry More and mathematician Isaac Barrow. More, who was himself a native of Lincolnshire, had been born at Grantham, the son of a staunch Calvinist. Though a mechanist and admirer of Descartes, he became the leader of the Cambridge-centered religious reaction to the acceptance of a full-blown Cartesianism. More shared Robert Boyle's fear that Descartes had come too close to divorcing God from the world order, a process that might eventually lead to the growth and spread of atheism. He thus rejected the concept that the phenomena of the world can be understood in mechanical terms alone. Mechanistic principles, More observed, are of great value to the natural philosopher when he wishes to discover the *immediate* cause of a phenomenon such as planetary motion; but the ultimate reason why the individual parts of the universe move rather than stand still, cannot be satisfactorily accounted for by envoking the mechanical doctrine. In order to maintain the concept of a providential order pervading the universe, More and his fellow thinkers rehabilitated the ancient Platonic idea of the *anima mundi* (the spirit of nature) at work in the world. This plastic, incorporeal spirit, they believed, permeates the entire physical matter of the universe and is the true cause of motion and other natural phenomena: it is the medium through which God, an active Agent, governs His divine creation. In taking this stand against the Cartesian demotion of God, once He had fashioned the universe, to the role of a passive spectator, More advocated an order of things that is direct and specific rather than indirect and general. Though Newton, while an undergraduate at Trinity, was deeply attracted to Descartes' mechanistic philosophy, his devotion to Cartesianism, thanks to More, always remained tempered. In his vision of the universe, Newton never seriously contemplated the abandonment of the interventionist Deity.

Isaac Barrow, the brilliant mathematician cum theologian, and a teacher with whom Newton became associated during his student days at Cambridge, occupied the newly created academic chair of Lucasian Professor. Born in London in 1630, and only twelve years Newton's senior, Barrow was educated at the famous Charterhouse. From there he went to Cambridge where he earned the Bachelor of Arts at Trinity in 1648, becoming a Fellow of the College a year later. Forced by the Independents into a European exile in 1655 because of his religious and political views, Barrow returned to Cambridge in 1659 on the eve of the Restoration. He was then ordained an Anglican minister and appointed professor of Greek. In 1662 he became professor of geometry at Gresham College, but he returned to Cambridge within a year, where he had been chosen to fill what eventually became one of the most distinguished chairs of science in the world. Barrow was a witty, outgoing, athletic man who traveled widely and relished the heat of combat, both physical and intellectual. Indeed, in many ways the contrasts between the two Isaacs could hardly have been more pronounced, as is illustrated by the following passage from Manuel:

> Newton was a Whig and a unitarian,* Barrow a Royalist and a devout Anglican in holy orders who became the King's Chaplain when he vacated the Lucasian Chair. Newton rarely used images in his writing and cultivated a sparse style; Barrow's prose is full of conceits and literary inventions, rich with classical allusions which he savored for their own sake. Newton, who was shy in public, was not noted for his oratory; Barrow was one of the great preachers of the Restoration, whose sermons were collected and published in voluminous editions. He was universally loved

---

* Newton remained a member of the Anglican church all his life. Privately, however, he rejected the doctrine of the Trinity on the Arian grounds that it was unreasonable to assume that Jesus was of the same substance as God. He believed that Christ was only the highest of created things.

in the College, unlike his successor, who was never too popular with his colleagues even when they elected him to Parliament. Newton hardly ever acknowledged the intellectual worth of others; Barrow could praise his fellows profusely and with conviction.[15]

Yet however great their differences of style and temperament, the two men were united by their shared gift of scientific genius. Barrow, though no match for Newton, was one of the most talented mathematicians of the seventeenth century—the supreme age of mathematics. His contribution to this revitalized field of study might have been far greater than it was had he not decided, in 1668, to resign the Lucasian Professorship, both in deference to his brilliant pupil, and to satisfy his deep personal desire to pursue the study of theology on a full-time basis.

Just how closely Barrow worked with Newton during the latter's undergraduate days is still a matter of debate among scholars. Despite D. T. Whiteside's view that there was little if any contact between the two until after 1666, I believe that it was Isaac Barrow who introduced Newton to the serious study of geometry, optics, and the emerging calculus, not to mention ancient chronology and church history. If so, this was Newton's first real opportunity to work with a scientist of the highest rank, and the latent intellectual processes that awaited only the proper stimulus burst forth in full flower. Barrow, though brilliant in his own right, must have been astounded and deeply moved, perhaps even somewhat frightened, by what he observed. Newton grasped the most abstruse mathematical and metaphysical concepts with an almost supernatural ease; his mind seemed to require no period of extended preparation or growth—inexplicably, full intellectual maturity came almost at once. No wonder Barrow recommended to his superiors that Newton, at the age of twenty-six, be appointed to the Lucasian Professorship!

To Isaac Barrow and Henry More, then, belongs much of the credit for assuring Newton's entrance into the exciting new world of thought. When he enrolled in Cambridge in 1661, Newton had encountered a curriculum that can be best described as nothing less than reactionary. The notes which he dutifully recorded as a student, in Westfall's words, "looked to the past instead of the future, and one will search them in vain to find a hint of their author's subsequent career." [16] Under Barrow's strong influence, however, Newton read Kepler's *Dioptrice*, which was the best textbook on geometrical optics available; Descartes' *Geometria*, which detailed the new analytical geometry; and Wallis' *Arithmetica Infinitorum*, a treatise which closely foreshadowed the calculus of Newton's own invention. Just as important, he attended several lectures on Copernican astronomy, studied Galileo's works on terrestrial mechanics, and waded deeply enough into the mystical waters of Kepler's esoteric tomes on celestial physics to assimilate the German astronomer's three crucial laws of planetary motion into his own thought processes. Thus, by 1665, when he took his B.A. and shortly before he was driven from Cambridge by the Great Plague, the young genius was already in command of the most recent advances in the fields of astronomy, physics, mathematics, and optics. Yet little did Newton dream that he was about to make a number of scientific discoveries so profound in their implications that they would revolutionize the study of each of these fields.

The New Moses

One fateful day toward the mid-fourteenth century, probably late in 1347, an Italian ship, laden with precious cargo from western Asia, reached its home port on the Mediterranean coast. Unbeknownst to its captain and crew, a stowaway lay hidden far below decks where the ballast and spare rigging were stored. Yet even had one of the sailors chanced to catch a fleeting glimpse of this unsolicited passenger, it seems doubtful that anything would have been done; for the intruder and his kind had been the constant, if unwelcome, companions of seamen almost since the dawn of sailing itself. Despite outward appearances, however, this was no ordinary ship's rat, because hidden beneath its dark fur was yet another and far more dangerous stowaway, a flea, whose tiny body carried deadly bacteria of the genus *Pasturella*. Not long after the ship had docked, a contagious and usually fatal disease swept northward out of Italy, quickly spreading over the entire European continent. The horrible sickness soon had a name—the Black Death—which derived from the hideous dark splotches it produced on the victims' skin: other symptoms included fever, chills, vomiting, diarrhea, and buboes, acute swellings of the lymphatic glands.* Before the disease had run its merciless course between one-fourth and one-third of the entire population of Europe had succumbed. Successive outbreaks were also commonplace during the fifteenth and sixteenth centuries, but these became less frequent and severe during the 1600s. In fact, the Great Plague of 1664–1665 made its appearance just when it seemed that Europe might be spared another major wave of the dreaded contagion.

It is ironic but true that this seemingly inexhaustible source of human suffering should have contributed, at least indirectly, to an improvement in the political and material well-being of those fortunate enough to elude its ubiquitous grasp: The wholesale loss of life, no less than the rise of towns, helped seal the fate of the repressive feudal system. The Black Death produced a severe shortage of labor and thereby enabled the remaining serfs to substantially strengthen their demands for greater

---

* Hence the modern appellation, bubonic plague.

freedom and economic rewards. With a free peasantry, the manorial system could no longer operate.

Strange and perverse as it may seem, the proponents of the new science also became the indirect beneficiaries of an unexpected recurrence of the plague in England. The disease started in comparably mild form in London in 1664, but by September of the following year it had reached epidemic proportions. Neither the exact cause nor its cure were yet known, but experience dictated that people were most susceptible when living in large numbers in close proximity to one another. In the summer of 1665, the decision was made to close Cambridge University; Newton, who was already vacationing at Woolsthorpe, did not return until March of 1666, when it was thought that the epidemic was over. However, this hope was quickly dashed by a new outbreak which forced him back to his Lincolnshire home in June; there he remained for almost another year, until April 22, 1667.

So magnificent were the twenty-four-year-old Newton's scientific achievements during his enforced retirement that historians usually refer to this crucial period in the history of science as the *annus mirabilis*, the marvelous year.* We will, of course, never know whether Newton would have made all of the discoveries credited to him during this startlingly brief period had not his plans to pursue advanced study been temporarily interrupted. Obviously, his intellectual endowment was too overpowering not to have asserted itself, no matter where he might have chosen to reside. Neither would I want to leave the impression that I believe Newton's contribution to the scientific revolution would have been less profound and far reaching had not a bubonic-infested rat, bored with life in the Orient, chosen to make its death-dealing way to Europe in 1347. On the other hand, true creativity in the sciences, no less than in the fields of art, music, and literature, comes when the individual is in an esthetically sensitive mood. Though I have some doubts as to whether Newton was trying to discover anything specific in 1665, his return to the familiar surroundings of his beloved Woolsthorpe and the company of his widowed mother must have created just such an atmosphere. Temporarily unburdened from the demands of academic life, he was free to let his mind wander where it would; to put together disparate and seemingly unrelated thoughts whose pieces were still little more than promising fragments of a giant intellectual puzzle; to meditate upon the apparently insignificant phenomenon of ripened apples falling from the branches of the windblown trees in Hannah's garden; and to watch the clouds race across the pitted coun-

---

* Strictly speaking, it would be more accurate to call this period the *anni mirabiles* because it encompassed two years rather than one. But this in no way detracts from the unparallelled magnitude of Newton's singular intellectual innovation: one year or two, it makes no difference.

tenance of the full autumn moon. And just like the moon-shadowing clouds which quickly disappeared from sight, Newton's thoughts raced on toward the distant horizon of discovery—toward the invention of the calculus, toward the law of the composition of light, and, most important of all, toward the principle of universal gravitation.

Since this essay is primarily concerned with the scientific revolution in astronomy, we will begin our examination of Newton's scientific accomplishments during the *annus mirabilis* with a discussion of his investigation of universal gravitation. On April 15, 1726, a year before Newton's death, Stukeley paid a visit to Sir Isaac at his home in what was then the village of Kensington, now a part of Greater London, where they dined together and spent the day in quiet conversation. "After dinner we went into the garden and drank tea, under the shade of appletrees, only he and myself," wrote Stukeley. "Amidst other discourse, he told me, he was just in the same situation, as when formerly, the notion of gravitation came into his mind. It was occasion'd by the fall of the apple, as he sat in a contemplative mood." Unlike many of the legends associated with the discoveries of famous scientists, Stukeley's account must be allowed considerable weight, I believe; not only because the good Doctor had no compelling reason to concoct such a story on his own, but also because it makes a certain amount of scientific sense when one pieces together Newton's thought processes at Woolsthorpe.

Suppose, for a moment, that Newton, as he told Stukeley, did observe ripened apples falling to the ground in his mother's garden. Suppose also that he began to ponder the cause of their falling and to what height the power of the earth's attraction might extend. Since from the bottom of the deepest crater to the summit of the highest mountain this power underwent no sensible change, might not its action be reasonably extended to the moon? If so, could what Newton called "gravity" be the force that kept the moon in its orbit? Did the moon, in fact, obey the same law as a moving body (the falling apple) on earth? Perhaps so, but the path of an apple to the ground is perpendicular while the path of the moon round the earth is roughly circular.

In an attempt to get a different handle on the problem Newton formulated one of those thought experiments for which Galileo became so famous. We know this because he later gave an account of it in the third book of his *System of the World*, the *Principia*, whose contents we will discuss in greater detail later in this chapter. Newton asked himself what would happen were one to shoot a projectile from a cannon situated atop a hill or mountain. The projectile, he knew, would describe a curve as it flew through the air and then fell back to the ground. Suppose, however, the charge were increased with each successive firing of the cannon. The velocity of the projectiles would thus be increased and they would describe larger and larger arcs, as illustrated in Figure 16. Newton concluded that if one had a large enough cannon (a rocket?) and sufficient powder (fuel?), it would be possible to fire a projectile with such force that it would no longer fall back to earth but go into orbit round the planet. In other words, it would become a satellite like the moon, its centrifugal tendency or outward pull exactly counterbalanced by its inward fall.* If this theory could be proven mathematically, then the moon could be thought of as a giant apple perpetually falling round the earth's surface. Newton now needed to work out the exact inward pull required to keep a circling body moving with a specific speed in a circle of a fixed radius.

In addition to his brilliant powers of abstraction, Newton, like Charles Darwin, possessed the rare capacity to master and build upon the work of others. He now discovered in the planetary laws of Johannes Kepler one of the most important clues to the riddle that confronted him. Disregarding completely Kepler's mystical attachment to the Platonic solids and the harmony of the spheres, he concentrated his attention on Kepler's Third Law which states: the square of the periods of revolution of any two planets are as the cubes of their mean distances from the sun. In other words, Kepler had shown that the lengths of the planetary

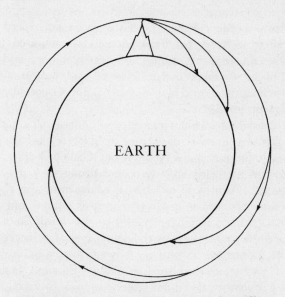

EARTH

FIGURE 16.   The making of an earth satellite.

* At this time in his career, Newton thought of circular motion in terms of an equilibrium of centrifugal and centripetal forces. He would later learn from his great rival Robert Hooke that a body revolving in orbit possesses rectilinear inertia and must therefore be continually diverted from its normally straight path by the inward pull of gravity.

"years" increase as one moves outward from the sun, not in direct proportion to their distances, but in a greater proportion. For example, a four-times-larger circular orbit has an eight-times-larger period and a two-times-smaller velocity; therefore, the centrifugal force is sixteen times smaller, or the inverse square of the distance. The gravitational force compensating for it must vary in the same ratio. At this point Newton concluded that the attraction that draws the planets to the sun must decrease inversely as the square of the distance separating them from the sun. Hence, a planet twice as far away from the sun would require only one-fourth the gravitational force to keep it in its orbit. He had formulated the inverse square law which is one of the most important among his many great discoveries. He next found that this law which governs the attraction between the sun and the planets also accounts for the intensity of attraction between the moon and the earth, as well as between the earth and an apple. To be more precise: Every body in the universe tends toward every other body directly in proportion to the product of their masses, and inversely in proportion to the square of the distance between them. Newton's "mathematical derivatives," wrote T. S. Kuhn, "were without precedent in the history of science. They transcend all other achievements that stem from the new perspective introduced by Copernicanism." With his inverse square law and his system of fluxions, the calculus, both the shape and speed of celestial and terrestrial trajectories could be accurately computed for the first time. "The resemblance of cannon ball, earth, moon, and planet was now seen, not in a vision, but in number and measurement." [17] The heavens and the earth were truly one at last: the Copernican Revolution had been accomplished!

This all seems so logical, which it was; indeed, so simple, which it was not. In this introductory essay, our inability to pursue the innumerable problems which not only Newton but Galileo, Kepler, and Copernicus had to wrestle with and overcome produces a grave distortion of one of the most profound intellectual achievements in the history of man. Had Newton's act of discovery been as simple and clear-cut as I have intimated, he could have published the result immediately and claimed his rightful place among the immortals of science. But this he did not do, could not do. Twenty years were to pass before he authored the *Principia*, and even then the astronomer Edmond Halley, playing the part of a seventeenth-century Rheticus, had to coax and cajole a reluctant Newton to share his discovery with the world.

Why, exactly, did Newton keep this product of his Woolsthorpe labors to himself, especially when he knew that it would reduce the last vestiges of Aristotelian cosmology to wormwood and ashes? Was it, as some have suggested, largely due to his ingrained distaste for the spotlight, coupled with the fear of arousing controversy within the scientific community? Or was it perhaps because Newton considered the act of discovery sufficient unto itself; he knew what he knew and therefore nothing else mattered? Or did he look upon his discovery in much the same way that the nineteenth-century monk Gregor Mendel looked upon his garden of peas, as beloved children of his own creation? Was Newton the overprotective parent who in his mind's eye conjured up terrible visions of what might happen to his delicate intellectual offspring should they be exposed to the rigors of a harsh and unappreciative world?

We would, I believe, be justified in answering all of the above questions in the affirmative had the *annus mirabilis* been postponed until a later stage of Newton's career. But as yet he knew few other scientists, had never been involved in a major intellectual controversy—scientific or otherwise—nor had any of his claims to the paternity of certain major discoveries been challenged by the likes of Hooke and Leibniz. No, the real reason for the protracted delay in the announcement of the inverse square law is that when Newton tested it, as in the case of the earth and moon, he was not completely satisfied with the result. He commented years later: "I thereby compared the force requisite to keep the Moon in her orb with the force of gravity at the surface of the earth, and found them answer pretty nearly." For most men "pretty nearly" would have been enough, but not for Newton.

The most important of several stumbling blocks arose over the question as to whether or not a body that is more or less homogeneous and spherical would act gravitationally toward another body as if all its mass were concentrated at its center. To illustrate this point, consider once again the case of the earth attracting the apple. Imagine, for a moment, that the entire earth is divided into small particles and that each interior particle of the planet attracts the particles of a body near its surface. The problem is that these attractions are not all of the same intensity because some particles of matter are at a greater distance from the outside than the others. Furthermore, the directions of these attractions vary. Generally speaking, however, those particles nearest the apple will pull immensely harder than those on the other side of the earth, but there will be far fewer of them. The question, in the words of Newton scholar E. N. da C. Andrade is this: "Will the extra number of the bits pulling at a distance just compensate for their greater distance, so that the pull of all the bits adds up to give the same results as if all the mass of the earth were at one point at the earth's center?" [18] The answer is "yes," but it is a subtle problem in the integral calculus. Though Newton had experimented with fluxions in 1665, a year before the formulation of the inverse square law, his mathematical method was not yet sophisticated enough to achieve the desired result. Newton engaged in further calculations in 1671 and again in 1672, but these results, too, showed only general agreement. "Pretty nearly" remained the operative phrase for almost two decades. The final solution to the problem was not forthcoming until 1684, when Robert Hooke, Newton's archrival, claiming the inverse square law as his own, faced Newton (through Halley) with the problem of what the orbit of a body would be if affected by the principle of variation of force with distance. Newton solved the problem within a relatively brief period, something Hooke could not do; the rest is history.

One sure measure of Newton's versatile genius was his ability to consistently devise methods for solving almost any scientific problem at hand. But because such solutions often came so easily, he seldom took the trouble to communicate his advances to the world—an irritating habit of character which, as in the case of Hooke, frequently led to sad and acrimonious controversy. A similar and just as ugly dispute erupted in later years between Newton and the German philosopher–mathematician, Gottfried Wilhelm von Leibniz. Their famous quarrel centered on another of the Woolsthorpe discoveries—the method of fluxions, or what is now known as the differential and integral calculus. Here again, the minds of several of Newton's contemporaries had been at work, including Wallis, Gregory, Fermat, and Barrow. Together they foreshadowed a method by which infinitesimal quantities might be computed and infinitesimal differences in varying quantities measured. Newton composed his first paper on the subject at Cambridge in May, 1665, shortly before the University closed its doors pending the abatement of the plague. It was concerned with the summation of infinitesimal arcs of curves. In it he initiated the practice of placing a dot over an algebraic symbol when it stood for an infinitesimal quantity. He completed a second paper at Woolsthorpe in November of the same year, and three others during the *annus mirabilis* of 1666. However, as was typical, he took no one into his confidence.

Yet, ironically, it was the invention of the fluxions that ultimately led Newton to reject the Cartesian belief, borrowed from the Pythagoreans, that the universe is fundamentally geometrical. Newton realized that concealed behind the geometrical forms that so appealed to Kepler and Descartes were such dynamic concepts as mass, force, and acceleration, none of which could be fully represented geometrically. Unlike algebra, which concerns itself with finding particular numerical values, calculus is based upon the idea of considering quantities and motions not as definite abrupt values but as in the process of originating, changing, or disappearing. It can be used, for example, to calculate the most minute changes in the acceleration of a body falling through a small amount of space, to compute the precise arc of a rapidly moving planet, or to measure the exact rate of deceleration of a ball as it slowly comes to a stop after being rolled across the ground.

All this can be briefly summed up by observing that the calculus is a highly effective tool for rapidly solving problems involving infinitesimal variations in the rates of motion. In a letter to the well-known scientific intermediary, John Collins, dated November 8, 1676, Newton wrote that there is no curved line expressed by any equation of three terms for which he could not, within a quarter of an hour, determine the simplest geometrical figure with which it could be compared. "This may be a bold assertion," he wrote, "but it is plain to me by the fountain that I have drawn from, though I will not undertake to prove it to others." As always, the proof for Newton was sufficient unto itself. Indeed it is difficult, though admittedly not impossible, to think of any changes in natural motion that are not mathematically reducible in Newton's terms. And though in his demonstrations and proofs he made some use of geometrical figures—straight lines, triangles, and rectangles of finite dimensions—he clearly realized that these were merely abstract representations of far more complex and continually changing forms. Thus, while the Cartesian scheme of the universe was pictorial and general, that of Newton was rigorously mathematical and precise.

In the history of science, the most brilliant experimental investigators—Robert Hooke, for example—have only rarely been known as outstanding theoreticians, while the greatest theoreticians, as in the case of Kepler and Einstein, have rarely been great experimental scientists. During the *annus mirabilis* Newton, ever the exception to the rule, demonstrated beyond any doubt that he was a consummate master of both facets of scientific activity. Having examined his prowess as a theoretician in the fields of mathematics and physics, we will now turn our attention to his innovative experimental work in the field of optics.

Strange as it may at first seem, Newton unquestionably devoted more time to his researches in optics during the Woolsthorpe period than to the calculus or the question of gravity. He had probably been introduced to the subject while attending lectures delivered by his mentor Isaac Barrow, who was also in the process of writing a book on optics—a volume which, because of Newton's innovative experimentation with the properties and behavior of light, was obsolete even before the printer's ink had dried upon its pages. Newton was most interested in what happened to light when it was refracted through pieces of ground glass. Ever since childhood he had derived deep satisfaction from working with his hands, and before retiring to Woolsthorpe to escape the plague he had taken up the art of grinding and polishing lenses with machines of his own construction and design. He seems to have been among the first to use pitch for optical polishing and was, apparently, an excellent technician. Like others before him, he observed that every lens, no matter how painstakingly ground, at its edges is a small prism. And in badly ground lenses this splitting up of white light into a spectrum of different colors resulted, as Galileo well knew, in the poor definition and visual distortion so common to the early refracting telescopes. This phenomenon, which is known as chromatic aberration, occurs because rays of light of different colors, i.e., of different wavelengths, on passing through a lens are not brought to the same focus. René Descartes, when a young man, had employed mathematics to determine the proper shape for the surface of lenses in an attempt to remedy this condition. The problem was that the hyperbolic and elliptical lenses he designed were so difficult to grind that neither Descartes nor those who followed his method succeeded in removing all the imperfections of the images. What Descartes did not know is that each ray of colored light has its own index of refraction. Thus, no single convex lens, no matter how perfectly ground, will eliminate all of the disturbing effects of chromatic aberration. The problem can be overcome only by using lenses of differing refrangibility in combination.

To Newton, the phenomena of color and aberration made a strong intellectual appeal, and he became convinced that there was more to learn about them than Descartes or anyone else had suspected. In order to test his theory, he undertook preparations for the famous experiment of 1666, the details of which he later communicated to the Royal Society.

> I procured a triangular glass prism, to try therewith the celebrated phenomena of colours. And for that purpose having darkened my chamber, and made a small hole in my window shuts, to let in a convenient quantity of the sun's light, I placed my prism at his entrance, that it might be thereby refracted to the opposite wall. It was at first a very pleasing diversion to view the vivid and intense colours produced thereby; but after a while applying myself to consider them more circumspectly, I was surprised to see them in an oblong form; which, according to the received laws of refraction, I expected would have been circular.[19]

Newton, of course, was not the first to view the spectral band; the knowledge that a prism transmits a beam of colored light dates back at least as far as Aristotle. But he was the first to undertake a systematic analysis of the phenomenon by performing what he called his *"experimentum crucis."* *

Newton was well aware of the mathematical law, discovered by Descartes, for calculating how much a ray of light will be bent when it passes through a prism at a given angle. According to this law, the

---

* Newton borrowed this concept from Francis Bacon. It refers to the specific experiment which verifies a novel hypothesis. See Ralph M. Blake, "Sir Isaac Newton's Theory of Scientific Method," *The Philosophical Review*, Vol. XLII, no. 5 (September, 1933), p. 483.

beam of light passing through Newton's prism ought to have made a circle of colors 2⅝ inches in diameter upon the opposite wall. The image, he found, was indeed exactly 2⅝ inches wide, but it was 13¼ inches high!" Comparing the length of this colored spectrum with its width, I found it about five times greater; a disproportion so extravagant, it excited me to a more than ordinary curiosity of examining from whence it might proceed." Obviously, if Descartes' law were valid, the top part of the beam had been bent far more than it ought to have been; the bottom part of the beam less, which was why the beam's height was elongated. There was something fundamentally wrong with the precise law of refraction.

Newton now took a second prism and placed it between the first prism and the wall in order to interrupt different parts of the refracted beam. He found, to his amazement, that each of the seven major colors—red, orange, yellow, green, blue, indigo, and violet—behaved much differently from the original white beam. For example, a narrow beam of red light before falling on the second prism was a narrow beam of red light after passing through it. He further discovered that the part of the beam that was bent too much the first time was also bent too much—by exactly the same amount—when refracted a second time. Likewise, the part of the beam that was bent too little was, when refracted again, bent too little by exactly the same amount. Finally, in an ingenious act of synthesis, he again started with a narrow beam of white light which he split up into a spectrum as before; he then directed the spectrum through another prism exactly like the first, but turned the other way up. There, on the opposite wall, he reproduced the narrow beam of white light. With that, the established view was shattered; for if light were modified by glass, as the conventional wisdom dictated, the second prism should have produced new colors by turning green to yellow or red to blue, but it did not.

Newton made two extremely important discoveries as a result of the *experimentum crucis*. First, he concluded that "the true cause of the length of that image was detected to be no other than that light consists of rays differently refrangible, which were, according to their degrees of refrangibility, transmitted toward divers parts of the wall." Second, he had observed that the "differently refrangible" rays were also differently colored. No amount of additional refraction could change any of the colors of the spectrum; blue remained blue, red remained red and so forth. Thus, wrote Newton, "Colors are not qualifications [i.e., modifications] of light derived from refractions or reflections of natural bodies as tis generally believed, but original and connate properties, which in divers rays are diverse. Some rays are disposed to exhibit a red color and no other; some a yellow and no other." The white light from the sun is not, as everybody thought, pure and homogeneous after all. It is a combination of many individual rays, each with its own degree of refrangibility and color. Newton had proved that these rays could be separated into primary colors and then reconstituted to form a beam of white light. He had, in the poet's words, "untwisted the shining robe of day" and deftly put it back together again.

It is little wonder that of all his remarkable discoveries Newton was most enthusiastic over the results of his experiments with light, for they were much more conclusive than the yet to be refined calculus or the still questionable inverse square law. Moreover, his work with prisms led to his first major success as an experimental scientist. Still, it was not in Newton's character to display great emotion, no matter how important the occasion. At his death he left a fragmentary memoir concerned with the months of unprecedented discovery which comprised the Woolsthorpe period: from reading it one might reasonably conclude that what was probably the most strongly ideational period in the life of any scientist, before or since, was for Newton an experience little out of the ordinary:

> In the beginning of the year 1665, I found the method of approximating series and the rule for reducing any dignity of a binomial into such a series. The same year in May I found the method of tangents of Gregory and Slusius, and in November had the direct method of fluxions, and the next year in January had the theory of colors, and in March following I had entrance into the inverse method of fluxions. And the same year I began to think of gravity extending to the orb of the moon, and having found out how to estimate the force with which a globe revolving within a sphere presses the surface of the sphere, from Kepler's Rule of the periodical times of the planets being in a sesquilaterate proportion of their distances from the centers of their orbs, I deduced that the forces which keep the planets in their orbs must be reciprocally as the squares of their distances from the centers about which they revolve; and thereby compared the force requisite to keep the moon in her orb with the force of gravity at the surface of the earth, and found them to answer pretty nearly. All this was in the two plague years of 1665 and 1666, for in those days I was in the prime of my age for invention, and minded mathematics and philosophy more than at any time since.[20]

This in a world turned upside down. For while Newton meditated in his mother's garden tens of thousands of his countrymen succumbed to the ravages of the Great Plague. Then, just as the worst seemed to be about over, the disease-ridden nation was shaken by yet another disaster of momentous consequences. In September, 1666, the Great Fire of London brought the already reeling capital to its knees. So widespread was the misery and destruction, many Christians predicted the coming of the Last Judgment while others, including prominent millennialists, spoke in exalted tones of the imminent appearance of the Messiah and the establishment of a kingdom of the righteous on earth. However, neither prophecy was destined to be fulfilled, at least not in 1666. Modern science, on the other hand, had at long last found the great lawgiver it had been waiting for, a genius capable of dispelling once and for all the old myths of the medieval Baals. For even as the physical world seemed to be falling apart at the seams, Newton was putting it back together in a more harmonious fashion than anyone since Pythagoras had dreamed.

Lucasian Professor

Newton returned to Cambridge in the spring of 1667 for the purpose of undertaking graduate study and was elected a Minor Fellow of Trinity College in October, an indication that his intellectual ability had earned him a position of respect among senior members of the faculty. It is of interest to note, however, that Newton's enrollment came at a time when a larger than usual number of fellowships stood vacant. No appointments had been made during the plague years of 1665 and 1666, and at least three other vacancies occurred as a result of rather bizarre circumstances. Two Trinity Fellows, probably disoriented by excessive drink, failed to properly negotiate the staircase leading to their quarters and fell to the bottom, sustaining injuries of such a serious nature they were forced to leave the University; a third was dismissed from the institution on the basis of insanity. Whether Newton would have been made a Fellow had he matriculated during a less unsettled year will never be known. We do know, however, that within six months he became a Major Fellow, and on July 7, 1668 he received the Master of Arts degree.

Newton's return to Cambridge led to a resumption of his relationship with Isaac Barrow, whose scholarly guidance had probably been instrumental in opening the door to the discoveries of the *annus mirabilis,* something for which Newton would forever remain in the great mathematician's debt. But Barrow's legacy to Newton went beyond that of a concerned and encouraging mentor. Convinced that the young man was possessed of genius, Barrow resigned the Lucasian Professorship in 1669, but only after recommending that Newton succeed him. Barrow's motives were not altogether altruistic, for he had long entertained the desire to devote himself more completely to the Anglican ministry. Moreover, as Whiteside has shown, Barrow may well have been in quest of an even bigger preferment, a mastership at Trinity, which he attained not long after his resignation. Yet conscious of his duty to the institution which had accorded him great honor, he was unwilling to resign his position if it might result in the appointment of a mediocre intellect. With Newton as his successor, he could be certain that the reputation of the endowed chair would not suffer. Working diplomatically behind the scenes, Barrow was successful in obtaining the prestigious appointment for his twenty-seven-year-old pupil, who in only eight years, two of them spent away from the University, had risen from the menial rank of Sizar to occupy one of the most respected professorships of the scientific world.

For the first time in his life, Newton's income was large enough to enable him to pursue his own inclinations and to fully develop his ideas. More estimates that in addition to his board and lodging, he received about £100 per year from his professorship and another £80 from rents collected on the land he inherited while still a boy. Add to this certain miscellaneous revenues and Newton's annual income from all sources averaged about £250, a tidy sum for a bachelor of conservative tastes. His position at the University was made even more attractive by the light duties prescribed by the Lucasian Professorship. Newton was required to give only a single lecture and hold two student conferences a week. Moreover, he could freely choose the topics of his lectures from among several fields in mathematics and natural philosophy including geometry, optics, mechanics, calculus, and astronomy. That Newton had become fascinated with the cosmologies of Copernicus, Kepler, Galileo, and Descartes is obvious, as evidenced by his discoveries at Woolsthorpe. Yet, so far as we know, he attempted few ongoing astronomical observations of a systematic kind. This seems rather odd for a man who built the first successful model of a reflecting telescope and devoted countless hours to the painstakingly difficult task of grinding and polishing optical lenses. However, like Kepler, Newton suffered from nearsightedness at birth, a condition which gradually improved until it finally disappeared altogether toward his middle years. But by this time his deep attraction to astronomy had long since been sacrificed to other nonscientific interests.

It was Newton's preoccupation with optics that commanded the greater share of his attention during this period. Shortly after he returned to Cambridge, but before his appointment as Lucasian Professor, Isaac Barrow, in yet another gesture of supreme confidence in the ability of his distinguished pupil, called upon Newton to consult with him on the preparation for publication of his *Lectiones Opticae (Optical Lectures)*. Historians of science have puzzled over the result of this collaboration ever since. Barrow's manuscript went to press in 1669, three years after Newton's revolutionary experiments on the composition of light. Yet the published work was totally devoid of even the slightest reference to the nature and origin of colors as described by Newton.

"Newton's failure to correct flagrant errors in the printed version of Barrow's course," writes Manuel, "seems a betrayal of confidence that makes him undeserving of the elegant praise lavished on him" by his mentor.[21] On the other hand, Manuel also speculates that Newton might have refrained from commenting on his discoveries out of deference to the man who had taught him so much. Perhaps. Yet at the very time the manuscript was at the printer's Newton, now Lucasian Professor, was delivering his own lectures on optics in which he disproved several of Barrow's major propositions. Is it possible, then, that Newton was hypocritically laughing up his sleeve at the work of the man who had done far more than any other to advance both his scholarship and his career? Historian of science I. Bernard Cohen thinks not. Rather, Cohen speculates that Newton did in fact inform Barrow of the *experimentum crucis*, but that Barrow was unwilling to embark upon a lengthy revision of his work, "especially at the time of quitting the field of science." [22] However, More asserts that, "It is quite inconceivable that Barrow would have permitted his book to be published if he had known about Newton's work. He was too able a scientist not to have recognized its importance and at least to have alluded to it." [23] In this particular instance the reader, no less than the expert, is free to choose from among a number of seemingly plausible, albeit highly contradictory, interpretations; for this is only one of several mystifying and probably never-to-be-explained happenings in the life of a truly enigmatic genius.

In the previous section of this chapter, we saw how Newton, in the course of his experiments with light, discovered what Descartes and others had not known—that the phenomenon of chromatic aberration was the major reason why the lenses of most refracting telescopes of the time produced distorted images. Having at last learned the root cause of the problem, the next logical question was: Could anything be done to correct it? By 1668, Newton was certain in his own mind that however carefully one might grind and polish lenses, it would still be impossible to construct a telescope capable of producing an image free from the distracting effects of colored rings. He was, as usual, absolutely correct, but only so long as one employs a certain kind of glass. Though by then Newton had developed a strong interest in chemical experiments, it seems not to have crossed his mind that different types of glass might separate out different amounts of color, and that by combining carefully selected lenses the prismatic effect produced by a single lens could be neutralized. It was only much later that the Englishman John Dollond and his son Peter developed a satisfactory process for the crafting of achromatic lenses, which in proper combination produce visual images completely free of chromatic aberration. Florian Cajori, the great Newton scholar, believes that this discovery might also have been Newton's had he only exercised his usual caution: "From very limited experimental evidence he drew a broad inference, to which he adhered with marvellous tenacity, but which was erroneous." [24]

This seemingly lost opportunity was not without its positive side, however. Devoid of all hope of making a refracting telescope free of distortion, Newton became preoccupied with the possibility of constructing a reflecting telescope. Unlike many of his ideas, this one was not original with him; it had already been suggested by several others including the respected Scottish mathematician and astronomer, James Gregory. But Gregory never completed the project, because he could not find an optician capable of successfully executing the demanding task of grinding the required mirrors. Unlike the refracting telescope which forms an image of a distant object with lenses, the reflector uses a concave mirror from which all colors are reflected at the same angle, so that the observer is not troubled by fuzzy rings of colored light. However, when it comes to the eyepieces the reflector and refractor are on the same footing, in that both are dependent upon lenses. Newton, after a considerable period of concentrated study, made a number of changes in Gregory's design, and in 1668 fashioned with his own skilled hands a small reflecting telescope—the first ever executed and directed to the heavens. Though a mere toy in comparison to its giant progeny, Newton used it to observe the moonlike phases of Venus and the four brightest satellites of Jupiter—the so-called Medicean stars made famous several decades earlier by Galileo in *The Starry Messenger*. It would be several decades, however, before the art of building reflecting telescopes reached a stage where they could compete with refracting models, the many technical defects of the latter type notwithstanding. The first major astronomical discovery made with a reflector came in 1781, when William Herschel became the first to observe the planet Uranus.

Word of important scientific achievements traveled quickly in the close-knit world of the virtuosi, and Newton was surprised, though obviously pleased, when he learned that knowledge of his invention had reached the membership of the Royal Society in London. Henry Oldenburg, Secretary of the Society, requested that Newton make his telescope available for examination by its members, which he did near the end of 1671. For this purpose he constructed a second and somewhat larger instrument, which on arrival in London was shown to an admiring Charles ii. Newton was elected a Fellow of the Royal Society on January 11, 1672, the day his telescope was formally discussed. Shortly thereafter, a detailed description of the invention appeared in the Society's official publication, the *Philosophical Transactions*. Newton, a lonely young professor with no close companions, must have thought that he had at last found a group of kindred spirits who would truly value his creations as a natural philosopher. Though he remained a Fellow of the Royal Society for the rest of his life, and would come to know a large number of the great scientists of the seventeenth century, this was perhaps as close as Newton ever came to a feeling of security and trust in his professional dealings with other men. Unfortunately it was not a mood destined to be long sustained.

Though deeply grateful for the attention and praise accorded him by the London virtuosi, Newton felt that the development of a reflecting telescope was of far less importance than were the results of his experiments with light. He did not wish to be known as a craftsman of popular renown, but as a scientific thinker of the first order. What better way, Newton thought, to show his gratitude for being elected to the Royal Society, and at the same time to further enhance his reputation as a natural philosopher, than openly to share with his peers the results of his research. Accordingly, he drafted a long letter to Oldenburg in which he offered to lay before the Society "an account of a philosophical discovery, which induced me to make the said telescope, and which I doubt not but will prove much more grateful than the communication of the instrument being in my judgment the oddest if not the most considerable detection which hath hitherto been made into the operations of nature." Newton, obviously in an expansive mood rarely associated with his normally suspicious and secretive character, was referring to his new theory of light and colors derived from the *experimentum crucis* of 1666.

The account of the famous experiment, which I have already described, was soon forwarded to the Society and duly published in the *Philosophical Transactions* of February 19, 1672. The paper aroused great interest and brought Newton considerable praise from several quarters, but it hardly evoked the universal admiration and acclaim that he had hoped it would. Objections were raised by a number of natural philosophers, both English and European, including Hooke, Huygens, Lucas, Linus, Pardies, and Gascoines. Though Newton, at the urging of Oldenburg, patiently answered each of his critics by letter, the pain which his sensitive nature suffered steadily intensified, and with it his irritability at having to sustain his arguments against many who had not even bothered to repeat his simple experiments with prisms. Yet he probably would have dismissed the criticism leveled at his work by lesser men had not such great masters of experimental science as Robert Hooke and Christian Huygens also rejected the product of his research, because it did not conform to their own preconceived hypotheses as to the nature and behavior of light. It was an alienated and despairing Newton who, in a letter to Oldenburg dated November 11, 1676, declared of natural philosophy: "I will resolutely bid adieu to it eternally, excepting what I do for my private satisfaction or leave to come out after me [meaning after his death]; for I see a man must either resolve to put out nothing new, or to become a slave to defend it."

Thus was the legend born. Newton, in the naive and confident manner of a trusting young professional, had been so certain of himself that he found it impossible to believe that men of ability could honestly take exception to what he had discovered. He was just now becoming cognizant of one of the less palatable sides of scholarly activity; namely, that a rise in the status of one individual often occurs at the expense of the professional reputation of another. Hence, his recently acquired confidence in his fellow virtuosi was soon damaged beyond repair. Newton came to view all questions and doubts pertaining to his work as nothing less than the product of deliberate persecution. Always a loner, at his best and worst, Isaac Newton now lived for only one man's approval—his own.

In response to the intense criticism which surrounded the published results of the *experimentum crucis,* Newton formulated his famous aphorism, *hypotheses non fingo,* meaning, "I do not feign hypotheses." * This brief statement of principle did not appear in print until 1713, however, when it was incorporated into the General Scholium of the second edition of the *Principia.* Four years later, in 1717, Newton discussed the concept in far greater detail in Query Twenty-Eight of the *Opticks,* one of thirty-one questions appended to the second edition of the brilliant work. "The main business of natural philosophy," he wrote, "is to argue from phenomena without feigning hypotheses, and to deduce causes from effects." Newton firmly believed that it was not the province of the scientist to formulate sweeping metaphysical principles regarding the nature of physical causation, as Descartes had done when he put forth his daring but unproven concept of vortices. To Newton, the only valid theory is a law based on indisputable facts which must be expressed in mathematical terms. Such a law may be dismissed only by the discovery of new facts or by proving, through the process of experimentation, that the accepted ones are false. If it were to remain otherwise, men would never achieve a degree of positive knowledge, for like the ancients and the medieval scholastics, they will be forever enslaved by the belief in occult forces and substances incapable of factual verification. And since one hypothesis (the Ptolemaic, for example) might well be as acceptable as another (the Copernican) in explaining the cause of certain natural phenomena, confusion and uncertainty are bound to proliferate. In effect, Newton was telling his detractors to "either put up or shut up," albeit in far more delicate terms. They must either disprove his laws on optics by formulating mathematical calculations, or else they must discard their preconceived notions and embrace the product of his labors, no matter how distasteful this might seem. The same later held true for his law of universal gravitation. Newton, for his part, was willing to be moved by the arguments of a rival, but only if those argu-

---

* It has also been translated as, "I do not make hypotheses."

ments eschewed *a priori* principles in favor of experimentation and mathematical demonstration. Francis Bacon's unqualified advocacy of the experimental method had clearly left a permanent mark on Newtonian methodology.

Newton's seemingly adamant rejection of hypothetical speculation notwithstanding, one of this century's most respected historians of science wrote not long ago: "Newton . . . was the most daringly speculative thinker about nature known to history, and the most fertile framer of hypotheses." [25] Indeed, the more one thinks about it, how else can one account for the *annus mirabilis*, if its accomplishments cannot be attributed to an explosion of Newton's incomparable speculative powers? Furthermore, Newton occasionally employed the term "hypothesis" in the *Principia* to designate certain propositions that formed an integral part of his theory, to say nothing of his frequent reference to the "Copernican Hypothesis," which he most certainly embraced even though it it had not been proven mathematically valid by any of his predecessors. Is this just another one of several seemingly irreconcilable contradictions in the thought processes of one of the most creative thinkers ever produced by the human race? The answer in this instance is, "no." The truth of the matter, observed intellectual historian Ralph Blake, "is that Newton's polemic against hypotheses is not directed against any and every use of hypotheses, but rather against certain current forms of their abuse." [26] To Newton, hypotheses served as fruitful aids in indicating new lines of investigation, and as a method of more clearly focusing one's inquiry on specific phenomena. Hypotheses might also suggest new experiments by which the cause of modern science can be advanced. But their use by natural philosophers must always be limited and tentative. Otherwise hypotheses will be presented, as in times past, as a kind of dogmatic truth from which no rational appeal can be made. Hence, Newton insists upon the priority of experimentation and mathematical deduction in all the natural sciences lest the legitimate results of rigorous empirical investigation be rejected on the basis of *a priori* principles.

Finally, Newton was made extremely uncomfortable by what he perceived as the hubris implicit in the formulation of broad and unproven principles of physical causation. Surprising though it may seem, he would have found alien the current notion that human understanding of nature can reach fulfillment in an absolutely certain and definitive science. From Newton's point of view the mathematical–mechanical laws of natural phenomena are probably the most that the scientist can ever hope to discover. The ultimate nature of things and their ultimate cause are not for man to know.

To be constantly engaged in studying and probing into God's actions was true worship and the fulfillment of the commandments of a Master. No mystical contemplation, no laying himself open to the assaults of devilish fantasies. . . . Working in God's vineyard staved off evil, and work meant investigating real things in nature and in Scripture, not fabricating metaphysical systems and abstractions, not indulging in the 'vaine babblings and oppositions of science falsly so called'. If God is our Master He wants servants who work and obey.[27]

Newton, who more than anything else perceived of himself as an obedient servant of God, retreated into the protective shell of his childhood. And once inside, the impenetrable barrier which separated him from other men would never again be breached. In 1673, he impetuously decided to sever totally his ties with the Royal Society on the pretext that he resided too far from London to take an active part in its weekly deliberations. But regular attendance at Society meetings was not required of its fellows, many of whom lived at a much greater distance from the capital than did Newton. The real reason for his decision to resign was his inordinate sensitivity to criticism of any kind. Because Newton's life was his work, he never accustomed himself to viewing scientific disputes on an abstract plane. Those who either questioned or objected to his conclusions, no matter how sincere in their intent, seemed to pose a direct threat to his very reason for being. Henry Oldenburg attempted to assuage him by drafting a flattering testimonial on the high esteem and deep affection in which he was held by the Society; the Secretary even offered to remit Newton's dues for the remainder of 1673, if only he would reconsider his action. Temporarily appeased, Newton withdrew his resignation, but he was more determined than ever to pursue his work in solitude and to keep secret the product of his future inquiries, no matter how significant. As in the case of Kepler and Galileo, scientific genius had once again become manifest in a man whose strongly unsocialized personality helped elevate him to the very pinnacle of revolutionary discovery, but which so cursed his private life that he was frequently plummeted into the very depths of inconsolable despair.

Though Newton lived in an age of genius, the extraordinary intellectual gifts showered upon him by the gods had been given to none of his fellow virtuosi, brilliant though many of them were. Still, Newton was not without his rivals, all of whom, no matter how respected for their intellectual integrity as individuals, seem to have been indistinguishable in his mind's eye from the hated Reverend Smith, the malicious interloper who had stolen Isaac's mother away during his childhood. He sincerely believed that his peers, like Smith, would either steal his scientific discoveries if given half a chance or, barring that, ridicule his work with the object of undermining its credibility. Of the many natural philosophers with whom Newton crossed intellectual swords, none so strongly aroused his enmity or more profoundly influenced the course of his research and contemplation than his gifted, tenacious, and equally petulant nemesis, Robert Hooke.

Hooke, seven years Newton's senior, was a many-sided genius cast in the same mold as Leonardo da Vinci. His thought process, like that of the great Florentine, was a curious mixture of sudden uprisings and agonizing downfalls, of crescendo and decrescendo. Such restless, peripatetic brilliance quickly burns itself out, then rises brightly from the ashes to attack a new problem, regardless of whether the old one has been satisfactorily resolved. Like a careening meteor entering the outer atmosphere of a planet, genius of this variety is both explosive and spectacular, but its product frequently disintegrates before reaching terra firma. Time and again Hooke intuited or hypothesized what only Newton or others were capable of establishing as scientific fact. Always painfully conscious of far more than he could prove, he watched in agonized distress while his rivals reaped the harvest for which he had prepared the ground. Indeed it was the personal misfortune of Robert Hooke, a very great genius in his own right, to have had as his contemporary Newton, an even greater one. Had Hooke lived at any other time, he would have doubtless shone like a star of the first magnitude; instead, his image was dimmed and blurred by that so brilliantly projected by Newton, the rising sun of modern science.

Hooke, like Newton, was an immensely gifted inventor of mechanical devices. While still a young boy he watched in quiet fascination as a clock was being dismantled for repairs; he then designed and built out of wood an exact working replica. Young Robert also constructed a model of a fully rigged man-of-war which not only sailed but fired a salvo from the miniature cannons mounted on its brightly painted deck. His father, John Hooke, was a clergyman who had high hopes of preparing his son for the ministry. But the elder Hooke soon abandoned the idea because persistent headaches made it impossible for his frail child to keep up with the rigorous program of study. Upon his father's death in 1648, Robert took his inheritance of £100 and left for London, where it was his good fortune to be befriended by Richard Busby, the master of Westminster School. Busby immediately recognized Hooke's special talents and became so concerned with the young man's welfare that he took him into his home. Having mastered the classical curriculum taught at Westminster, Robert moved on to Oxford where he enrolled in Christ Church in 1653.

Hooke's arrival at the great center of learning could not have been better timed. Oxford was the home of that brilliant group of intellectuals who, on the collapse of the Protectorate and the restoration of the Stuarts in 1660, removed to London where they founded the Royal Society under the patronage of Charles II. Hooke soon earned a place in this select circle which included William Petty, John Wallis, Seth Ward, John Wilkins, Thomas Willis, Christopher Wren, and Robert Boyle. They acknowledged and drew upon his talent in mechanics; Hooke, in return, received from them his introduction to the new world of scientific thought, a fair trade indeed. It was Thomas Willis who introduced Hooke to the illustrious virtuoso Robert Boyle. A bond formed between the two men, and Hooke accepted Boyle's offer to become the chemist's paid assistant, a position he held until his appointment as Curator of Experiments for the Royal Society late in 1662.

It is difficult to believe that anyone could have been better suited for this latter task than Robert Hooke, though most men would have found the demands of the position far too excessive in relation to its meagre financial rewards. The Society charged Hooke to furnish each of its weekly meetings "with three or four considerable Experiments" in addition to any others proposed by the members. The pace was indeed hectic but Hooke thrived on the intense pressure, for it nourished his inner compulsion to skip from one idea to another without pause. Moreover, the innumerable responsibilities associated with his position brought a kind of disciplined order into the life of the natural philosopher that otherwise would have been lacking. As long as he worked at a fever pitch, the brilliant ideas poured forth in a seemingly endless stream. It was only many years later, when the demands placed on him by his peers finally slackened, that Hooke's creative powers perceptibly waned; his mind then became flaccid, his ability to innovate permanently impaired.

Shortly after Newton submitted his paper on light and colors to the Royal Society in 1671, a committee was appointed to undertake a study of the document's scientific findings. However, Hooke, a member of the committee, wrote the final report, which contained only his personal and highly critical evaluation of Newton's conclusions. The Curator of Experiments considered himself an expert in the field of optics and had discussed his own theory of light in his *Micrographia*, the first great work devoted to microscopic observations. An advocate of the mechanical philosophy of nature, Hooke believed that light also behaves according to mechanical principles. Unlike Descartes, however, he rejected the idea that light is a pressure transmitted instantaneously through space from a cosmic vortex. Instead, Hooke held that light is composed of pulses of motion transmitted through an invisible material medium of ether, an hypothesis that makes him one of the forebears of the wave theory of light. Newton himself would later borrow this idea from Hooke when he formulated his own theory of light.

What Hooke failed to understand, however, was that Newton had not put forth an hypothesis regarding the material composition of light. Rather, he had only described the results of his experiments to show how white light and colored light behaved and, by so doing, had demonstrated their measurable properties. Hooke remained steadfast in his belief that a change in the color of light is caused by the glass or other medium through which it passes. Thus, he completely overlooked the fact that Newton was replacing the old concept of modification with the new and revolutionary principle of analysis. In fact, Hooke's evaluation of Newton's paper was so wide of the mark he mistakenly concluded that Newton had obtained his main ideas from reading the *Micrographia*.

The superficial and highly disparaging nature of Hooke's critique would have doubtless tried the patience of any normal man, let alone a highly suspicious and temperamental individual like Newton. Upon learning of his fellow virtuoso's misplaced criticism, Newton was literally consumed by rage. And to make matters worse, the fires of hatred were fanned to even greater heights by Henry Oldenburg, who had quarreled with Hooke many times in the past and now decided to use Newton to gain a measure of revenge. Stung to the quick and egged on by Oldenburg, Newton launched a bitter public attack on the unsuspecting Hooke, which found its way into the *Philosophical Transactions*. Hooke's critique of Newton, on the other hand, for all its irritable vanity, had never been published. Hence, the major embarrassment in this instance was suffered by Hooke. Though during the next several years their friends made a number of attempts to reconcile the differences separating the quarrelsome rivals, the best that could ever be accomplished was a series of uneasy truces which were inevitably broken by one or the other.

The antipathy nurtured by the two men for each other flared into the open a second time in 1675, and was again concerned with the question of the composition and behavior of light. During a rare visit to London, for Newton traveled hardly at all, he mistakenly thought he heard Hooke accept his theory of colors over which they had clashed in 1672. This encouraged Newton to submit a second paper to the Royal Society, the contents of which make up something like three-quarters of Book Two of the *Opticks*, a work whose publication he withheld for obvious reasons until after Hooke's death. The primary purpose of the paper was to explain the phenomena of colors on solid bodies which are seen by reflected light, and secondarily to explain colors produced by light when it is reflected from transparent films such as mica, soap bubbles, or other similarly thin surfaces. There is no question that in the latter instance Newton's investigation was inspired by Hooke, who, in the *Micrographia*, recognized that the colors are periodical, and that the spectrum repeats itself as the thickness of the film increases. Hooke had correctly concluded that the succession of colors which he observed was produced by a combination of a beam of light being partially reflected from the upper surface of the film and also penetrating the film to be reflected from its lower surface. Thus, to Hooke belongs the credit for suggesting the concept of periodicity in modern optics.

Yet, as so frequently happened with the mercurial genius, he failed to take the final and decisive step of submitting his observations to exact measurement, and it was at this point that Newton entered the picture. The latter superimposed a plano-convex lens on a flat sheet of glass, trapping a thin film of air between the two surfaces. Within this film he discovered a series of colored concentric rings (Newton's rings), and found that the distance between them is directly dependent upon the increasing thickness of the film of air. Newton carefully measured this relationship and succeeded in quantitatively demonstrating the concept of periodicity. Hooke's failure to probe the depths of a phenomenon which he unquestionably elucidated had again cost him dearly, for a far greater share of the credit for this discovery went to Newton, as indeed it should have.

Still, the primary source of the new controversy can be traced to the contents of a second paper which Newton, in 1675, had sent to the Royal Society along with the first. Titled "An Hypothesis Explaining the Properties of Light," it was in fact a general system of nature which Hooke believed Newton had stolen from the *Micrographia*. This claim does not bear up under critical examination, although it again illustrates the overall compatibility of the two scientists' fundamental ideas on nature. The hypothesis presented in Newton's second paper was of importance not only because it revealed the direction of his thought on the physical composition of light, but also because it presented a theory of matter which deeply affected his speculations on the nature and cause of gravity.

Though a confirmed mechanist, Newton, like Hooke, rejected Descartes' theory that light is a pressure resulting from the cosmic circulation of matter. The phenomena of color, Descartes had argued, occurs when the tiny, whirling globules of particulate material undergo modification as they pass through glass or are bounced off the opaque surfaces of other objects. However, as we observed earlier in this chapter, the peculiar qualities of refraction which Descartes believed resulted in the creation of colored light were not borne out by Newton's *experimentum crucis*.

During the seventeenth century another mechanist, Pierre Gassendi (1592–1655), who opposed Descartes' conception of space as a plenum, revived the doctrine of the ancient atomists of the Hellenistic Age. Whereas Descartes claimed that matter occupies all of space, the atomists held that it occupies only a part of the great void. They postulated that between the invisible particles of matter are equally invisible pockets or pores which contain nothing of a corporeal nature. They also maintained that all bodies are composed of an indefinitely large number of indivisible atoms of various shapes and sizes. Material objects differ from one another because of the number, density, and motions of these tiny constituent parts. But even the most solid bodies, the planets included, are honeycombed by minute interstices of empty space. Thus, Newton, like the other natural philosophers of his day, was faced with a choice between two theories: the plenum of Descartes or the atoms of Gassendi. He decided for atoms and by so doing, in Westfall's words, "set his course down a particular road from which he would never turn."[28] It should be kept in mind, however, that the choice open to Newton was more one of detail than of radically divergent systems, for he still "embraced a mechanical conception of nature as the only possible system, as the only feasible basis for scientific discussion. . . . Invisible mechanisms must exist whereby particles in motion cause what appear to be attractions and repulsions." [29] Newton and his fellow mechanists viewed the possibility of action at a distance without the impact of one body upon another as a chimerical product of the mystical imagination.

During his earlier experiments with prisms, Newton had proved that light is complex and heterogeneous; hence, the phenomena of colors can only be explained by reducing light to its constituent parts. This Newton proceeded to do, and he immediately discovered that the individual rays of colored light which make up the spectrum have immutable properties, for, try as he might, Newton could not alter the composition or the behavior of a single beam of colored light. And though he possessed no absolute proof, he had become convinced that the immutable properties displayed by individual rays of colored light imply the existence of immutable particles of matter, or what were commonly called corpuscles (atoms). Newton further postulated that these tiny particles fly through space with great velocity and cause vision by their impact on the retina. Color results because particles, according to their various sizes, reflect rays of one color and transmit those of another—the color that meets the eye being the one reflected while the other rays are either transmitted or absorbed.

Central to Newton's mechanistic theory of light was his concept of an etherial medium, much like air but far rarer and more strongly elastic. In a modification of Gassendi's atomic theory, Newton postulated that ether, a fluid of homogeneous matter, permeates the pores of all bodies and is expanded through all of infinite space. He also speculated that when corpuscles of light pass through the ether, its varying zones of density alter the direction in which the corpuscles move. Thus, certain phenomena associated with light—reflection, refraction, and diffraction, for example—arise from the influence of ether on the motion of the particles of which light is composed.

As far as optical phenomena are concerned, Newton's theory proved most important in explaining the periodicity of the thin films which he had been the first to examine in detail. How is it, he asked himself, that on the surface of a transparent medium like glass, part of the light is refracted and part of it is reflected? Or, to put it another way, how can an impinging particle at one time be refused admittance and be thrown back, and at another time be permitted passage into the interior? As his answer, Newton advanced his famous "theory of fits of easy reflection and easy transmission." Every thin film, such as a film of air trapped between two pieces of glass, he conjectured, is also a thin film of ether. The procession of flying corpuscles of light sets the ether near the surface of the glass into a spasmodic or wavelike motion, which results in a succession of compressions and rarefactions of the ether. When a particle of light reaches the surface at the moment the ether is compressed it is thrown back; but if it arrives at the moment of rarefaction it will pass through. Thus, the optical phenomena of thin films are periodic, the product of "fits" or spasms in the ethereal medium.

As was noted above, Newton's "Hypothesis of Light" presented a general system of nature, and as such dealt with a broad range of physical phenomena. A problem that increasingly occupied his thought during this period, and one which he dealt with in his paper as well as in his later works, was that of the cohesion of particles of matter to form bodies, and the natural attraction of those bodies to one another. By this time, Newton was deeply immersed in the study of alchemy, and it now seems clear that he was hopeful of discovering at the atomic level some universal substance or law of chemical action by which he could explain the principle of attraction. He had even gone so far as to speculate whether the force which held together the innumerable particles of individual bodies might be the same force which bound the planets into a solar system, and the countless solar systems into a seemingly infinite universe. However, the hundreds of chemical experiments Newton undertook never yielded the results he so ardently pursued, never enabled him to unite the microcosm with the macrocosm in a truly satisfactory manner. Like other mortals, the greatest lawgiver in the history of science also experienced the bitter taste of failure, although at an intellectual level achieved by only a handful of immensely gifted men. In the end, Newton returned to the theory of matter presented in the "Hypothesis of Light" to explain the mystifying phenomenon of physical attraction.*

As we have observed, the ethereal medium propounded by Newton was far rarer and much more elastic than air. And, though all-pervasive, Newton's ether was rarer within the "dense bodies of the sun, stars, planets and comets, than in the empty celestial spaces between them." Indeed, he "saw no reason why this increase of density should stop any where, and not rather be continued through all distances from the Sun to Saturn, and beyond." He then enquired into the possibility whether the ether could, by its presence, account for the dual phenomena of attraction and repulsion.

* Newton's alchemical studies have recently been analyzed in a most illuminating work. See Betty Jo Teeter Dobbs, *The Foundation of Newton's Alchemy* (Cambridge, 1975).

Newton, in his mind's eye, pictured the earth much like a great sponge, steadily soaking up a stream of subtle ethereal matter which continually presses down upon its surface. This same stream of ether, by its impact on bodies above the earth, causes them to descend, thus accounting, he believed, for the principle of attraction. On the other hand, what was to prevent the earth from becoming larger and larger as the ether continued to accumulate? With an explanation too long and intricate to pursue here, Newton hypothesized that the ether, after falling to earth, and penetrating its surface, changed form, and then ascended once more into space. By so doing it would produce the repulsion which experimenters had discovered whenever two bodies closely approach one another. According to Newton, this circular process might also account for the "generation [of] fluids out of solids, fixed things out of volatile, and volatile out of fixed . . . some things to ascend, and make the upper terestriall juices, Rivers, and the Atmosphere; and by consequence others to descend for a Requitall to the former." He further speculated that "as the Earth, so perhaps may the Sun imbibe this Spirit copiously to conserve his Shineing, and keep the Planets from receding from him. And . . . that this Spirit affords or carreys with it thither the solary fewell and materiall Principle of Light." As with Boyle, Newton's ethereal medium had come to occupy two distinct functions: it propagated the motions of matter across vast distances of interstellar space, and it possessed qualities which accounted for extramechanical phenomena like growth, decay, and combustion.

The mechanical theory of gravity advanced by Newton clearly shows that his mind, like Descartes', was fully capable of sustained flights of scientific imagination. But here the similarity between their methodological approach ends, illustrating once again the contrast between the Baconian, practical-minded English virtuoso, and his theoretically minded Continental counterpart. For while Descartes was continually preoccupied with the question concerning from what abstract truth his scientific theory derived, Newton was content with a theory only if it yielded *quantifiable* results. Hence, Newton took his fluctuating opinions on ether for exactly what they were—metaphysical hypotheses uncorroborated by mathematically demonstrable principles. He formulated the law of universal gravitation without being obliged to give an explanation of the real forces which produce it, forces which remain as mystifying today as they were to men of Newton's generation. "To us," he wrote in the *Principia*, "it is enough that gravity does really exist, and acts according to the laws which we have explained, and abundantly serves to account for all motions of the celestial bodies, and of our sea." Unlike so many of the famous in history who have disregarded the advice they have freely dispensed to ordinary mortals, Newton himself was the most faithful practitioner of his own motto: *hypotheses non fingo*. It really

amounted to saying that his task, as that of all natural philosophers, was
to explain under what law things happened in nature, rather than to attempt to deal with the problem of ultimate causation in the universe
which, in the final analysis, is the domain of God.

Robert Hooke was also an advocate of the corpuscular theory of
matter and of ether; and, as he scanned his rival's latest scientific paper,
he again took it for granted that Newton had borrowed heavily from
his work without giving him due credit. Newton, it is true, did build
upon certain fundamental concepts advanced by Hooke in the *Micrographia,* but there were major differences in the two scientists' explanations of the natural phenomena discussed above. Furthermore, Hooke
would have realized this had he only bothered to give to Newton's work
the detailed attention which it merited. Newton, for his part, lost all
patience with those who had dared to raise serious objections to his
findings. In 1676, he all but severed his few remaining ties with his
fellow virtuosi, withdrawing into the security and solitude of his private
intellectual world.

The Principia

Isaac Newton had always preferred to deal with things and ideas
and to avoid persons and entangling emotional commitments. Now, re-
pulsed and angered over the criticism engendered by the publication of
his scientific papers, he permitted even fewer outsiders than before to
invade the privacy of his university chambers, much less his secret world
of thought. Indeed, as Manuel suggests, his life at Cambridge during the
1680s and early 1690s is somewhat reminiscent of the austerity normally
associated with the monastic cloister, except that in this instance the
"abbot" established his own demanding rule of order. It was, in fact, a
more exacting regimen than prevailed in many of medieval Christendom's
most illustrious monasteries.

As so frequently happens with the solitary and the gifted, Newton
took an inordinate interest in his health and, like Kepler, became somewhat of a hypochondriac. Yet despite his dangerously premature birth
and delicate infancy, he possessed an extraordinary constitution for a
man of the late seventeenth century, especially when one considers that
he consciously shunned rugged physical activity during childhood and
almost never exercised as an adult. One of the major indicators of his
physical well-being is that at a time when the extraction of teeth was the
most common way of treating any and all dental problems, Newton,
though he died an octogenarian, lost but one permanent tooth. Neither,
in spite of his tendency toward hypochondria, did he pamper himself
when he became deeply immersed in the solution to an important scien-

tific problem. During such periods Newton would often forget to eat his meals, sleep for only a few hours a night, and remain within the narrow confines of his rooms for days on end. Yet on only two occasions did he suffer serious attacks of illness, and in both instances they were mental rather than physical. Newton seems to have suffered a nervous breakdown in 1664 from the exacting work and late hours he kept while observing the comet of that year. In 1693 he was again stricken by the same nervous disorder which resulted from his almost superhuman labors on the *Principia*. In the latter instance, rumors circulated of his being mentally deranged and incapable of undertaking further scientific studies. The crisis soon passed, however, and Newton recovered his mental stability. Yet his enormous powers of creativity do appear to have suffered, for only briefly did he ever return to sustained scientific work.

Newton's biographers have long lamented the fact that few of those personally acquainted with the scientist during the height of his career bothered to record their impressions of him for the benefit of posterity. And of those who did, none delved very deeply below the surface in an attempt to capture the spirit of the inner man. No doubt this is at least partially the result of Newton's unceasing attempt to keep even the men whom he most respected at arm's length. But it is also important to recognize that biography, particularly in the sciences, had not yet taken on the importance that it has in the twentieth century. Most of the few intimate facts we possess of Newton's manner of living during the Cambridge period were recorded by Humphey Newton, a distant relative from Grantham, who served as the great man's assistant and amanuensis* during the writing of the *Principia*. "His carriage then was very meek, sedate, and humble," wrote Humphrey Newton, "never seemingly angry, of profound thought, his countenance mild, pleasant, and comely." Newton's generally serious nature is attested to by his secretary's observation that he saw his employer laugh only once. The scientist had supposedly lent a friend a copy of Euclid to read. When the book was returned, Newton inquired how his friend liked the ancient geometrician. "The friend answered by desiring to know what use and benefit in life that study would be to him. Upon which Sir Isaac was very merry."

It was unusual for Newton to leave the confines of the university, and, while there, according to his secretary, "he always kept close to his studies . . . and had few visitors. . . . I never knew him to take any recreation or pastime either in riding out to take air, walking, bowling, or any other exercise whatever, thinking all hours lost that was [sic] not spent in his studies." So intense were Newton's unexplained powers of concentration that "oftentimes he has forgot to eat at all, so that, going into his chambers, I have found his mess untouched, of which,

* Manuscript copyist and secretary.

when I have reminded him, he would reply—'Have I?' and then making to the table, would eat a bit or two standing, for I cannot say I ever saw him sit at table by himself." The old woman who made Newton's bed and delivered his meals truly delighted in this seemingly providential aspect of her employer's strange behavior: "she sometimes found both dinner and supper scarcely tasted of" and "has very pleasantly and mumpingly gone away with [them]."

It is also from Humphrey Newton that we learn of his employer's exacting work schedule. "He very rarely went to bed til *two* or *three* of the clock, sometimes not until *five* or *six*, lying about *four* or *five* hours, especially at spring and fall leaf, at which time he used to employ about six weeks in his laboratory, the fire scarcely going out either night or day." During these periods, scientist and assistant would take turns passing the night by the furnaces "till he had finished his chemical experiments, in the performance of which he was the most accurate, strict and exact." Nor did the secretive genius divulge the purpose behind his exhausting labors. "What his aim might be I was not able to penetrate into, but his pains, his diligence at these set times made me think he aimed at something beyond the reach of human art and industry." The singularity of purpose with which Newton worked is further testified to by the fact that when, on some special occasion, he did venture out into public he "would go very carelessly, with shoes down at heels, stockings untied, surplice on, and his head scarcely combed." [30] And despite his strong attachment to institutionalized religion, when deeply absorbed in a problem Newton seldom attended chapel. However, since he looked upon the pursuit of science as a religious activity in its own right, he may well have reasoned that what took place in his laboratory was as fundamental a part of religious experience as any late afternoon prayers he might offer to the Creator from a pew in St. Mary's Church.

In 1677, Robert Hooke was elected Secretary of the Royal Society, and Newton, who had remained a Fellow despite his earlier threats to resign, sent the great experimentalist a brief letter of congratulation. During the next few years the great rivals seem to have adoptd a more conciliatory attitude toward one another, but in retrospect this only proved to be a deceptive lull before the final, acrimonious rupture. The opening move came in 1679, when Hooke wrote Newton about various scientific matters, and asked him for a paper on any subject of his choice. Among the various propositions put forth in Hooke's letter was a revolutionary theory of planetary dynamics. We observed earlier in this chapter how Newton had explained the orbit of the moon in terms of an exact balance between centripetal and centrifugal forces. However, Hooke suggested to Newton that the moon, or a planet, would move steadily through space in a straight line in the absence of a central force drawing it aside, a force decreasing with the increase of the distance. Most notable in Hooke's analysis of the problem is the absence of any reference to the concept of centrifugal force as previously employed by Newton. Thus it was Hooke, rather than Newton, who was the first to clearly see the elements of planetary motion as we continue to accept them. Many years were to pass before Newton acknowledged that his correspondence with Hooke modified his thinking on celestial dynamics and stimulated him to attack the question of gravity anew. For he had not as yet satisfactorily resolved the central problem which prompted him to set the subject aside in the late 1660s: whether it is legitimate to assume that the mass of a spherical body may be considered as concentrated at its center.

Newton, in his reply to Hooke, did not comment on the latter's speculation that planetary motion is the result of a continuous diversion from a rectilinear path by a central attraction. He excused himself by referring to the death of his mother earlier in the year and went on to explain that the press of family business made it impossible for him to concentrate on matters of natural philosophy. Moreover, wrote Newton, "I had for some years past been endeavoring to bend myself from philosophy to other studies in so much that I have long grutched the time spent in that study unless it be perhaps at idle hours sometimes for a diversion." Though he had not exactly lied, Newton was intentionally putting Hooke off, and he admitted as much to astronomer Edmond Halley a decade later. Newton simply did not wish to correspond with the person who he believed had cast serious doubts on his integrity both as a man and as a dedicated scientist.

Nevertheless, as a partial sop to Hooke, Newton went on to discuss an experiment designed to demonstrate the diurnal rotation of the earth. He pointed out that a heavy ball let fall from a high place should strike the earth slightly east of the perpendicular. Hooke replied that the ball should fall to the south as well as to the east and pictured the path of its fall as part of an ellipse, rather than the spiral which Newton had proposed. In this instance Hooke was correct and he knew it. Notwithstanding his explicit promise that Newton might "be assured that whatever shall be soe communicated shall be noe otherwise further imparted or disposed of than you yourself shall praescribe," Hooke tactlessly demonstrated Newton's error before a meeting of the Royal Society, thus gaining a measure of revenge for his public humiliation at Newton's hands in 1672. Newton, who hated to be corrected, especially by Hooke, and in public, was so enraged that he refused to answer a number of his rival's future letters. Still, Newton later conceded that Hooke's correspondence enabled him to demonstrate that an elliptical orbit entails an inverse square attraction to one focus—one of the two fundamental principles on which the law of universal gravitation rests. Moreover, Hooke's definition of orbital dynamics, though limited to certain kinds of matter, suggested to Newton a cosmic application. Hooke, by excluding comets and certain other celestial bodies from his concept of gravitation, had failed to raise it to the level of a universal principle. This Newton did by subsequently applying the law of gravitation to every particle of matter in the universe. Once again Hooke's remarkable intuitive powers contributed immensely to Newton's eventual solution of a fundamental problem of nature. Unfortunately for Hooke, his brilliance of insight was not matched by equally strong powers of analysis. It is little wonder that he found it impossible to suppress his mortification and outrage at the success of the man whose thinking he so profoundly influenced. To know and yet not to know was the hapless Hooke's tragic fate, a truly agonizing form of mental torture which pursued him to the grave.

The scene now shifts to London, where the question of planetary motion had begun to occupy the minds of several prominent members of the Royal Society. Early in January 1684, Robert Hooke, Edmond Halley, and Christopher Wren took part in an animated discussion on the question of orbital dynamics. Hooke argued that the attraction between a planet and the sun decreases as the square of the distance, a proposition he had earlier communicated to Newton, but one which rested on intuitive grounds alone. What Hooke did not know, of course, is that Newton had discovered the inverse square law in 1666, but had characteristically kept it a secret. Wren and Halley were skeptical of their friend's claim to say the least, and Hooke, whose irritating boastfulness often got him into trouble when he felt threatened, made a serious blunder by claiming that he had proved the law to his own satisfaction. Wren, who knew better than to take Hooke at his word, told the great experimenter that he would give him two months' time to provide a convincing demonstration; if Hooke proved successful, Wren promised to present him with a book worth forty shillings. Hooke, backed into a corner of his own making, began to temporize. He continued to insist that he had proved the law, Halley later recalled, "but [said] he would conceal it for some time, that others trying and failing might know how to value it, when he should make it public."

At this point Halley, who had grown impatient with Hooke's dilatory tactics, took the initiative. Fourteen years younger than Newton and the son of a successful soap manufacturer, Edmond Halley had been a student at Queens College at Oxford. Enthralled by astronomy since childhood, he had sailed at the age of 20 to the island of St. Helena to observe the positions of the fixed stars in the southern hemisphere. Halley had read Newton's scientific papers and had come to admire the marvelous deductive powers of their secretive author. A few months after his discussion with Wren and Hooke, he mustered up sufficient courage to approach the reclusive Newton, whom he had never met. Like Rheticus a century and one-half earlier, Halley came as a supplicant seeking the indulgence of the master. Newton, who approved of the sincere and unaffected manner of the brilliant young astronomer, immediately took him into his confidence. Halley now posed the crucial question: "What would be the curve described by planets, supposing the force of attraction towards the sun to be reciprocal to the square of their distance from it?" Newton was able to answer without hesitation that the curve would be an ellipse, for, in 1679, this same question had arisen in his correspondence with Hooke, and Newton had sat down to find the answer for himself. Using the mathematics of nascent and ultimate ratios, a concept very similar to the differential method, Newton brought the law of gravitation a mighty step closer to reality by demonstrating that a force, varying inversely with the square of the distance, would "bend" a moving planet into an elliptical orbit. Thus, Newton had confirmed Kepler's law that the ellipse is the true path of the planets. In so doing he had again triumphed over Hooke, whom Newton derisively, though accurately, characterized as "a mere smatterer in mathematics."

Halley, who was almost as awe-struck by Newton's matter-of-fact manner as the magnitude of his revelation, pressed Newton for his mathematical proof. The scientist searched among his papers but could not find it. However, Newton promised the young astronomer that he would repeat his calculations and forward the results to him within a short time. Three months later Newton, true to his word, sent Halley a short tract entitled "De Motu" ("On Motion"), which contained a mathematical demonstration of the proposition that motion in an ellipse about an attracting body at one focus entails an inverse square law of attraction. During this same period Newton solved a number of other important problems on motion and devoted his autumn lecture course at Cambridge to the subject. In November 1684, a second and more detailed version of "De Motu" was sent to London and entered in the Register Book of the Royal Society on February 23, 1685, as "Propositiones de Motu." This treatise constituted the germ of the *Principia*, and was clearly meant to be a short account of the greatest scientific book ever written.

By taking the problem of orbital dynamics to Newton, Halley had thoroughly aroused the natural philosopher's unparalleled analytical powers. What is more, the tactful stranger convinced the reluctant physicist to divulge all that he knew about universal gravitation and its effects upon the planets, stars, comets and tides. The years 1685 and 1686 have, with justification, been frequently compared with the *annus mirabilis* of 1666, for Newton composed almost the whole of the *Principia* within the remarkably brief period of eighteen months. The undertaking subjected him to the severest mental strain of his life, and the numerous anecdotes, many of them true, about his absentmindedness and other idiosyncratic traits of character are largely related to this period.

On April 28, 1686, Newton presented and dedicated to the Royal Society the first of three volumes which together make up his most illustrious work. He chose to title it *Philosophiae Naturalis Principia Mathematica (Mathematical Principles of Natural Philosophy)*, but it is almost always referred to as the *Principia*. The history of the production of the treatise was anything but smooth; almost immediately two crises treatened the book's publication—one financial, the other professional. At their meeting of May 19, the members of the Royal Society agreed that the importance of Newton's work dictated that the Society should underwrite the cost of its publication; but on June 2, only two weeks later, this decision was reversed by a resolution that "Mr. Newton's book be printed, and that Mr. Halley undertake the business of looking after it and printing it at his own charge." Between its May and June meetings the Society, much to the embarrassment of its Fellows, had learned that there was no money in the treasury with which to issue Newton's treatise. It had, some time before, published a long-since forgotten volume by one Willoughby, titled *Historia de Piscium (A History of Fish)*. It is hardly an exaggeration of the facts to state that Willoughby's offering was something less than a best seller, and the Society was forced to adopt a number of rather unpalatable measures in an attempt to cut its considerable losses. For example, when the salaries of its officers fell into arrears the Council voted to pay them off in books of fishes." Halley, who served as second secretary, was agreeable; but Hooke, who was financially less well off, wisely asked for six months to consider his decision, hoping, no doubt, that the fortunes of the organization would reverse themselves in the interim. Halley, to his lasting credit, accepted the charge of his peers despite the fact that he had a family to support and was less prosperous financially than he had been a few years earlier. In fact, it is quite conceivable that without Halley the entire project might have been abandoned—and not for want of money alone.

No sooner had the financial question been resolved than a second and far more controversial issue came to the fore. Almost immediately after Newton's first installment had been presented, Robert Hooke raised his now familiar cry of plagiarism. That Hooke sincerely felt he had been wronged is indisputable: He based his claim to priority for the discovery of the inverse square law, announced by Newton in Book I of the *Principia*, on the contents of the letter he had sent to Newton some six years earlier. Hooke now charged that Newton intended to publish the solution to the problem of planetary motion along the very lines he himself had proposed.

This, for Newton, was the last straw, the ultimate defy! Aroused beyond control, he dashed off an angry letter to Halley in which he threatened to suppress Book III of the *Principia*, which contained the very capstone of his lifework. "Philosophy [science] is such an impertinently litigious Lady," he wrote, "that a man had as good be engaged in lawsuits, as have to do with her. I found it so formerly, and now I am no sooner come near her again, but she gives me warning." Unless the whole world was willing to receive the tablets containing his laws, and without reservation, Newton, the Moses of modern science, would smash them to bits in the manner of the ancient Hebrew lawgiver, who could not contain his rage when confronted with the infidelity of Israel's idolatrous children.

Halley, who had developed a far more subtle understanding of Newton's character than anyone else, acted quickly to avert disaster. He wrote Newton that Hooke only "seems to expect you should make some mention of him in the preface, which it is possible you may see reason to prefix." In other words, Hooke, for all his bluster, sought nothing more than a public acknowledgment of his assumed role in the solution of the problem. It would have been a magnanimous gesture on Newton's part, costing him nothing, for Hooke still did not know that Newton had formulated the inverse square law over a decade before he broached it in the letter of 1679. Instead, Newton went over his entire manuscript and deleted nearly every reference to his rival. Fortunately, however, the tact and devotion of Halley ultimately saved the day; for he succeeded in pacifying Newton to the extent that the publication of the work was allowed to proceed. Still, Newton's hatred for Hooke continued to feed upon itself. From this time onward he would have nothing further to do with the haunted, sickly man whose intellectual powers were rapidly slipping away. Not until Hooke died did Newton publish the *Opticks* or assume the presidency of the Royal Society. Yet Newton, though unwilling to admit it, owed as much or more to Robert Hooke than to any other natural philosopher of his day.

The *Principia* is far from an easy book to read. Its proofs are given in the form of equations in geometry even though it would have been simpler to obtain the same results by utilizing the calculus of Newton's own invention. Contrary to Whiteside's convincing counterargument, many historians of science still believe that Newton did in fact employ the calculus to arrive at his conclusions, but that he then reworked the problems using the mathematical system whose roots reach back to the ancients whom he so deeply admired. Be that as it may, Newton intentionally made the *Principia* abstruse in an attempt to keep unfounded criticism to a minimum, and, as he informed his friend the Reverend Dr. Derham, "to avoid being bated by little smatterers in mathematics . . . but yet so as to be understood by able mathematicians." Such hypersensitivity is strongly reminiscent of Copernicus' blunt pronouncement to potential readers at the outset of the long-delayed *De revolutionibus:* "for mathematicians only." Newton, moreover, shunned Galileo's use of the vernacular, choosing instead to compose his revolutionary work in classical Latin, still the international language of the scientific community.

The *Principia*, as noted above, consists of three books, the first of which deals generally with various problems of motion under the idealized conditions of no friction and no resistance. Book II is primarily concerned with the motions of fluids, and the effect of friction on the motions of solid bodies suspended in a fluid medium. It is in this book that Newton, by mathematical demonstration, succeeded in demonstrating that not only was Descartes' hypothesis of vortices nonessential to a scientific analysis of planetary motion, it was incapable of mathematical demonstration. Still, it is Book III, titled *System of the World*, which most concerns us, for it deals directly with the motions and mutual attractions of the celestial bodies with special reference to the moon, the nature and distances of the stars and comets, and the phenomena of the tides.

The essential problem faced by Newton was that of formulating a quantitative mechanics capable of unifying earthly and celestial motion into a single coherent system. Of the many pieces to the giant puzzle confronting him, the most important were Kepler's three laws of planetary motion and Galileo's laws of the motions of bodies on earth. Yet despite certain areas of affinity, these major components did not seem to fit together. For example, the forces which drove the planets in their Keplerian orbits appeared not to apply on earth. And vice versa, Galileo's laws of motion and falling bodies on earth apparently had little if any bearing on the movements of the planets, the moon, or the comets. According to Kepler, planets moved in ellipses while Galileo said they moved in circles; Kepler had said that the planets are driven along by "spokes" of forces issuing from the sun (the *anima motrix*), but Galileo, harkening back to the ancients, believed they were not driven at all because circular motion was self-perpetuating. To connect the work of his two illustrious predecessors Newton formulated his three laws of motion which appear in Book I of the *Principia*, but are expanded in Book III.

Newton's first law states that every body continues in its state of rest, or of uniform motion in a straight line, unless it is compelled to change that state by forces impressed upon it. Galileo, it will be recalled, had been the first to formulate this principle, which Newton, taking up where the Italian had left off, restated and integrated into his own system of mechanics. Newton's second law holds that the change of momentum (or the change of velocity times the mass of a body) is proportional to the force impressed. In this instance Newton borrowed from Galileo the idea that the acceleration in a body's motion could be taken as a measure of the effect produced by one body on another, or the measure of forces acting upon such a body. But Newton, as was characteristic of his genius, carried this a giant step further by introducing the concept of "mass." In experimenting with pendulums, Newton had been able to show that the mass of a body is different from but proportional to its weight. Furthermore, mass is independent of the body's position but dependent on its motion with respect to other bodies. This constituted a fundamental breakthrough because Newton clearly recognized mass as a property of all bodies, a property dynamically fundamental and measurable.

Unlike his first two laws of motion, Newton's third law was totally of his own making and involved the concept of force, a principle little understood by his contemporaries. The third law states that to every action there is always opposed an equal and opposite reaction. This principle applies when we are concerned with the motion of two or more bodies subject to their mutual attractions: if one body acts upon another at a distance, the second also reacts on the first with an equal and oppositely directed force. Hence, the moon pulls the earth with the same force with which the earth pulls the moon. The same is true of the earth and an apple, except that in this instance the force exerted causes the apple to visibly change its position while the earth, because of its far greater mass, seems totally unaffected. With his laws of motion Newton exactly defined for the first time the concepts of mass, inertia, and force, and their relation to acceleration and velocity. He thus founded that branch of modern physics now known as dynamics.

We have observed that Newton, long before he wrote the *Principia*, formulated the inverse square law, applying it with a considerable measure of success to the attraction of the earth for the moon and of the sun for the planets. But, according to his third law of motion, gravitation could no longer be thought of as a property peculiar to the central body of a solar system; it must apply to every planet, star, comet, and moon in the universe! Yet before he could clearly establish the validity of the sweeping principle of universal gravitation, one further point needed to be settled; namely, that one spherical body would act gravitationally toward another body as if all its mass were concentrated at its center. This Newton triumphantly demonstrated in Book I of the *Principia*, and by so doing, he was able to establish that gravity is indeed a universal property of *all* bodies, solely dependent upon the amount of matter each contains. In Proposition VII of Book III Newton announced the most celebrated of all his scientific discoveries, the law of universal gravitation which states: every particle of matter attracts every other particle with a force proportional to the product of the masses and inversely proportional to the square of the distances between them. With this single law of physics Newton "democratized" the universe as it were by laying to rest once and for all the belief in the concept of hierarchical dominance among the celestial bodies, an idea which appealed as strongly to the mystical Kepler as it had to the practical-minded Aristotle. In the infinite universe envisioned by Newton, no one body is more important than any other; all matter—from the largest of the stars to the smallest of invisible atoms—obeys the same invariant principle. This being established, Newton now found himself in a position to explain how his extended conception of gravitation tied together astronomical phenomena which for countless centuries had perplexed the finest minds in the history of scientific thought.

Ironically, the very law which promised to clarify man's understanding of the cosmos through the unification of all matter presented its discoverer with innumerable difficulties when he came to the calculation of planetary, lunar, and cometary motions. To speak or to write in the abstract of the law of universal gravitation is one thing, to demonstrate it in concrete physical terms is quite another. For contradictory though it may seem, the very strength of Newton's law—its universal applicability—also constituted the greatest drawback to its clearcut demonstration and acceptance by men of the late seventeenth century. Consider, for example, the problems encountered by Newton when he attempted to calculate the orbit of Saturn round the sun. Had it only been a matter of dealing with the relatively simple problem of two mutually attracting bodies, Newton could have rather easily deduced the planet's orbital motion on the basis of the studies he undertook between 1666 and 1672. But, as Newton well knew, the problem is made far more complex because while Saturn revolves around the sun it is also influenced by the attraction of other bodies, most notably its neighboring planet Jupiter, the largest in our solar system. Although such secondary attractions are minor when compared with the pull of the sun, which contains about a thousand times more matter than all the planets and their satellites put together, they nevertheless produce small disturbances or perturbations from the true orbit, deviations which must be taken into consideration by the modern physicist. Even Newton, armed as he was with the calculus and blessed with unmatched analytical powers, could not resolve such problems in more than a preliminary way. Yet by considering the gravitational effect of a third body, Newton was able to achieve results of a far more accurate nature than any suggested by his many gifted contemporaries. The pioneering study of planetary perturbation begun by Newton became increasingly sophisticated during the eighteenth and nineteenth centuries. In 1846 the planet Neptune was discovered by means of it.

Of much greater concern to Newton than the relatively minor disturbances in the orbits of the distant planets were the irregularities in the moon's motion, which had been known to astronomers for centuries. Even though the moon is controlled by the pull of the earth, the enormous mass of the sun, even at its far greater distance, visibly disturbs

Sun exerts x 2 times force of Earth on Moon

the normal lunar orbit. And unlike the perturbations of the planets, those of the moon are far more numerous and much larger, making lunar theory one of the most complicated and difficult in astronomy. Through an extremely complex system of calculations Newton succeeded in accounting for most of these major disturbances. He also discovered other irregularities which had not been previously observed, and indicated the existence of similar aberrations in the motions of the moons of Jupiter and Saturn. And, as a corollary to his calculations of the motions of the planets and their satellites, Newton demonstrated how to estimate the masses of the planets and of the sun from the mass of the earth. His results, quite understandably, were not as reliable as those computed by the latest methods, but they were remarkably accurate for the time. For example, Newton computed the earth's density to be between five and six times that of water: the accepted figure is almost exactly five and one-half!

One of the most interesting deductions offered by Newton in the *Principia* concerns his belief that the earth is an oblate body. In other words, the planet is somewhat flattened at the poles and bulges slightly at the equator, much like a balloon gently compressed between one's hands. He suspected this because a rapidly spinning body composed of any kind of yielding material tends to swell at its circumference and shrink along its axis. Thus, a plastic sphere the size of the earth, held together by its own gravity, and rotating once a day, can be shown to have an equatorial diameter a few miles greater than its polar diameter. Newton's basic proof, however, was gathered from experiments with pendulums and columns of water, the details of which need not be related here. He also calculated the oblateness of Jupiter which, owing to its more rapid revolution, is considerably greater than that of earth. This latter result was confirmed telescopically by the Italian astronomer Giovanni Domenico Cassini only a few years after the publication of the *Principia*.

Based on his study of the nonspherical shape of the earth, Newton made an additional and far more significant discovery. A sphere, as he had demonstrated, attracts another body as if its mass were all concentrated at its center, but a spheroid does not. This means that the intensity of the earth's gravitational field will not be the same everywhere. As a spheroid, it pulls and is in turn pulled by the moon with a slightly uncentric attraction, meaning that the line of pull does not pass directly through the earth's exact center. What we are dealing with, in effect, is a giant top slightly overloaded on one side, so that gravity acts on it asymmetrically. What is the result? The axis of the earth begins to very slowly change its angle of rotation so as to trace out the shape of a cone in the heavens, a movement known to astronomers as "the precession of the equinoxes." This phenomenon was first observed by the Greek

Or by the Hindus several hundred years before.

astronomer Hipparchus in the late second century B.C., but its significance had continued to elude astronomers ever since. Copernicus devoted a good deal of attention to it in *De revolutionibus* but, like his puzzled predecessors, he failed to shed much practical light on the phenomenon. It is little wonder, for astronomical observation had not been going on long enough to measure more than a small fraction of the complete axial cycle.

Newton undertook the calculation of this conical motion, which he correctly ascribed to the slightly asymmetrical attraction of the moon. He found that it takes about 26,000 years for the earth's axis to complete the cone, or exactly what was wanted to explain the precession of the equinoxes. Newton had discovered both the physical cause of the precession and calculated the extended time frame in which it occurs. As E. N. da C. Andrade, in a classic example of British understatement, remarked: "A calculation of this kind is not the same kind of thing as suggesting that an apple falls because the earth pulls it!"

Of all the mysteries which confounded astronomers through the ages none produced greater consternation and less illumination than the continuous fluctuation of the world's great oceans. According to legend, Aristotle became so despondent over not being able to explain the phenomenon of the tides that he committed suicide by flinging himself into the sea. Kepler came the closest to a correct explanation of tidal motion by attributing it to magnetic forces generated by the sun, earth, and moon, a concept he borrowed from the English natural philosopher William Gilbert. But Galileo dismissed this idea as simply another example of Kepler's superstitious preoccupation with occult forces. He explained the tides as a result of the combined effects of the earth's rotations on its axis and its movement around the sun, thus linking the phenomenon to purely mechanical principles without having to appeal to any such force as the attraction of the moon on the waters, which Galileo considered nothing less than an occult property. Descartes, too, had his own view on the subject: he believed that when the moon approached earth it pressed down upon the planet's waters by the centrifugal force of its vortex and hence produced a low tide under it. Conversely, a high tide resulted on those areas of the earth's surface which were relatively free of the moon's vortical pressure.

Newton, in one of the *Principia's* most brilliant yet simply stated propositions, wrote: "That the flux and reflux of the sea arise from the actions of the sun and moon." By applying the law of gravitation to the problem Newton found that the power of attraction is greater on the waters facing the attracting body than on the earth as a whole, and greater on the earth as a whole than on the waters on the opposite side. Since in this instance the moon is the more powerful of the two attracting bodies, the main tides are due to the earth's satellite. Its chief effect is to cause a pair of waves or ocean humps, of tremendous area, to travel round the earth once in a lunar day, or a little less than twenty-five hours. The sun also causes a similar but lower pair of waves to circle the earth once in a solar day of twenty-four hours. The effect of these two pairs of waves periodically overtaking each other accounts for the spring and neap tides. The spring tides, the more marked of the two, occur when the sun, moon, and earth are in a line, thus exerting a maximum gravitational pull on the earth's surface. The neap tides result when the sun and the moon, both attracting the waters of the seas, pull at right angles to one another. These alternating alignments regularly occur every two weeks. Though Newton's theory was not sufficiently refined to furnish predictions of the exact time or height of a tide at any given place, it did provide a satisfactory explanation of the main details of tidal action and of the general characteristics of the tides themselves. Yet another tremendous advance in scientific knowledge was his to claim.

Equally as unsuccessful as the many attempts of astronomers to provide a sound explanation for the rise and fall of the seas were their efforts to account for the sudden appearance and disappearance of the comets. Until shortly before Newton's time, these mysterious "visitors" had been regarded as nothing more than vaporous exhalations from the earth launched in the higher regions of the terrestrial atmosphere. Although generally recognized by the natural philosophers of the seventeenth century as celestial bodies, no one had as yet been able to account for their seemingly irregular movement across the vast heavens. Since Newton supposed comets to be composed of matter, he reasoned that they must be acted upon by gravitational forces in exactly the same way as the planets. Yet when he employed observational data collected in part by Astronomer Royal John Flamsteed, with whom Newton quarreled almost as relentlessly as he did with Hooke, he found that the motions of the comets were far more complex than those of the planetary bodies. As a result, Newton set out to demonstrate that besides elliptical orbits, parabolic and hyperbolic orbits also lead to the law of gravitational attraction. Newton soon gave substance to this conjecture by stating "that the comets move in some of the ⌐onic sections, having their foci in the center of the sun, and by radii drawn to the sun describe areas proportional to the times." In other words, he assumed cometary orbits to be widely extended ellipses of great eccentricity, which at their tops were so nearly parabolas that parabolas could be substituted for them.

Suddenly, comets were no longer the new creations or celestial disturbances the ancients thought them to be; nor were they phenomena which suddenly disappeared, never to be seen again. Moreover, every comet must return with the same regularity of a planet unless seized by the gravitational force of another star more powerful than the sun. Edmond Halley, employing Newton's complex method of calculation, carefully plotted the orbit of the brilliant comet of 1682. Then, through a painstaking search of past records, he determined that there had been reports of similar sightings in 1531 and 1607, or about once every seventy-five years. Could not these sightings, Halley conjectured, indicate successive appearances of the same object, rather than the regularly spaced arrival of three unrelated comets? He calculated the orbit on the assumption that if, in fact, this was one and the same comet, it would next appear in 1758, give or take a year. The huge, fiery body, which now bears Halley's name, was first observed on Christmas Day, 1758, by an amateur astronomer named George Palitsch. Since then it has appeared twice more, in 1835 and again in 1910. Halley's confirmation of Newton's reduction to rule of the complex movements of the comets, and their inclusion with the planets in the category of bodies orbiting the sun, may be justly regarded as one of the most profound of the many physical discoveries first announced in the *Principia*.

It will perhaps be recalled that one of the central objections raised by those who in the beginning strongly opposed Copernican theory was that it required a universe of such immense proportions as to challenge the credulity of rational men, not to mention the teachings of the Church. Though what Koyré calls "the world-bubble" swelled considerably during the century and one-half since the death of the reluctant revolutionary, Newton was the first natural philosopher to establish a true idea of the distances separating the celestial bodies, especially the stars. Like Copernicus, Newton noted that stellar parallax was too inconsequential to measure, but this in itself proved relatively little.* However, from the principal of universal gravitation Newton calculated that the stars must be hundreds of times more remote than Saturn, then the most distant planet known. Were it otherwise, they would either fall into the sun or swing into orbit around it. Moreover, at such great distances the stars would not be visible by reflected lights as are the planets; they must be self-luminous bodies like the sun. And if indeed the stars are suns like our own, they too must act as centers of other planetary systems.

Newton further explained his views on the extension of matter and of the universe in a letter sent to his friend the Reverend Dr. Richard Bentley in 1692:

---

* It was not until 1837 that the phenomenon of stellar parallax, whose existence had been crucial for the heliocentric theory since Aristarchus, was finally established by Friedrich Wilhelm Bessel in Germany and Wilhelm Struve in Russia.

It seems to me that if the matter of our sun and planets, and all the matter of the universe, were evenly scattered throughout all the heavens, and every particle had an innate gravity towards all the rest, and the whole space throughout which this matter was scattered was but finite, the matter on the outside of this space would, by its gravity . . . fall down into the middle of the whole space, and there compose one great spherical mass. But if matter was evenly disposed throughout an infinite space, it could never convene into one mass; but some of it would convene into one mass, and some into another, so as to make an infinite number of masses, scattered of great distances from one to another throughout all that infinite space.[31]

Thus, Newton's universe is an infinite void of which only an infinitesimal part is occupied by unattached material bodies moving freely through the boundless and bottomless abyss, a colossal machine made up of components whose only attributes are position, extension, and mass. Life and the sensate world have no effect upon it, and are banished from its rigorously mechanical operations. Man's only contact with the objective world is limited to that of observing and interpreting its manifold phenomena through what Louis More describes as "mechanical actions impressed on his nerves and by them transmitted to a sensorium in his brain." Man's soul, assumed to be a rational entity apart from the brain, "then interprets these nerve stimuli as sight, sound, taste, etc.; interpretations which are purely subjective and incapable of mechanical formulaion." [32] Newton accepted the now familiar concept of primary and secondary qualities; hence, for him, as for Galileo and Descartes, the gap between the mechanistic outer world and the inner world of sensations, emotions, and ideas was unbridgeable.

Yet for all its lack of feeling, Newton's universe is a precise, harmonious, and rationally ordered whole. Chaos and disunity have been barred through the application of a mathematical law which binds each particle of matter to every other particle on the basis of universal attraction. By flinging gravity across the infinite void, he was able to unite physics and astronomy in a single science of matter in motion, fulfilling a dream of astronomers from Pythagoras to Kepler. And even though Newton was unable to discover a demonstrable principle with which to actually explain the phenomenon of gravitation, the laws he formulated provided convincing proof that man inhabits a preeminently orderly world. We honor Newton today not because he provided us with ultimate answers to extremely complex scientific problems—a proposition that he himself would have rejected, and one which Einstein, among others, has disproved—but because Newton, in apprehending the Pythagorean power by which number holds sway above the flux, contributed more than any other individual of the modern age to the establishment and acceptance of a rational world view.

Newton and God

Isaac Newton—even among those who know relatively little about his work—is most commonly remembered as the scientific genius who pictured the cosmos as a great machine, running according to immutable mathematical principles. Frequently overlooked is the fact that Newton was a profoundly religious man, who sought to develop a theory of the universe which would accommodate both an interventionist God and the new science. For, like Boyle, Newton stopped far short of accepting Descartes' total mechanization of the natural world. And though he resembled his fellow virtuosi in that he emphasized a rational approach to knowledge as the best means of exploring the works of the Creator, Newton went considerably beyond the others in his attempt to bring Christianity into conformity with science.

All during his life Newton sought to determine the extent of God's role in physical causation, a complex question he never resolved to his complete satisfaction. This much can be said with certainty, however: though Newton was a confirmed mechanist, he rejected the Cartesian belief that mechanical laws alone are sufficient to explain the phenomena of the universe. Newton, in the same letter to Bentley quoted from above, confessed to his inability as a scientist to explain how—on the basis of mechanical causation—matter originally divided itself to compose shining bodies like the sun and opaque bodies like the moon. "I do not think [this phenomenon] explicable by mere natural causes," he wrote, "but am forced to ascribe it to the council and contrivance of a voluntary Agent." Nor, Newton continued, "is there any natural cause which could give the planets those just degrees of velocity, in proportion to their distances from the sun and other celestial bodies, which were requisite to make them move in concentric orbs about those bodies."

On this point Newton seemingly differed little from Descartes (or for that matter from Aristotle) who pictured God as the First Cause, the creator of the world. But whereas Descartes perceived of God as an impersonal and uninterested master mechanic, who once having accomplished his task became little more than an absentee deity, Newton struggled to maintain a role for God by arguing that He was left with essential operations to perform. Indeed, Newton saw science as of little intrinsic value except as a means of collecting further evidence of the laws and attributes of God. And even though he employed a modified form of Bacon's experimental method to discover those laws, Newton was essentially unmoved by the Baconian belief that science should or could be used as a tool to insure man's domination of nature. Newton, like Kepler, worked for the greater glory of God, rather than for the more mundane purpose of aiding mankind in any material sense. For both men, science was an integral part of their religious experience, and was always subject to the moral sanctions and Divine revelations of the Bible. Moreover, Newton, as was noted earlier, was completely free of the notion that an understanding of nature can ever be fully embodied in an absolutely certain and definitive science. Rather, he sensed that no matter how rational the world may appear to us, there is still nothing below a certain depth which is truly explicable in human terms. Furthermore, Newton's view of the cosmos was cyclical as opposed to Descartes' belief in a self-perpetuating world order. In the 1717 Queries to the *Opticks*, Newton advanced the theory that nature was gradually running down, as evidenced by the increasing irregularities in the motions of the planets. This tendency, Newton believed, if not reversed, would eventually lead to the cessation of all movement and an end to life on earth. Just as God was responsible for the creation, it was also His place to initiate a renewal of the motion necessary to reform the planetary orbits. He had doubtless done so before, and there was no reason to believe that He would not do so in the future, unless, of course, He was ready to render His final judgment of man as promised in the Apocalypse. And for God to produce activity, Newton did not question the traditional Christian belief that He would have to be where He produced it. To think otherwise, one would have to accept Descartes' assertion that matter was independent of God, a conclusion that Newton rejected because of its atheistic connotations. For if one separated God from the universe of His creation, one might just as easily abolish the Creator altogether. Hence, matter, motion, and space, as Newton defined them, could not be conceived of without supposing God's existence and perpetually guiding presence.

In order to explain how God could constantly be the active agent in a world which but for His presence would revert to the state of an amorphous and motionless matter suspended throughout an infinite space, Newton advanced the idea of the *sensorium*, which he borrowed from Henry More and the Cambridge Platonists. This concept holds that God is an infinitely extended intelligence who is everywhere and in everything. The whole material universe, created by God and endowed by Him with special properties, is contained in the vastness of God Himself: it is the crucible in which all physical movement ordained by Him takes place. We would do well to let Newton speak for himself on his concept of the Almighty, which he does in the *General Scholium* to the second edition of the *Principia*.

He [God] is eternal and infinite, omnipotent and omniscient; that is, his duration reaches from eternity to eternity; his presence from infinity to infinity; he governs all things, and knows all things that are or can be done. He is not eternity and infinity, but eternal and infinite; he is not duration or space, but he endures and is present. He endures forever, and is everywhere present; and, by existing always and everywhere, he constitutes duration and space. . . . In him are all things contained and moved; yet neither affects the other: God suffers nothing from the motion of bodies; bodies find no resistance from the omnipresence of God. It is allowed by all that the Supreme God exists necessarily; and by the same necessity he exists *always* and *everywhere*. Whence also he is all similar, all eye, all ear, all brain, all arm, all power to perceive, to understand, and to act; but in a manner not at all human, in a manner not at all corporeal, in a manner utterly unknown to us. As a blind man has no idea of colors, so have we no idea of the manner by which the all-wise God perceives and understands things. He is utterly void of all body and bodily figure, and can therefore neither be seen, nor heard, nor touched; nor ought he to be worshipped under the representation of any corporeal thing. We have ideas of his attributes, but what the real substance of anything is we know not. In bodies, we see only their figures and colors, we hear only the sounds, we touch only their outward surfaces, we smell only the smells, and taste the savors, but their inward substances are not to be known either by our senses, or by any reflex of our minds: much less, than, have we any idea of the substance of God.[33]

Just as man cannot know the ultimate nature of God, neither can he know the ultimate nature of the universe created by Him. The power and wisdom of the Almighty are beyond man's inconsequential grasp. For this reason, were Newton alive today, I seriously doubt that he would be considered a good scientist. He was not sufficiently proud or confident of the powers of the human intellect. There were many times when, like the great Charles Darwin, Newton wobbled and held back, for neither man possessed the supreme confidence in their respective systems that their followers had. It is too easily forgotten that those who bring forth new ideas are seldom ruled by them. Thus to Newton the doctrine of mechanical causation was only one minor aspect of God's unfathomable method of overseeing His stupendous creation, a shrug of the Divine shoulders which man, with his severely limited understanding, had somehow succeeded in translating into mathematical principles. Nothing would have surprised or dismayed Newton more than to have seen his scientific laws triumph at the expense of the religious principles which he held so dear. The modern doctrine of scientific progress, whose formulation was largely made possible by Newton's unprecedented discoveries, would not have appealed to him in the least, much less the perversion by contemporary thinkers of his mechanical principles in a facile attempt to explain the very basis of human behavior itself.

The Parting

The *Principia* was finally published during the summer of 1687, and its appearance immediately raised Newton to a position of international prominence. Even the Continental scientists, who in their unwavering loyalty to the mechanical ideal rejected the concept of action at a distance, were unable to contain their awe for the technical expertise revealed by the book. Meanwhile, the brightest young British scientists collectively adopted Newton's system as their model.* Though the publication of the great work did not mark the end of Newton's brilliant career as a scientist, it left him exhausted and anxious to turn to other things. He was doubtless aware that in establishing the laws of the universe he had reached the pinnacle of his genius, that the flood of revolutionary ideas had been permanently checked. What, after all, was there left for him to achieve after he had given birth to a new system of the world? He could only amend, reform, and restate principles which had been formulated during his younger days.

Moreover, there is a distinct possibility that the very severe intellectual and physical strain involved in the production of the *Principia* permanently impaired Newton's creative powers in physics and mathematics. I am reminded of Bertrand Russell's autobiographical account of his own struggle, between 1902 and 1910, to finish the *Principia Mathematica*, which he coauthored with the mathematician–philosopher Alfred North Whitehead. "I persisted, and in the end the work was finished," Russell wrote, "but my intellect never quite recovered from the strain. I have been ever since definitely less capable of dealing with difficult abstractions than I was before. This is part, though by no means the whole of the reason for the change in the nature of my work." [34] If Manuel is correct, Newton's growing preoccupation with Biblical history and chronology may have also been a refuge, for "it represents a diminution of his generalizing capacity." [35]

Yet this is not to assert that Newton's intellectual powers precipitously declined; quite clearly they did not, as illustrated by the following example. In June, 1696, the Swiss mathematician Jean Bernoulli challenged "the acutest mathematicians of the world" to solve a particularly difficult problem requiring the use of the calculus, which by this time had been independently discovered by Leibniz. Bernoulli allowed one year for the solution in order that all contestants should have an equal chance. Newton learned of the problem on January 29, 1697, and he submitted the correct answer to the President of the Royal Society the very next day. Though his calculations were published anonymously in the *Transactions*, Bernoulli immediately recognized the lion of mathematics by what he referred to as the unmistakable mark of his claw; *tanquam ex ungue leonem.*

Whatever the condition of Newton's mental faculties, the fact remains that once the *Principia* had been completed he found it increas-

---

* For a general discussion of the popularization of the Newtonian system see David Layton, "Motions of the Planets: Newton's Effect on English Thought," *History Today*, Vol. 7 (June, 1957), pp. 388–395.

ingly difficult to settle back into the scholarly routine of the academic
community. One reason is that for the first time in his life he became
deeply embroiled in political affairs. Newton, a dedicated if somewhat
unorthodox Protestant, assumed a role of leadership in the resistance of
Cambridge to James II's attempt to Catholicize the University by making
faculty appointments favorable to the crown. As a result, Newton was
elected to Parliament by his grateful colleagues, and thus became a
member of the historic convention which arranged the revolutionary
settlement following James' ignoble flight to the court of Louis XIV in
1688. Newton, now forty-seven, began to take a genuine interest in the
broader world which he had so adamantly rejected as a young man.
In London he made the acquaintance of an impressive group of talented
men, including the powerful Whig leader Charles Montague (later Lord
Halifax), philosopher John Locke, and Samuel Pepys, the famed diarist
and President of the Royal Society. About the same time, Newton
became friends with Fatio de Duillier, a young Swiss mathematician,
who had taken up residence in the capital. Fatio, according to his own
ingratiating account, had studied the *Principia* in greater depth and with
more intensity than any other natural philosopher. Newton, flattered by
Fatio's attentions, soon admitted him to his inner circle of devoted followers. So strong was the bond between them that Westfall characterizes
Newton's affection for Fatio as the most profound experience of his
adult life.

Despite his prestige as a scientist, Newton had no real prospect of
obtaining a mastership at Cambridge because his anti-Trinitarian views
precluded him from taking holy orders. Certainly, he was not badly off
financially as Lucasian Professor, but he had become increasingly dissatisfied with both his status and his way of life. Spurred on by Fatio's
suggestion that he find a position in London, Newton turned to his
influential friends for assistance in obtaining an administrative post. First,
an effort was made to have him appointed Provost of King's College,
Cambridge, but it proved unsuccessful. He then asked Locke to aid him
in securing the post of Comptroller of the Mint, again without success.
Locke next sought to secure him the Mastership of the Charterhouse,
an appointment Newton refused on the grounds that it was far too
demanding for the modest annual salary of £200. Newton had also
solicited the help of Charles Montague, who, like Locke, did the best
he could for his friend, but for the time being nothing came of it. Ever a
victim of his own suspicions, Newton gradually slipped into one of his
darkest moods; his friends, he thought, must be conspiring against him.
Matters were made worse by the failure of his alchemical experiments
which, though pursued at a fever pitch between 1691 and 1693, yielded
nothing of major scientific value. Moreover, his plans to have Fatio
join him for an extended visit at Cambridge never materialized. Dejec-

tion gradually gave way to delusion, until finally in 1693, Newton suffered a second nervous breakdown, one similar to but far more severe than that of 1664.

The first intimation of the impending mental crisis was contained in a letter to John Locke in which Newton blamed Charles Montague for his unsuccessful efforts to obtain a position. "Being convinced that Mr. Montague is false to me," wrote Newton, "I have done with him." Several months later an astounded Samuel Pepys received the following message dated September 13, 1693.

> I am extremely troubled at the embroilment I am in, and have neither ate nor slept well this twelve-month, nor have my former consistency of mind. I never designed to get any thing by your interest, nor by King James's favor, but am now sensible that I must withdraw from your acquaintance, and see neither you nor the rest of my friends any more, if I may leave them quietly. I beg your pardon for saying I would see you again, and rest your most humble and most obedient servant,
>
> Is. Newton.[36]

Only three days later John Locke, now a co-conspirator in Newton's imagined plot, was equally as stunned as Pepys when a letter arrived from Newton with the following charge: "being of [the] opinion that you endeavoured to embroil me with women and by other means, I was so much affected with it, as that when one told me you were sickly and would not live, I answered, 'twere better if you were dead.'" [37]

Alarmed for their friend's sanity, both men overlooked Newton's accusations and answered his letters in a kindly and dignified fashion. In a subsequent reply to Locke, Newton stated that, "When I wrote to you, I had not slept for an hour a night for a fortnight together, and for five nights together not a wink." And though he recalled writing to the political philosopher, Newton could not remember what he had said. Fortunately, his recuperative powers and resilience were almost as extraordinary as his mental capacity, and the delusion, sleeplessness, and despondency soon passed. By the end of 1693—the Black Year—Newton seems to have returned to his normal self. Yet only briefly did he ever again undertake sustained scientific work; 1693 marks the effective conclusion of the most creative period in the life of any scientist the world has yet seen.

Earlier in this chapter, I observed that no survey can do justice to the contributions made by the astronomers discussed in these pages, least of all the work of Sir Isaac Newton. For what are obvious reasons, I have focused upon the first fifty years of his long and always fascinating life. The reader who wishes to follow Newton's career as the Master of the Mint, chronologist of Biblical history, as the virulent opponent of Flamsteed and Leibniz, and as the autocratic President of the Royal Society is encouraged to consult other more comprehensive works. However, before concluding this chapter, one further observation seems to be in order.

Time and again, during the course of my research and writing, I have asked myself the imponderable question: How does one explain the genius of a Newton—or for that matter the genius of any of the extraordinary men we have come to know? Obviously, one must assume that each of them was born with a wealth of intellectual powers denied we ordinary mortals. Furthermore, Newton, in addition to his surfeit of natural talent, possessed an unmatched capacity to compartmentalize his activity and studies to the point of totally excluding any intrusions from the outside world. When he was once asked by a friend how he made his discoveries, Newton tersely replied, "By always thinking unto them." On another occasion he reiterated this view, "I keep the subject constantly before me and wait till the first dawnings open little by little into the full light." Andrade, More, and other biographers have concluded that Newton, unlike most great thinkers who have found it impossible to fully concentrate on a problem for more than an hour or two, sat permanently transfixed, oblivious to his surroundings or the passage of time until a solution was forthcoming. It is a view which is difficult to argue with given the prodigious amount of work he was able to complete during his two most creative periods. Unquestionably, Newton experienced ecstasy in the truest sense of the word—a state of exalted delight in which he was continuously able to transcend the boundaries of normal understanding.

Still, this is only a description of how his mind worked rather than an explanation of why it functioned as it did. In this respect, it is much like Newton's inability to explain the true cause of universal gravitation even though he was satisfied that it exists. Neither do we know what genius is and must consequently satisfy ourselves with the fact that, like gravity, it is there for the benefit of all. The truth is, that despite the intensive research conducted on man and his behavior, we do not yet know whether the mind can truly understand the mind; unless and until we do, human genius, by virtue of the awe and wonder it commands, is bound to remain an elusive and somewhat mystical quality.*

Isaac Newton, knighted by the Queen in 1705, was 85 years old when he died on March 20, 1727. His body, after lying in state in Jerusalem Chamber of Westminster Abbey, where it had been brought from his home in Kensington, was buried in a place of high honor in front of the beautiful choir. Voltaire was present at the funeral, as were the leading members of the Royal Society and numerous peers of the

---

* For a more detailed analysis of this question see Gerald Holton, "On Trying to Understand Scientific Genius," *The American Scholar*, Vol. XXXXI (Winter, 1971–72), pp. 95–110.

realm. Newton, I think, would have been deeply pleased to know that, in death, he had at last attained the position of prominence and respect to which he had aspired ever since he enrolled at Cambridge as a callow Sizar some sixty-six years earlier.

Strange indeed are the twists and turns of history, especially when reflected in the poet's eye. On learning of the new Copernican astronomy, John Donne was literally driven to the edge of despair as he contemplated the ultimate collapse of the medieval order:

> 'Tis all in pieces, all cohaerence gone;
> All just supply and all Relation.

But little more than a century later his countryman, Alexander Pope, exulted in the power and majesty of the Newtonian world view in the epitaph which he composed for the fallen giant:

> Nature and nature's laws lay hid in night:
> God said, let Newton be! and all was light.

Truly, it was not a bad end for a country lad from Lincolnshire, who passed many a pleasant hour of his solitary childhood daydreaming—of what we shall never know—as his homemade kite swam, like a shimmering skyfish, into the gentle summer wind.

9. The Copernican Revolution: The Nature of Scientific Discovery

The Fairy Ring

IF WE THINK OF SCIENCE as that procedure through which the human mind develops consistent and comprehensive theories about the natural world, we are at once struck by its discontinuity as a historical process: only twice in mankind's history has science flourished vigorously as a major intellectual activity. The first great burst of critical speculation began in classical antiquity during the sixth century B.C., when the brilliant philosophers of Ionia started to investigate the form and substance, the structure and laws, the origins and evolution of the universe. This heroic quest ended, a few centuries later, in the wake of the Macedonian conquest of Greece during the Alexandrian Age. The search for simple, ultimate principles underlying all diversity was abandoned as an unattainable goal; and the central philosophical and intellectual tradition that supported scientific inquiry was replaced by a growing preoccupation with Oriental religions on the one hand, and with technology on the other. Even Ptolemy's constructions of the heavens, brilliant though they were from a technical point of view, reflected the permutations of the age. Indeed, Copernicus was so put off by the lack of unity and of esthetic sensitivity in the Alexandrian's picture of the world that he complained in his prefatory letter affixed to *De revolutionibus:* "It is as though an artist were to gather the hands, feet, head, and other members for his images from diverse models, each part excellently drawn, but not related to a single body, and since they in no way match each other, the result would be monster rather than man." The Heroic Age had taken as its guiding example Prometheus, who had dared to steal fire from the gods; but "the philosophers of the Hellenistic period," writes Arthur Koestler, "dwelt in Plato's cave, drawing epicycles on the wall, their backs turned to the daylight of reality." [1]

A millennium and more passed before Western man again found

himself back on the scientific road opened up by Thales, Pythagoras, and Aristotle. Only with Copernicus' declaration that the result of Hellenistic astromony is more "monster rather than man" was the way prepared for the establishment of scientific discovery in the modern sense. *De revolutionibus*, which after a lapse of seventeen centuries reintroduced astronomers to the long-neglected heliocentric planetary system of Aristarchus, inaugurated the second heroic age of science—the Copernican Age, the age in which we still live. Having discussed the lives and work of the colossi who laid the scientific foundations of the modern era—Copernicus, Tycho, Kepler, Galileo, Bacon, Boyle, Descartes, Hooke, and Newton—the time has come to attempt an analysis, however tentative, of the process of scientific discovery and change, particularly as it relates to the Copernican Revolution.

The most comprehensive analysis of scientific revolutions to date is that undertaken by Copernican scholar Thomas S. Kuhn. In his pioneering work titled *The Structure of Scientific Revolutions*, Professor Kuhn has developed a method with which he attempts to interpret the nature of change in science, and to determine what factors separate the various sciences from the study of history, philosophy, music, art, literature, and the many other academic disciplines. Though Kuhn's thesis, as we shall see, is not without its detractors, it is a tribute to the vitality of his stimulating analysis that many of the finest minds in the history of science have felt his work worthy of lengthy comment, and not only in the classroom but in published form as well. The very least that can be said of *The Structure of Scientific Revolutions* is that it has served as a powerful catalyst in refocusing the attention of scholars on the concepts and doctrines which are fundamental to an interpretation of the development and structure of modern science. For those who have demonstrated their loyalty by following this narrative from its beginning, an analysis of Kuhn's work will provide a final opportunity to pause and reflect even more deeply on what is truly one of the most profound chapters in the intellectual history of Western man. Thus, we shall examine Kuhn's thesis in some detail, keeping in mind (as has its originator) that his explanation of the nature of scientific revolutions will doubtless require further modification and revision as historians continue to learn more about this always fascinating process.*

It will be helpful, I think, if we go back in time for a moment to

---

* Since Professor Kuhn first advanced his thesis some two decades ago, he has subsequently revised it both in response to those critical of certain aspects of his methodology, and in light of further thought and research of his own. The most recent edition of his work was used in writing this chapter: Thomas S. Kuhn, *The Structure of Scientific Revolutions*, 2nd ed. (Chicago, 1970). Still, it is only fair to caution the student unfamiliar with Kuhn's thesis, a circumstance which doubtless applies to most readers of this essay, that it is a complex and often very subtle presentation of highly complicated issues. Only when the reader compares the first and second editions of Kuhn's work will the differences become apparent. Such comparisons are, for the most part, beyond the scope of this, a general presentation.

the early sixteenth century, to the period just prior to the publication by Copernicus of *De revolutionibus*. The vision of the cosmos accepted by most thinking men in Western Europe was a combination of the thought of Aristotle, Ptolemy, and the medieval church. The universe was thought to be earth-centered, finite, and operating according to two sets of principles, one celestial and the other terrestrial. Then, as now, the student of nature constructed a conceptual model of the world, and with this model attempted to explain the structure and behavior of all natural phenomena. The practice of model building, as we observed earlier in this essay, originated with Thales and his Ionian followers during the sixth century B.C. It proved itself a most practical device for picturing nature as a rationally ordered whole, particularly in such fields as astronomy, and for providing promising young scholars with a conceptual framework to aid them in mastering the teachings of the particular school of thought with which they were to become associated.

For upwards of 2000 years the Aristotelian model of the universe, modified by Claudius Ptolemy during the second century A.D., held sway over the thought of the Christian and Moslem worlds, though it was lost to the West during the long Dark Age following the demise of ancient Rome. By Copernicus' time, however, Aristotelianism was again firmly entrenched as the main body of thought in the scholastic-dominated universities of Renaissance Europe. But with the publication of *De revolutionibus*, something remarkable in the history of thought occurred. The scientific teachings of Aristotle and Ptolemy were openly challenged for the first time; a period of protracted debate, intellectual conflict, and dissension set in, and it continued, in one form or another, up to the development and acceptance of Newtonian physics in the eighteenth century. Not since the pre-Aristotelian period had so many differing views of nature openly competed with one another.

Of course, a majority of thinkers, bolstered by the authority of the Church, clung tenaciously to the teachings of the Stagirite. However, others, sympathetic to the work of an obscure canon who had lived on the outskirts of Christian civilization, now believed the sun to be the center (or at least near the center) of the universe. Still others, like progressive conservatives Tycho Brahe and Ursus, thought the planets revolved round the sun and, with the sun, round the earth. There were yet other models of the universe, but they need not concern us here. The important thing to keep in mind is that the consensus so long associated with Aristotelian science had vanished. The closed circle of thought had been broken, and many a concerned thinker wondered aloud whether the uncertainty and doubt that poured in to fill the gaping intellectual and psychological vacuum would ever give way to a new universally acceptable orthodoxy. Eventually it did, but the process was some two centuries in duration. Moreover, its successful completion required a genius of the very highest order, Isaac Newton, whose only true peer in the history of Western thought was probably Aristotle himself. Within fifty years of the publication of the *Principia*, the Copernican system ruled unchallenged. Aristotle's magnificent model was replaced by another of even greater grandeur. Copernicus, Kepler, and Newton had constructed a model of the universe so superior to those of their rivals in explaining the positions and motions of the planets and their satellites that all others became obsolete by comparison. Hence, the competing schools of thought disappeared and were not to be heard from again.

As Kuhn points out, we now have names for the conceptual models of nature first constructed by the Ionians and later employed by Aristotle, Tycho, Newton, and all other scientists, no matter what their particular field of specialization. They are called paradigms. "These," writes Kuhn, "I take to be universally recognized scientific achievements that for a time provide models and solutions to a community of practitioners." [2] Paradigms "are the source of the methods, problem-field, and standards of solution accepted by any mature scientific community at any given time." [3] They are what give coherence to various schools of science, to " 'Ptolemaic astronomy' (or 'Copernican'), 'Aristotelian dynamics' (or 'Newtonian'), 'corpuscular optics' (or 'wave optics'), and so on." [4] In sum, paradigms constitute the conceptual framework which provides any community of like scientists with its particular view of nature. It tells the scientist "what the world is like" and establishes the requisite foundation without which research could hardly proceed. And today, as in classical Greece, the mastery of the accepted paradigm "is what mainly prepares the student for membership in the particular scientific community with which he will later practice." [5]

Not only are paradigms important because they provide scientists with a specific framework from which to view the world and conduct further research, a matter we shall explore in more detail below, but because all scientists in a given field usually accept the *same* paradigm, thus committing themselves to a *single* set of rules and standards of scientific practice. Before such unanimity can be achieved, however, there occurs what Kuhn calls a preparadigm stage, which is usually followed by what might be termed a multiparadigm stage.* To begin with, researchers in any new field of scientific inquiry have at their disposal only a large number of facts, all of which are likely to seem equally relevant. "Being able to take no common body of belief for granted," each scientist is "forced to build his field anew from its foundations." [6] Gradually the early fact-gathering stage is eclipsed by a period of consolidation, during which the facts are woven into a

---

* Admittedly, Kuhn does not speak specifically of a "multiparadigmatic" period, but it seems to be implicit in his analysis.

conceptual framework or paradigm. Quite often several paradigms arise simultaneously to challenge one another, as with astronomy in the pre-Aristotelian period, and again during the sixteenth and seventeenth centuries.

The preparadigm and multiparadigm stages are "regularly marked by frequent and deep debates over legitimate methods, problems, and standards of solution." But as more and more data is gathered, the number of acceptable paradigms is drastically reduced. Finally, a single paradigm, the Aristotelian or Copernican, for example, triumphs over its competitors. This is so because it has proven itself significantly better than its rivals at solving the basic problems which confront it. A given paradigm is "particularly likely to succeed" if it "displays a quantitative precision strikingly better than its older competitor." Copernicus himself did not possess such quantitative precision, but both Kepler and Newton did. Hence, Newtonian physics overwhelmed the last resistance to Copernican theory and guaranteed its survival as the only generally accepted paradigm in Western astronomy for the next 200 years. Yet quantitative precision alone, though of primary significance to scientific revolutions, is only one of several criteria which determine whether a new paradigm wins acceptance. How well a paradigm aids the scientist in solving puzzles, its promise for making successful predictions, and its esthetic appeal also have a direct bearing on its claim to superiority.

With the acceptance of a single paradigm, the period of debate and crisis attending the preparadigm and multiparadigm stages gives way to what Kuhn describes as "normal science." Now that the scientists in a given field see "the world" in pretty much the same way, they are ready to undertake the task of further articulating the paradigm itself. "Few people who are not actually practitioners of a mature science realize how much mop-up work . . . a paradigm leaves to be done. Mopping-up operations are what engage most scientists throughout their careers. They constitute what I am here calling normal science." [7] The object of normal science is to provide the factual evidence required to support the theoretical claims implicit in the accepted paradigm. Or to put it another way, the paradigm is the promise while normal science is the vehicle for fulfilling that promise.

In certain respects, this process can be likened to that of a master painter of the Renaissance. Usually a genius, he drew the broad outlines of the work to be undertaken, after which he painted in the major figures. The remaining work, the sketching in of details, was left to apprentices. Under no circumstances was the apprentice allowed to innovate or deviate from the model designed by his master. Contrary to popular belief, most scientists pursue a similar course. Having inherited the sweeping work of a great genius like Copernicus, Darwin, Newton, or Einstein, they too fill in the details with a degree of precision and a general absence of spontaneity lacking in any other field of study.

If it is not the scientist's aim to produce major novelties, conceptual or phenomenal, then exactly what is his goal and function? According to Kuhn, he is motivated by the desire to solve puzzles designed to further elucidate and strengthen the paradigm to which he is committed. "Many of the greatest scientific minds have devoted all of their professional attention to demanding puzzles of this sort. On most occasions any particular field of specialization offers nothing else to do, a fact that makes it no less fascinating to the proper sort of addict." [8] The fact is, the outcome of most research problems can almost always be anticipated within very narrow limits. Thus the scientist is as much, if not more, concerned with the best *method* of achieving a solution as with determining the expected result. This "requires the solution of all sorts of complex instrumental, conceptual, and mathematical puzzles." The individual who achieves the most accurate results "proves himself an expert puzzle-solver, and the challenge of the puzzle is an important part of what usually drives him on." [9] The fascination of normal science lies in the ingenuity of making the accepted rules work, not in discovering fundamental paradigmatic novelties. The purpose of normal science is to better articulate the paradigmatic theory and to apply that theory ever more precisely to nature.

Kuhn asserts, in effect, that normal science rarely if ever innovates. The great majority of scientists not only shy away from formulating new theories, "they are often intolerant of those invented by others." [10] Francis Bacon once spoke of those drawn into a powerful circle of thought as "dancing in little rings like persons bewitched." If Kuhn is to be believed, and he marshals considerable evidence both historical and contemporary to support his view, the models employed by scientists constitute a kind of fairy ring or magic circle which, once it has encompassed its proponents, makes it virtually impossible for them to view the world from any other perspective.* The Aristotelian saw the world differently from a Copernican, as did Einstein from Newton.

If I interpret Kuhn correctly, he uses the term "seeing" in two separate but by no means mutually exclusive ways. Galileo, it will be recalled, experienced great difficulty when he attempted to convince those Aristotelians open-minded enough to look through his telescope that certain of the phenomena they observed were similar to those common on earth; for example, the mountains and craters of the moon. The question is: Did Galileo's opponents actually see what Galileo saw? Kuhn suggests that they did not. Using arguments derived from the modern psychology of

---

* This and a number of related issues are developed by Arthur Koestler in *The Case of the Midwife Toad* (New York, 1971). Also see Bernard Barber, "Resistance by Scientists to Scientific Discovery," *Science*, vol. 134 (1961), pp. 596–602.

perception, he cites the well-known shifts of gestalt in which an observer fluctuates back and forth between seeing a duck or a rabbit in some appropriate design. In other words, the habitual expectation of what the person expects to see may well define what in fact he does see. Perceptually the universe of Copernicus and that of Aristotle are two different things. "The very ease and rapidity with which astronomers saw new things when looking at old objects with old instruments may make us wish to say that, after Copernicus, astronomers lived in a different world. In any case, their research responded as though that were the case." [11]

Kuhn also uses "seeing" in the sense of conceiving or conceptualizing the world around us. Normal science is nothing less than a "strenuous and devoted attempt to force nature into conceptual boxes." Though this can be and frequently is an arbitrary process, "we shall wonder whether research could proceed without such boxes." [12] Only with the acceptance of a paradigm which is underpinned by normal scientific research is the progress so commonly associated with scientific activity possible. The very possibility of interpreting nature in a manner different from the conventional wisdom of a given school of thought is usually closed off to most scientists; their vision is limited to a single direction only. They, much like the rest of us, seem to dance contentedly inside Bacon's "little rings," while sketching in the details of their cherished paradigms.

The Questions of Children

The single most important reason why one paradigm (the Copernican, for example) triumphs over another (the Ptolemaic) is that it has proven itself superior in solving the problems which confounded those scientists committed to its predecessor. Yet it is equally important to keep in mind that no single paradigm provides a totally satisfactory explanation of all the diverse natural phenomena which confront its adherents. There are always some discrepancies which refuse to yield to the pursuit of normal scientific practice. Kuhn refers to such discrepancies as "anomalies"— properties of nature which stubbornly defy repeated attempts to fit them into a neat conceptual box.

The question is, how do scientists respond to the appearance of an anomaly when it cannot readily be fitted between paradigmatic theory on the one hand, and nature on the other? We are told that quite often fundamental novelties are intentionally suppressed because they are "necessarily subversive" of the basic principles according to which a scientific community operates. Usually, however, even the most stubborn of anomalies responds to normal scientific practice, though it may take years or even generations to accomplish this task. It took a century and a half, for example, before the consciousness of the breakdown of Hellenistic astronomy announced in *De revolutionibus* was resolved by the mathematical solution of a number of major anomalies in the *Principia*. Moreover, the ability to achieve a solution to certain particularly difficult problems within the established rules—the precession of the equinoxes, comets, and the tides, for example—serves to deepen the commitment to the accepted paradigm, even as it confers great prestige on the scientist who accomplishes it. In such instances normal science has fulfilled its primary function of bringing theory and fact into ever closer agreement. Yet even in those exceptional instances when an anomaly resists integration with the established pattern of scientific research, the paradigm usually remains intact. This is most important because, "By ensuring that the paradigm will not be too easily surrendered, resistance guarantees that scientists will not be lightly distracted and that anomalies that lead to paradigm change will penetrate existing knowledge to the core." [13] Jefferson's pronouncement that political institutions should not be challenged or overthrown for light and transient reasons applies equally well to the institutionalized paradigm of modern science.

But what happens in those rare instances when the fit between nature and theory fails, when anomaly becomes the rule rather than the exception? Almost immediately normal scientific activity, its paradigm undermined, ceases. A period of intellectual crisis ensues during which a number of new theories are advanced to challenge the old one. Kuhn believes that Ptolemaic astronomy, with its ever growing list of anomalies, was in just such a period of crisis by 1500. Copernicus first turned against the Alexandrian's model for its lack of unity; Tycho, though not a heliocentrist, discovered that comets are a superlunar rather than a sublunar phenomenon; Galileo, with telescope in hand, demonstrated that the heavens are potentially boundless and that physical change penetrates what was once thought of as the immutable celestial realm; Kepler rejected the ancient doctrine of circular planetary motion; and Newton supplied the *coup de grâce* by developing a revolutionary dynamics based on his concept of force. In sum, the old paradigm and its standard beliefs and procedures was destroyed and a new one adopted in its stead, a scientific revolution had occurred.*

"All crises," writes Kuhn, "begin with a blurring of a paradigm and the consequent loosening of the rules for normal research." With normal scientific activity at a virtual standstill, there is a transition to what is termed "extraordinary science." Research during the crisis period closely

---

* Several noted historians of science have questioned Kuhn's use of the term "crisis," believing it to be suggestive of far greater discontinuity than actually occurs when scientists seek to reconcile anomaly with established fact. See, for example, the articles by Owen Gingerich, Dudley Shapere, Thomas B. Settle, and Martin J. Klein in *Copernicus: Yesterday and Today,* ed. by Arthur Beer & K. A. A Strand (Oxford and New York, 1975), pp. 85–133. For Kuhn's response to this criticism see *The Structure of Scientific Revolutions,* p. 181.

resembles research during the preparadigm period when no single model has as yet been established as the universally accepted vehicle for further study. Extraordinary scientific activity is not nearly so structured as normal science, for it leads to a "proliferation of competing articulations, the willingness to try anything, the expression of discontent," and "recourse to philosophy and to debate over fundamentals." [14] It is during periods of breakdown such as this that a genius turned revolutionary, a Newton, Darwin, or Einstein, faced with a fundamental and irreconcilable anomaly in theory, will advance a new paradigm to replace its faltering predecessor. Though he is the rarest of all scientific types, he represents the public's most commonly accepted stereotype of a scientist. The revolutionary genius "will often seem a man searching at random, trying experiments just to see what will happen, looking for an effect whose nature he cannot quite guess." Yet however spontaneous and unorthodox his methods may seem, he too operates within a delimited conceptual framework, albeit one at considerable variance with that being rejected by his disoriented colleagues. If successful, his yet unarticulated conceptual framework may well become the new paradigm, or what Dudley Shapere aptly defines as "a single overarching *Weltanschauung* [world view], a disciplinary *Zeitgeist*, that determines the way scientists of a given tradition view and deal with the world, that determines what they would consider to be a legitimate problem, a piece of evidence, a good reason, an acceptable solution, and so on." [15]

This is not to assert that all geniuses in the sciences automatically become revolutionaries. The majority of them, like their less gifted peers, work within the boundaries of normal science. A scientific genius usually turns revolutionary only in response to great provocation and with considerable reluctance, and only then when his birth and intellectual maturity happen to coincide with a period of crisis and breakdown in his chosen field. Such periods have been very rare; in the entire history of astronomy only three paradigms have succeeded in capturing the universal admiration of practicing astronomers: the Aristotelian–Ptolemaic, which survived for about 1800 years; the Copernican, lasting for about four centuries; and the Einsteinian, which only came into its own during the 1920s and 1930s. In astronomy, as in related fields, there will doubtless be others, for we have discovered no evidence of immutability in the sciences. But scientific revolutions of the magnitude of those brought about by Copernicus, Lavoisier, and Darwin are most unusual and will not easily be duplicated in the future, the ever accelerating pace of current scientific activity notwithstanding.

Indeed, the more specialized the various sciences become, the more difficult it is for any single individual to come to grips with all of the current knowledge and research in his field. Were Newton alive today, he would find it virtually impossible to conduct major research in the many branches of science which commanded his attention between 1665 and 1695: physics, chemistry, mathematics, optics, and astronomy. He would more than likely find himself a member of some professional sub-specialty, say particle physics, focusing his attention on a far narrower range of natural phenomena and scientific problems than he ever thought possible.

This rapid multiplication of subspecialties suggests to Kuhn the emergence of a second and more common type of scientific revolution in the future, one limited to a very small scientific community, numbering anywhere between twenty and one hundred individuals. "There can be small revolutions as well as large ones," revolutions which "affect only the members of a professional subspecialty, and . . . for such groups even the discovery of a new and unexpected phenomenon may be revolutionary." [16] Members of these subgroups share their own special paradigms in addition to being committed to the more general paradigm which unites them with the other subspecialists in their chosen field. Given this new set of conditions, normal science no longer functions as "a single monolithic and unified enterprise that must stand or fall with any one of its paradigms as well as with all of them together," as in the past. Paradigm-shattering research undertaken by one group of specialists will sometimes scarcely affect the view of nature held by those in another, if at all.

What comes to mind is the familiar model of an atom, which one encounters in almost every classroom where introductory chemistry and physics courses are taught. First think of the atom as representing a particular field of science, say physics. Then think of its nucleus as the paradigm around which circle minute subparticles of matter, each representative of a specific professional subgroup. Though all of the many subgroups are linked to a central mass, each is a semi-independent entity in its own right. A major paradigmatic shift in one subspecialty may go completely unheeded by those aligned with the others.* Only if the overarching paradigm, which loosely links the various subspecialties together, were found inadequate to the task of fitting theory with scientific practice would the entire field be seriously affected. Then, as in the past, normal science would be held in abeyance until someone advanced a new paradigm which could adequately account for the anomalies deemed insoluble under the terms of the old one.

One of the most interesting and well-documented facets of scientific revolutions has to do with the fact that the individuals responsible for these fundamental innovations are almost always very young or very new to

---

* In situations of this type, normal and extraordinary science do not seem the distinct and mutually exclusive entities described by Kuhn. Rather, they can both occur at the same time, serving as complementary components of ongoing scientific change. For a more detailed discussion of this point, see Thomas B. Settle, "On Normal and Extraordinary Science," in *Copernicus: Yesterday and Today,* pp. 105–111.

the field whose paradigm they overthrow. Though Copernicus withheld publication of *De revolutionibus* until his seventieth year, there is sufficient evidence to conclude that his conversion to heliocentric astronomy occurred during his student days in Renaissance Italy, probably when he was in his early or mid-twenties. Johannes Kepler became a convert to Copernican astronomy while Maestlin's student at Tübingen, and at the age of twenty-six published the *Mysterium Cosmographicum*, the first of his great iconoclastic treatises. Newton, who at twenty-four had already developed his revolutionary theory of colors, the calculus, and the inverse square law, observed that "in those days I was in the prime of my age for invention, and minded mathematics and philosophy more than at any time since." Einstein, whose theory of relativity was published in 1905, when he was just twenty-six, once observed: "A person who has not made his great contribution to science before the age of thirty will never do so." Even Darwin, whose legendary caution caused him to delay publication of *The Origin of Species* until relatively late in his career, had begun to fit the innumerable pieces of the great mosaic of evolutionary theory together while serving as a naturalist aboard the H.M.S. *Beagle.* "When on board the *Beagle*," Darwin wrote to a friend, "I believed in the permanence of species, but as far as I can remember vague doubts occasionally flitted across my mind." The great naturalist was only twenty-two when the little ship left Devonport in south-western England on December 27, 1831 to begin its historic five-year cruise round the world. When the *Beagle* finally docked again in England, it was a twenty-seven-year-old revolutionary, dressed in the garb of a prosperous country gentleman, who walked uneasily down its oaken gangplank and into the pages of history. These and many similar examples provide indisputable evidence that scientists build their reputations and careers upon the intellectual innovations of their early adulthood.

How does one account for the fact that scientific revolutions have almost always been initiated by the young? Obviously youth, despite its natural tendency to think otherwise, has no monopoly on brains; older scientists are no less intellectually competent as a group than those they train to one day take their place. Indeed, the answer to the question has nothing whatsoever to do with the intelligence of one generation as opposed to that of another, but with the impact of aging upon the world view of the individual. As one grows older, it becomes increasingly difficult to relinquish one set of values or institutions in favor of another. This is especially true when it involves the sacrifice of what constitutes a major part of one's lifework. Young men, on the other hand, "are so new to the crisis-ridden field that practice has committed them less deeply than most of their contemporaries to the world view and rules determined by the old paradigm." [17] In other words, a highly innovative newcomer has relatively little to lose, as compared to those whose careers are at a more advanced stage. We see here a direct parallel with political revolution, the success of which usually depends heavily upon the actions of the young, whose committment to traditional institutions is frequently more tenuous than the generation of their fathers. Thus the dictum "Revolutionary when young, conservative when old," applies as much to innovation in the sciences as to sudden and radical changes in the political arena.

Still, there is a much deeper principle in operation here than youth versus age alone. Most of the time, normal scientific activities bind the new members a particular school of thought even more tightly to the prevailing paradigm than the generation of its founder: witness, for example, the Aristotelians of the fifteenth and sixteenth centuries, the Darwinists and neo-Darwinists of the late nineteenth and twentieth. Committed at an early age, as most scientists are by a rather narrow system of education, to a particular paradigmatic theory, they cannot easily tolerate the possibility that they may be in error. In this sense, modern science has taken on a major characteristic of the medieval Church. It excludes the heterodox from its inner circle, denying them access to its most important activities through a particularly effective form of professional ban and interdict. The Lamarckians were treated little differently by those committed to Darwin's theory of natural selection and Mendelian genetics than was Galileo after his brush with the Holy Tribunal, in 1633. Of course, the Lamarckians did not undergo physical confinement, but they suffered a form of professional ostracism bordering on intellectual imprisonment.

The schisms which occur occasionally between young men and those who have labored in the vineyards for many years are related directly, I believe, to a point discussed in the previous section of this chapter. I refer here to the concept of gestalt, the manner in which scientists from divergent fields view the natural world. We have observed that so long as normal scientific research based on a given paradigm satisfactorily melds fact with theory, there is no cause for alarm. Only when a major anomaly stubbornly refuses to conform to the accepted standards of scientific practice does the malaise associated with paradigmatic breakdown set in. Almost inevitably, it is the new convert or youthful initiate to the field who sees things differently.* To put it simply, a young genius of the caliber of a Newton or an Einstein brings with him to his field of study a special kind of perception (gestalt), which enables him to plumb the depths of a crisis-creating problem in a manner quite alien to his colleagues. His vision "can hardly be communicated to others," writes Gerald Holton, and often leads the individual who possesses it to the conclusion "that he is, in some sense, one of the 'chosen' ones."[18] This apparently

---

* For an excellent first-hand account of how a crucial scientific innovation can result when scientists in one field shift to another, see James D. Watson, *The Double Helix* (New York, 1968).

inherent inability to deal with the world on conventional terms is well illustrated in Lewis S. Feuer's comparison of scientific revolutionaries Wilhelm Conrad Roentgen and Albert Einstein:

> Both men as youngsters were expelled from their high schools, both were filled with dislike for examinations and the classical curriculum, and both encountered the active hostility of some of their teachers. Both were a source of extreme concern to their fathers because of their inability to adjust to the world's ways, found their way to the Federal Polytechnic at Zurich, and discovered friendship in circles where a student revolutionary culture prevailed in act or spirit.[19]

Thus it would seem that youthful genius alone, even when manifested during a period of scientific upheaval, is not sufficient in and of itself to initiate a major paradigmatic shift. The great scientific revolutionaries, from Kepler through Einstein, have almost always been among the unsocialized personalities of their generation. Perhaps further research will one day confirm the hypothesis that major innovators in the science, because of their alienation from society at large, have used science, consciously or otherwise, as a substitute outlet for the radical changes sought but denied them in the nonscientific world.

From the still limited evidence available, it would also seem that the youthful genius' mode of seeing things is quite often suffused with an innocence and naiveté commonly associated with that of a child. The poet William Cowper referred to Newton as a "childlike sage," a characterization which strikes us as being rather odd based upon our earlier discussion of the great man's life and work. Yet consider this; how many adults would bother, let alone think, to ask themselves the profoundly simple question: "Why do apples fall from the branches of trees?" Einstein, even more than Newton, continuously addressed the questions of children to nature. Doubtless the most simple, yet remarkable, of his queries came when he was a student in Aarau, at the age of sixteen: "What would the world look like if I rode on a beam of light?" [20] Less than a decade later Einstein had his answer, and the world of science shuddered from the shock waves generated by the resounding footfalls of his transcendent genius.

Interestingly, both Newton and Einstein were keenly aware of the affinity between their respective methods of focusing on a scientific problem and the child's way of coming to grips with the world. Einstein once commented that he was led to the formulation of relativity theory largely because he kept positing questions concerning time and space that only a child would think about. And in a rare moment of deep personal insight, Newton, in one of his more famous statements, said much the same thing: "I do not know what I may appear to the world; but to myself I seem to have been only like a boy playing on the sea-shore, and diverting myself in now and then finding a smoother pebble or a prettier shell than ordinary, while the great ocean of truth lay all undiscovered before me." Newton and Einstein somehow brought with them to their work in science that precious secret normally forgotten in the waning moments of childhood's magic hour, that pertinent answers quite often result from asking impertinent questions.

The Edge of Objectivity

Isaac Newton, in his oft-quoted tribute to the men of genius whose singular contributions to modern science helped make possible his own, once wrote: "If I have seen farther than other men, it is because I have stood on the shoulders of giants." Thus, despite his fervent anti-Baconian warning that natural philosophers must never permit themselves to be tricked into thinking that nature will one day be theirs to completely understand, and thereby to totally dominate, Newton envisioned science as at least a partially cumulative process, by which a degree of progress is the logical result when one keen mind, in an effort to add to man's knowledge of the natural world, utilizes and expands upon the work of others.

For the most part, the theological limitations imposed by Newton on the efficacy of his scientific work and that of other natural philosophers were greatly overshadowed by the essentially materialistic conception of his mechanical science, which won a profound victory first in England, then on the Continent. So far has this emphasis on materialism been carried in the two and one-half centuries since Newton's death that many, including both scientists and laymen, now fully embrace what can only be described from their point of view as a self-evident principle; namely, that "progress" in the sciences is the only legitimate yardstick by which the progress of a society can be measured, that science is the only field in which man can discover the "truth." Without doubt, the unprecedented growth in the belief in scientific progress is one of the most profound consequences of the Copernican Revolution, whether intended or not by its makers. In this section, we will undertake a brief analysis of Thomas Kuhn's view of the relationship between the concept of progress and the process of discovery in the sciences.

Kuhn believes that one of the most important reasons why science and progress seem to go hand in hand stems from the special method of education peculiar to the training of young scientists. To an extent unprecedented in other fields, textbooks in the sciences serve as the major pedagogic vehicle for the perpetuation of the discipline. This is not a recent development by any means, for such was the case long before the rise of our present system of higher education, when certain classic works were regarded as the "textbooks" of their time. "Aristotle's *Physics,*

Ptolemy's *Almagest*, Newton's *Principia* and *Optics*, Franklin's *Electricity*, Lavoisier's *Chemistry*, and Lyell's *Geography* [all] served for a time implicitly to define the legitimate problems and methods of a research field for succeeding generations of practitioners." [21]

The student pursuing a science curriculum relies almost exclusively on textbooks throughout his undergraduate days and, if he chooses to undertake advanced work, until his third or fourth year of graduate study, when he normally begins independent research of his own. Textbooks are of course assigned to students in other academic disciplines by those responsible for the training of future practitioners, but the farther the nonscientist progresses in his undergraduate program, the less he will rely on such secondary tools. It is largely through the mastery of his textbooks that the future scientist is prepared for membership in the particular community of specialists with which he will later practice. For this is the device by which the newcomer becomes committed to the same rules and standards of practice—the same paradigm—as those who are already full-fledged members of his chosen field. When the individual scientist can take a paradigm for granted, he no longer need attempt to build his field anew. That is, he does not have to start from first principles and justify the use of each concept or rule which he employs. Since he "joins men who learned the basis of their field from the same concrete models, his subsequent practice will seldom evoke overt disagreement over fundamentals." [22] This frees members of any given scientific community to concentrate almost exclusively upon the subtlest and most minute phenomena that concern it. Since fundamental disagreements are quite rare, progress seems the inevitable result.

Contrast this situation with the nature of education and professional activity in such fields as music, art, literature, philsosophy, and history. Though textbooks are frequently employed in introductory courses, the student acquires the main body of his education by being exposed to the major works of the masters or great thinkers in his field. For example, no teacher of Shakespeare would dream of offering a course on the master's work without having his students read certain of the great Elizabethan's poetry or plays. Nor would any competent historian offer an advanced course without introducing students to some part of the important primary source material in his field. Yet few, if any, physics students are ever required to read the works of Newton, Planck, Heisenberg, or Einstein. For everything they need to know about the discoveries of these men is presented in a more precise and far briefer form in the textbooks assigned by their professors. The basic difference between these two approaches is that the student pursuing a discipline outside of the sciences has no universally accepted paradigm on which to base his studies. Rather "he has constantly before him a number of competing and incommensurable solutions" to the disparate problems which he encounters along the way.

He must ultimately evaluate those problems for himself and provide his own solutions, knowing full well that there will be many whose conclusions will be diametrically opposed to his own. At most, he can expect to become a member of one of a number of competing schools, each of which constantly questions the very fundamentals of its rivals. Unlike the scientist, the historian, the philosopher, and the artist must constantly reexamine first principles for himself. He is in no position to leave such an important aspect of his work to the writers of textbooks. In the absence of universally accepted paradigms, it is difficult to conceive of progress in these fields in the same way that we do in the sciences.

From what has just been said, it is clear that the education of a student in the sciences is, as Kuhn suggests, both a narrow and rigid undertaking. But it is equally apparent that a scientific community is a highly effective instrument for producing solutions to the problems defined by its paradigms. While those in the nonscientific disciplines argue interminably over fundamentals, the puzzle-solving activities of scientists proceed apace. Only during rare periods of paradigmatic breakdown is the forward momentum generated by ever more intensive scientific research markedly slowed. But so long as someone presents a new paradigm for acceptance by his peers, the assault on nature is quickly renewed, the forward momentum regained.

Still, there is much about the education of scientists that presents a false image of the doctrine of progress, both to scientists themselves and to the public at large. However efficient their method of education may be, scientists are frequently the victims of the very narrowness of their isolated academic experiences. Unlike the student of art, music, or philosophy, who is constantly broadening his perspective as he simultaneously comes to grips with several competing schools of thought, the student of science thrives in a rather authoritarian environment, which is clearly circumscribed by an already articulated body of data, problems and theory derived from the paradigm which he must master. Furthermore, as Kuhn, a trained physicist, points out, "Textbooks [in the sciences] begin by truncating the scientist's sense of his discipline's history and then proceed to supply a substitute for what they have eliminated." [23] At best, science textbooks contain only a smattering of history, in addition to scattered references to the major figures of the past. Almost never is mention made of a scientist, however wise, or a school of thought, however perceptive, which took major exception to the paradigms which ultimately prevailed in the frequently intense struggle for ascendancy. Rarely, in fact, are such struggles even mentioned. This problem is compounded by the fact that textbooks used in the training of scientists are simply rewritten whenever a previously accepted scientific theory is replaced by another, which is believed to be more compatible with the results of ongoing research. This distortion of historical tradition, no matter how compelling the reason, strongly reinforces the widely held view that progress in the sciences is a straight line, cumulative process unaccompanied by protracted debate, confusion, or periodic breakdown. Yet a closer look at the history of science reveals something quite different—that scientific advance is not a continuously linear process, but almost always takes a zigzag path. It is both continuous and discontinuous at the same time.

The tendency to write history from the viewpoint of hindsight alone is a dangerous one. To uphold Copernicus, Kepler, and Galileo as wise men simply because they were "right" automatically implies that those who opposed heliocentric astronomy were "wrong" and therefore foolish. Yet when *De revolutionibus* was first published, its supporters were far outnumbered by its detractors within the scientific community, because Copernicus could not provide sufficient physical evidence for the rejection of geocentric astronomy. Einstein's relativity theory suffered a similar fate until the perihelion advance of the planet Mercury was definitely proven. Yet who would be so brash as to call the physicists who opposed Einstein's theory fools, or to disregard their carefully reasoned arguments in any balanced history of modern physics. Nevertheless, "the depreciation of historical fact is deeply, and probably functionally, ingrained in the ideology of the scientific profession," according to Kuhn, "the same profession that places the highest of all values upon factual debate of other sorts." [24]

The educated lay public, no less than scientists themselves, also tends to view science as the only field in which progress is made. This belief had its genesis during the eighteenth century, when the historical materialism of the Enlightenment *philosophes* emerged out of the creation of the new science of Descartes and Newton. Sidney Pollard in *The Idea of Progress* observes:

> When Descartes built up a new mathematical world of confident certainty as great as that of religion, and when Newton began to encompass complex, contradictory reality itself within a few universal laws, it had become clear that here was a field of endeavor in which modern man had overtaken the classics and was set on a course which would leave the past far behind. Scientists, advancing by action rather than by contemplation, foreshadowed the attitude of modern man. At the same time, science and its methods were acquiring a prestige which would lead other fields of thought to pay it the compliment of imitation. Social inquiries, history itself, sought to buttress themselves by becoming scientific, and historians, looking for determinate regularities which they could designate as "laws," were soon to light upon one for which science itself had given the first hints: the law of human progress. [25]

Bacon's "engine of progress" continued to gain momentum during the eighteenth century, spurred on by the spirit of capitalistic enterprise and the ever expanding desire to establish man's mastery of the world. The

Industrial Revolution of the nineteenth century, by wedding science a[...] technology, produced the seemingly unshakable belief in the inevitability of material progress. And though the ranks of the doubters and pessimists have grown appreciably since the mid-twentieth century, science is still viewed by a majority in the West as the only major field in which a marked advancement of the human condition is yet possible.

This is not, Kuhn asserts, simply the result of the public's equation of science with materialism, and of materialism with progress. Scientists, as no other professional community, experience an unparalleled degree of insulation (some would say isolation) from society as a whole. Unlike the poet, theologian, artist, or novelist, who must be constantly concerned with the layman's reaction to his creative endeavors, the work of the individual scientist is almost exclusively addressed to and judged by fellow members of his profession. His insulation from society permits him to select and concentrate his attention upon only those problems which he believes can be solved. "Unlike the engineer, and many doctors, and most theologians, the scientist need not choose problems because they urgently need solution and without regard for the tools available to solve them." [26] This naturally gives the scientist a decided advantage when it comes to achieving concrete results. With their freedom to pick and choose, we would expect scientists to solve problems at a considerably more rapid rate than other professional groups, which they in fact do. And even if a scientist proves unsuccessful in attaining the results he had hoped for, that failure is rarely publicly exposed, for he answers only to an inner circle composed of his peers. Neither must we forget that science has become such a complex field of study that the layman, no matter how well educated he may be, is usually unable to follow its highly technical developments, except in a most cursory manner. Its virtual monopoly on knowledge surrounds the scientific community, like the priests of ancient Egypt and Delphi, with a kind of mystique, an impenetrable aura which carries the implication of professional infallibility. Thus, while progress is assured during periods of normal scientific research, it is equally important to remember that much of what scientists and the public take for progress lies, like beauty, in the eyes of the beholder.

Scientific progress in the direction of increased material well being is one thing, progress towards absolute truth about nature is quite another. "Modern science," writes historian Charles Coulston Gillispie, "is impersonal and objective. It takes its starting points outside the mind in nature and winnows observations of events which it gathers under concepts, to be expressed mathematically if possible and tested experimentally by their success in predicting new events and suggesting new concepts." [27] With this most of us would certainly agree. And yet there is something clearly unsettling to many thinkers about the much broader conclusion commonly derived from Gillispie's observation: that modern man's deepening penetration of nature moves him ever closer to the development of a right and perfect science, ever nearer to what Gillispie calls the "edge of objectivity."

It is hardly disputable that normal science, the puzzle-solving activity discussed above, is a highly cumulative process, markedly successful in its single major aim, the constant expansion of the body and precision of scientific knowledge. One need only momentarily look back upon those scientific developments most commonly associated with the Copernican Revolution to see that this is so. Kepler could never have discovered his three laws without access to Tycho's observations of the planets and stars; Newton could never have given birth to a new physics without Kepler's Laws and Galileo's kinematics. Science as we know it simply could not exist without handing down its body of conceptual and instrumental techniques from one generation to another. Newton was by no means the only man of his profession to stand on the shoulders of giants.

This being said, we are still confronted by the unavoidable question of whether modern science is moving toward a truly objective understanding of the natural world. Kuhn, for one, would argue that it is not. Scientific revolutions on the scale of the one instituted by Copernicus result in major paradigmatic change; in this instance the rejection of the Aristotelian–Ptolemaic world view in favor of a sun-centered cosmology. This "transition from a paradigm in crisis to a new one from which a new tradition of normal science can emerge," writes Kuhn, "is far from a cumulative process, one achieved by an articulation or extension of the old paradigm. Rather, it is a reconstruction of the field from new fundamentals, a reconstruction that changes some of the field's most elementary theoretical generalizations as well as many of its paradigm methods and applications."[28] This does not, of course, mean that every time one paradigm is replaced by another all scientific knowledge associated with the old paradigm is automatically rejected. We know that Copernicus borrowed heavily from the observational data of the Aristotelians, while Einstein incorporated certain important elements of Newtonian dynamics in his radically changed vision of the universe.

On the other hand, in the process of being assimilated, the discredited paradigm is inevitably displaced by another, for two paradigms cannot exist side by side; the successful new model must permit predictions and yield results that are fundamentally different from those derived from its predecessor. This not only entails a reordering of the scientific facts, it involves a major shift in the conceptual scheme by which a scientific community embraces the same phenomena dealt with by its predecessor, but explained in a different way. Though the factual content of science is highly cumulative, the concepts that lead to the discovery of facts and the development of scientific laws are not.

To better illustrate this point let us turn once again to an example from the field of art. Classical Greek sculpture of the Periclean Age was based on the Olympian ideal of unblemished physical perfection, and therefore displayed little if any concern for the expression of individual human emotion. In fact, when viewing a piece of sculpture executed during the fifth century B.C., it is virtually impossible to differentiate between the idealized mortal and the deities of the Greek pantheon. During the Hellenistic period, however, a new conceptual model of art based on the portrayal of individuality replaced the classical Greek view. Hellenistic sculptors drew upon their predecessors' extensive technical knowledge of physical anatomy, but they added to it the dynamic element of human emotion. By so doing Hellenistic man employed a much different standard of what to him constituted a masterwork of art from that applied by the Periclean sculptors Myron and Praxiteles. This is equally true of the history of science, where one school of thought often borrowed heavily from another. But none has proven itself correct for *all* time, for *all* people, or for *all* societies. Yet each was correct for what it sought to achieve. Scientific revolutions, no less than revolutions in the arts, reflect changes of world views, changes which are indigenous to the societies and the communities of scientists that produce them. However, in the sciences we continue to make the serious mistake of confusing knowledge with certainty, of attempting to convince ourselves that the latest discoveries are unerringly right, and for all time, while believing discarded principles based on seemingly outmoded paradigms to be wrong. But Kuhn reminds us that

> as science progresses, its concepts are repeatedly destroyed and replaced, and today Newtonian concepts seem no exception. Like Aristotelianism before it, Newtonianism at last evolved . . . problems and research techniques which could not be reconciled with the world view that produced them. For half a century we have been in the midst of the resulting conceptual revolution, a revolution that is once again changing the scientist's (though not yet the layman's) conception of space, matter, force, and the structure of the universe. Because they provide an economical summary of a vast quantity of information, Newtonian concepts are still in use. But increasingly they are employed for their economy alone, just as the ancient two-sphere universe is employed by the modern navigator and surveyor.[29]

The Aristotelians are long gone, Newton's physics is under siege, and the conceptual world in which their respective followers lived is part of the unrecoverable past. Only certain parts of the still useful factual content of their work are employed by scientists today. Yet both Aristotle and Newton were as "right" for their respective times as Einstein is for ours. Furthermore, there is every reason to believe, based on the study of history, that Einstein's conceptual model of a vast and expanding but finite universe will one day give way to a new one, and so on *ad infinitum*.

I share Kuhn's view that there is nothing absolute in what is termed scientific truth. A system is "true" only so long as it satisfactorily explains the phenomena as a given scientific community defines them. That is precisely what geocentric astronomy did prior to 1600 and what heliocentric astronomy did after Copernicus. Kuhn would have us deal with the rise of modern science by placing it in the context of Darwinian historiography; that is, by treating it as a process that moves steadily away from primitive beginnings but *toward* no objectively defined goal. This, it seems to me, is very sound advice. For if we are to survive the vicissitudes of an increasingly unstable world, we would do well to begin by relinquishing, as Newton did some three centuries ago, the hubristic Baconian notion that scientific discovery carries those who participate in the process closer and closer to absolute truth.

Historical time is a tricky thing; it flows at an ever accelerating speed, like a river approaching a great waterfall. We must soon learn to cope with the awesome power given to us by the heirs of Copernicus or our species, and all others on this planet, are doomed to a painful and purposeless extinction. Were this to happen, only the stars would remain; and what are the stars, after all, without the eyes of man to gaze upon them or the human mind to contemplate the vastness of their wonders?

Appendix: Some Important Dates in the History of Modern Astronomy: 1473-1727

1473   Nicolas Copernicus born on February 19 at Torun in Prussian Poland.

1496-1503   Copernicus undertakes advanced study in Italy where he is attracted to the heliocentric cosmology of the ancient Greek Aristarchus of Samos.

1512-1514   Copernicus composes and privately circulates the *Commentariolus* in which he ascribes the apparent motion of the sun to the earth.

1539   Rheticus travels to Frauenburg for his first meeting with Copernicus.

1540   *Narratio prima*, authored by Rheticus, published. Marks the first time that the Copernican theory has appeared in print.

1543   Copernicus' *De revolutionibus* published. Copernicus dies at Frauenburg on May 24.

1546   Tycho Brahe born on December 14 at Knudstrup, then a part of Denmark.

1564   Galileo Galilei born at Pisa, Italy, on February 16.

1571   Johannes Kepler born on December 27 in Weil-der-Stadt, Germany.

1572   A supernova (Tycho's nova) becomes visible to European astronomers in early November.

1573   Tycho concludes that the new celestial object is not a comet but a previously invisible star. Publishes *De nova stella*.

1576   King Frederick II grants Tycho the island of Hveen for life. Cornerstone of Uraniborg, Tycho's observatory, laid on August 8.

1577   The Great Comet of 1577 appears in November.

1573   Tycho proves that the comet of 1577 is not a sublunar body, lays the basis for his own model of the universe, and takes up residence at Uraniborg.

1588–1590   Kepler accepts the Copernican theory while studying at Tübingen University under Michael Maestlin.

1592   Galileo appointed to the faculty of the University of Padua. Conversion to Copernican astronomy dates from this period.

1594   Kepler assumes duties as the District Mathematician of Gratz.

1597   Publication of the *Mysterium Cosmographicum*, Kepler's first pro-Copernican treatise.

1599   Tycho appointed Imperial Mathematician to Holy Roman Emperor Rudolph II in Prague.

1600   Kepler joins Tycho at Benatky Castle outside Prague.

1601   Tycho dies on October 24 in Prague. Kepler appointed Imperial Mathematician on November 6 and acquires Tycho's astronomical data.

1604   Appearance of a second supernova (Kepler's nova) in October.

1606   Kepler publishes *De Stella Nova*.

1609   Kepler's *Astronomia Nova* published announcing his first and second laws. Galileo constructs his first model of the telescope and undertakes the first telescopic observations of the stars. Galileo appointed Court Mathematician to the Grand Duke of Tuscany.

1610   Galileo publishes *The Starry Messenger*. Kepler publishes the *Dioptrice* and his *Conversation With the Starry Messenger*.

1612   Kepler becomes District Mathematician of Linz.

1613   Galileo's *Letters on Sunspots* published.

1615   *Letter to the Grand Duchess Christina* published by Galileo.

1616   Galileo called to Rome by the Holy Office to explain his position on Copernican astronomy. *De Revolutionibus* placed on the Index of Prohibited Books until it can be "corrected."

1619   Kepler's masterwork, the *Harmonice Mundi*, published containing his third law.

1620   *De revolutionibus* revised and removed from the Index.

1623   *The Assayer* published by Galileo.

1627   Kepler publishes the *Rudolphine Tables*.

1630   Kepler dies in Regensburg on November 15.

1632   Galileo's *Dialogue Concerning the Two Chief World Systems* published and subsequently banned by the Holy Office. Galileo ordered to appear before the Inquisition in Rome.

1633   Trial, condemnation, and recantation of Galileo. He is confined to house arrest for the remainder of his life. The works of Copernicus, Kepler, and Galileo placed on the Index where they remain until 1822.

1638 *Discourse Concerning Two New Sciences* secretly completed by Galileo, smuggled out of Italy, and published in Leyden.

1642 Galileo dies on January 8 in his villa at Arcetri. Isaac Newton born on Christmas day at Woolsthorpe, England.

1661 Newton enters Trinity College, Cambridge.

1665–1666 Newton lays the groundwork for his greatest discoveries at Woolsthorpe: the calculus (fluxions), the theory of light, and the inverse square law.

1668 Newton constructs the first working model of the reflecting telescope.

1669 Newton appointed Lucasian Professor of Mathematics at Cambridge.

1672 Newton elected a Fellow of the Royal Society in London. Publishes a detailed description of his telescope in the *Philosophical Transactions*.

1680 Comet of 1680, later to be named Halley's comet, appears.

1687 The *Principia* published announcing Newton's law of universal gravitation.

1703 Newton elected President of the Royal Society.

1704 Newton's *Opticks* published.

1727 Newton dies on March 20 at his home in Kensington.

Notes

Chapter 1

1. Loren Eiseley, *Darwin's Century* (New York: 1958), p. 59.
2. Lewis Mumford, *The City in History* (New York: 1961), p. 17.
3. William McNeil, *The Rise of the West* (Chicago and London: 1963), p. 15.
4. Mumford, *The City in History*, p. 18.
5. Antonie Pannekoek, *A History of Astronomy* (New York: 1961), p. 25.
6. *Ibid.*, p. 38.
7. *Ibid.*, p. 48.
8. Willy Ley, *Watchers of the Skies* (New York: 1969), p. 14.
9. Karl A. Wittfogel, *Oriental Despotism: A Comparative Study of Total Power* (New Haven: 1957), p. 3.
10. *Ibid.*, p. 4.
11. *Ibid.*, p. 28.
12. *Ibid.*, pp. 26–27.
13. Lewis Mumford, *The Myth of the Machine: Technics and Human Development*, vol. I (New York: 1967), p. 191.
14. *Ibid.*, p. 194.
15. *Ibid.*, p. 198.

Chapter 2

1. H. D. F. Kitto, *The Greeks* (Baltimore: 1957), p. 39.
2. *Ibid.*, p. 8.
3. Stephen Toulmin and Jane Goodfield, *The Fabric of the Heavens: The Development of Astronomy and Dynamics* (New York: 1961), pp. 59–60.
4. The most definitive account of ancient Greek and medieval astronomy is still J. L. E. Dreyer's, *A History of Astronomy from Thales to Kepler*, 2nd ed. (New York: 1953). The work was originally published under the

title of *History of the Planetary Systems from Thales to Kepler* by the Cambridge University Press, 1905. See also Sir Thomas Heath, *Aristarchus of Samos: The Ancient Copernicus* (Oxford: 1959).

5. Heath, *Aristarchus of Samos: The Ancient Copernicus*, p. 8.

6. Arthur Koestler, *The Sleepwalkers* (New York: 1959), pp. 21–25.

7. Toulmin and Goodfield, *The Fabric of the Heavens: The Development of Astronomy and Dynamics*, pp. 64–65.

8. Aristotle, *On the Heavens*, trans. by W. K. C. Guthrie, 4th ed. (Cambridge and London: 1960), p. 225.

9. *Ibid.*, p. 235.

10. Dreyer, *A History of Astronomy from Thales to Kepler*, p. 13.

11. Koestler, *The Sleepwalkers*, p. 23.

12. Dreyer, *A History of Astronomy from Thales to Kepler*, pp. 36–37.

13. Aristotle, *On the Heavens*, p. 217.

14. Dreyer, *A History of Astronomy from Thales to Kepler*, p. 42.

15. E. J. Dijksterhuis, *The Mechanization of the World Picture* (Oxford: 1961), p. 7.

16. Koestler, *The Sleepwalkers*, p. 29.

17. Benjamin Farrington, *Greek Science* (London: 1953), p. 45.

18. Koestler, *The Sleepwalkers*, p. 62.

19. Aristotle, *On the Heavens*, p. 253.

20. *Ibid.*

21. *Ibid.*

22. *Ibid.*, p. 25.

23. *Ibid.*, p. 95.

24. *Ibid.*, p. 91.

25. *Ibid.*, p. 93.

26. *Ibid.*, p. 189.

27. *Ibid.*, p. 179.

28. *Ibid.*, p. 31.

29. *Ibid.*, p. 149.

30. Dijksterhuis, *The Mechanization of the World Picture*, p. 35.

31. Dreyer, *A History of Astronomy from Thales to Kepler*, p. 90.

32. Quoted in Heath, *Aristarchus of Samos: The Ancient Copernicus*, p. 302.

33. *Ibid.*, p. 304.

34. Koestler, *The Sleepwalkers*, p. 50.

35. Toulmin and Goodfield, *The Fabric of the Heavens: The Development of Astronomy and Dynamics*, p. 125.

36. The only acceptable English translation of the *Almagest* by Ptolemy was done by R. Catesby Taliafero and appears in *The Great Books of the Western World*, vol. 16, published by the Encyclopedia Britannica, Inc., 1952.

37. *Ibid.*, p. 7.
38. *Ibid.*, p. 8 .
39. Koestler, *The Sleepwalkers*, p. 69.
40. Ptolemy, *Almagest*, pp. 5–6.

Chapter 3

1. Saint Augustine, *Confessions*, trans. with an introduction by R. S. Pine-Coffin (Baltimore: 1968), p. 89.
2. *Basic Writings of St. Augustine*, vol. II., ed. by Whitney J. Oates (New York: 1948), pp. 275–276.
3. Mumford, *The City in History*, p. 243.
4. "Old Ireland, Her Scribes and Scholars," in *Old Ireland*, ed. by Robert McNally (New York: 1965), p. 125.
5. Mumford, *The Myth of the Machine: Technics and Human Development*, p. 265.
6. Quoted in Mumford, *The City in History*, pp. 258–259.
7. Dreyer, *A History of Astronomy from Thales to Kepler*, p. 208.
8. *Ibid.*, p. 214.
9. Stephen Toulmin and Jane Goodfield, *The Discovery of Time* (New York: 1966), p. 46.
10. *Basic Writings of St. Augustine*, p. 202.
11. Saint Augustine, *Confessions*, p. 298.
12. *Ibid.*, p. 278.
13. Alfred North Whitehead, *Science and the Modern World* (Cambridge: 1953), p. 6.
14. Toulmin and Goodfield, *The Fabric of the Heavens: The Development of Astronomy and Dynamics*, pp. 159–160.
15. *Basic Writings of St. Thomas Aquinas*, vol. II, ed. by Anton C. Pegis (New York: 1945), p. 35.
16. S. Pines, "The Semantic Distinction Between the Terms Astronomy and Astrology According to Al-Biruni," *ISIS*, vol. 55 (1964), p. 344.
17. *Basic Writings of St. Thomas Aquinas*, p. 35.

Chapter 4

1. Angus Armitage, *Sun, Stand Thou Still* (New York: 1947), p. 40.
2. *Three Copernican Treatises*, 3rd ed., edited with an Introduction by Edward Rosen (New York: 1971), p. 342.
3. See Koestler, *The Sleepwalkers*, pp. 121–122.
4. Nicolas Copernicus, *On the Revolutions of the Heavenly Spheres*, trans.

by Charles Glenn Wallis in *The Great Books of the Western World*, vol. 16 (Chicago: 1952), p. 508.

5. *Ibid.*, p. 4.

6. For a modern translation of the *Commentariolus* see *Three Copernican Treatises*.

7. Copernicus, *On the Revolutions of the Heavenly Spheres*, p. 507.

8. *Ibid.*, p. 506.

9. *Ibid.*

10. *Ibid.*

11. *Ibid.*, p. 505.

12. *Ibid.*, pp. 506–507.

13. *Ibid.*, p. 513.

14. *Ibid.*, p. 516.

15. *Ibid.*, pp. 525–526.

16. *Ibid.*, p. 514.

17. I suggest Thomas S. Kuhn's, *The Copernican Revolution* (Boston: 1957) or Angus Armitage, *Copernicus* (New York: 1957).

18. Copernicus, *On the Revolutions of the Heavenly Spheres*, p. 508.

19. *Ibid.*, pp. 526–528.

20. *Ibid.*, p. 526.

21. Alexandre Koyré, *From the Closed World to the Infinite Universe* (Baltimore: 1957), pp. 33–34.

22. *Ibid.*, pp. 34–35.

23. Grant McColley, "The Eighth Sphere of Copernicus," *Popular Astronomy*, vol. X (March, 1942), p. 135. See also Grant McColley, "Nicolas Copernicus and an Infinite Universe," *Popular Astronomy*, vol. XLIV (December, 1936), pp. 529–533.

24. Copernicus, *On the Revolutions of the Heavenly Spheres*, p. 519.

25. *Ibid.*

26. Quoted in Koyré, *From the Closed World to the Infinite Universe*, p. 40.

Chapter 5

1. Quoted in John Allyne Gade, *The Life and Times of Tycho Brahe* (Princeton: 1947), p. 12.

2. J. L. E. Dreyer, *Tycho Brahe, A Picture of Scientific Life and Work in the Sixteenth Century*, 2nd ed. (New York: 1963), p. 12.

3. Gade, *The Life and Times of Tycho Brahe*, pp. 13–14.

4. *Ibid.*, p. 30.

5. Quoted in Antonie Pannekoek, *A History of Astronomy*, pp. 205–207.

6. Arthur Koestler, *The Sleepwalkers*, p. 290.

7. Dreyer, *Tycho Brahe, A Picture of Scientific Life and Work in the Sixteenth Century*, pp. 86–87.

8. T. S. Kuhn, *The Copernican Revolution*, p. 201.

9. *Ibid.*, p. 204.

10. Dreyer, *Tycho Brahe, A Picture of Scientific Life and Work in the Sixteenth Century*, p. 128.

11. Gade, *The Life and Times of Tycho Brahe*, p. 157.

Chapter 6

1. Gerald Holton, *Thematic Origins of Scientific Thought: Kepler to Einstein* (Cambridge: 1973), pp. 69–70.

2. Quoted in Arthur Koestler, *The Sleepwalkers*, pp. 236–7. Koestler's provocative but highly controversial biography of Kepler originally contained in *The Sleepwalkers* has since been published independently of the longer work. See Arthur Koestler, *The Watershed* (New York: 1960). The definitive biography of Kepler is Max Caspar's, *Kepler*, trans. and ed. by C. Doris Hellman (London and New York: 1959).

3. Caspar, *Kepler*, p. 368.

4. See Gale E. Christianson, "Kepler's *Somnium:* Science Fiction and the Renaissance Scientist," *Science-Fiction Studies*, vol. 3, no. 8 (March, 1976), pp. 79–90.

5. Quoted in Koestler, *The Sleepwalkers*, p. 241.

6. Caspar, *Kepler*, p. 57.

7. Quoted in Koestler, *The Sleepwalkers*, pp. 242–243.

8. E. A. Burtt, *The Metaphysical Foundations of Modern Physical Science*, rev. ed. (New York: 1952), p. 67. For a more detailed discussion and analysis of Kepler's belief in archetypal ideas see C. G. Jung and W. Pauli, *The Interpretation of Nature and the Psyche* (New York: 1951), pp. 151–211.

9. Carola Baumgardt, *Johannes Kepler: Life and Letters* (New York: 1951).

10. *Ibid.*, p. 38.

11. Quoted in Caspar, *Kepler*, p. 136.

12. Holton, *Thematic Origins of Scientific Thought: Kepler to Einstein*, p. 78 (emphasis Holton's).

13. Baumgardt, *Johannes Kepler: Life and Letters*, p. 95.

14. *Ibid.*, pp. 80–81.

15. Koestler, *The Sleepwalkers*, p. 349.

16. Baumgardt, *Johannes Kepler: Life and Letters*, pp. 83–84.

17. Koestler, *The Sleepwalkers*, p. 363.

18. Baumgardt, *Johannes Kepler: Life and Letters*, pp. 99–100.

19. *Ibid.*, p. 102.

20. *Ibid.*, p. 103.

21. Caspar, *Kepler*, p. 220.

22. Johannes Kepler, *Harmonice Mundi*, trans. by Charles Glenn Wallis in *The Great Books of the Western World*, vol. 16 (Chicago: 1952), p. 1010.

23. Caspar, *Kepler*, p. 359.

Chapter 7

1. Lewis S. Feuer, *Einstein and the Generations of Science* (New York: 1974), p. 240.

2. *Ibid.*, p. 290.

3. T. S. Kuhn, *The Structure of Scientific Revolutions*, 2nd ed. (Chicago: 1970), p. 6.

4. Quoted in Mary-Allan Olney, *The Private Life of Galileo* (London: 1870), p. 2.

5. For an excellent background analysis of Galileo's attitude toward religion see Giorgio Spini, "The Rationale of Galileo's Religiousness," in *Galileo Reappraised*, ed. by Carlo L. Golino (Berkeley & Los Angeles: 1966), pp. 44–66.

6. Ludovico Geymonat, *Galileo Galilei: A Biography and Inquiry Into His Philosophy of Science*, trans. with additional notes and appendix by Stillman Drake (New York: 1965), p. 7.

7. Alexandre Koyré, "Galileo and Plato," *Journal of the History of Ideas*, vol. IV, no. 4 (October, 1943), p. 406.

8. For a modern translation of *The Little Balance* see Laura Fermi and Gilberto Bernardini, *Galileo and the Scientific Revolution* (New York: 1961), pp. 133–143.

9. Herbert Butterfield, *The Origins of Modern Science, 1300–1800*, rev. ed. (New York: 1967), p. 14. I am indebted to Professor Butterfield for his cogent analysis of pre-Galileian mechanics, elements of which have been incorporated into my discussion of the subject.

10. *Ibid.*, p. 23.

11. Leonardo Olschki, "Galileo's Philosophy of Science," *The Philosophical Review*, vol. LII, no. 4 (July, 1943), p. 352.

12. Quoted in Olney, *The Private Life of Galileo*, p. 14.

13. Baumgardt, *Johannes Kepler: Life and Letters*, p. 41.

14. See, for example, Edwin Panofsky, *Galileo as a Critic of the Arts* (The Hague: 1954).

15. Jacob Bronowski, *The Ascent of Man* (Boston: 1973), p. 200.

16. Koestler, *The Sleepwalkers*, p. 362.

17. Galileo Galilei, *The Starry Messenger*, in *Discoveries and Opinions of Galileo*, trans. with an introduction and notes by Stillman Drake (New York: 1957), p. 29.

18. *Ibid.*, p. 28.

19. *Ibid.*, p. 49.

20. *Ibid.*, pp. 47–48.

21. *Ibid.*, pp. 25–26.

22. Giorgio de Santillana, *The Crime of Galileo* (Chicago: 1955), p. 17.

23. See Galileo Galilei, *Letters on Sunspots* in *Discoveries and Opinions of Galileo,* trans. by Stillman Drake, pp. 59–144. Both the personal and scientific details of the Galileo–Scheiner controversy are well analyzed.

24. William R. Shea, "Galileo, Scheiner and the Interpretation of Sunspots," *ISIS,* vol. LXV, no. 209 (Winter, 1970), pp. 498–519.

25. Quoted in *Discoveries and Opinions of Galileo,* pp. 146–147.

26. *Ibid.*, p. 152.

27. Santillana, *The Crime of Galileo,* p. 44.

28. *Ibid.*, pp. 45–46.

29. Quoted in Koestler, *The Sleepwalkers,* p. 447.

30. Santillana, *The Crime of Galileo,* p. 126.

31. *Ibid.*, p. 128.

32. *Ibid.*, p. 132.

33. Galileo Galilei, *Dialogue Concerning the Two Chief World Systems— Ptolemaic and Copernican,* trans. by Stillman Drake, foreword by Albert Einstein, 2nd ed. (Berkeley & Los Angeles: 1967), p. 50.

34. *Ibid.*, p. 11.

35. Galileo Galilei, *Letter to the Grand Duchess Christina* in *Discoveries and Opinions of Galileo,* pp. 182–183.

36. Ernst Cassirer, "Galileo: A New Science and a New Spirit," *The American Scholar,* vol. 12, no. 1 (Winter, 1942–43), p. 7.

37. Geymonat, *Galileo Galilei: A Biography and Inquiry Into His Philosophy of Science,* p. 170.

38. Burtt, *The Metaphysical Foundations of Modern Physical Science,* p. 81.

39. Harold L. Burstyn, "Galileo's Attempt to Prove That the Earth Moves," *ISIS,* vol. LIII (June, 1962), p. 181.

40. Galileo Galilei, *The Assayer* in *Discoveries and Opinions of Galileo,* p. 272.

41. Galilei, *Dialogue Concerning the Two Chief World Systems—Ptolemaic and Copernican,* p. 464.

42. Quoted in Santillana, *The Crime of Galileo,* p. 252.

43. *Ibid.*, pp. 259–260.

44. *Ibid.*, pp. 312–313.

45. Quoted in Koestler, *The Sleepwalkers,* pp. 490–491.

Chapter 8

1. Frank E. Manuel, *A Portrait of Isaac Newton* (Cambridge: 1968), p. 28.

2. Frank E. Manuel, *The Religion of Isaac Newton* (Oxford: 1974), p. 17.

3. Louis Trenchard More, *Isaac Newton, A Biography* (New York: 1934), p. 3.

4. *Ibid.*, p. 4.

5. Manuel, *A Portrait of Issac Newton*, p. 26.

6. More, *Isaac Newton, A Biography*, p. 11.

7. *Ibid.*, p. 12.

8. Manuel, *A Portrait of Isaac Newton*, p. 44.

9. More, *Isaac Newton, A Biography*, p. 19.

10. Dijksterhuis, *The Mechanization of the World Picture*, p. 397.

11. Richard S. Westfall, *Science and Religion in Seventeenth-Century England* (New Haven & London: 1958), p. 14.

12. *Ibid.*, pp. 72–73.

13. *Ibid.*, p. 5.

14. Eiseley, *Darwin's Century*, p. 177.

15. Manuel, *A Portrait of Isaac Newton*, p. 93.

16. Richard S. Westfall, *Force in Newton's Physics* (London and New York: 1971), p. 325.

17. Kuhn, *The Copernican Revolution*, p. 256.

18. E. N. da C. Andrade, *Sir Isaac Newton: His Life and Work* (New York: 1954), pp. 40–41.

19. More, *Isaac Newton, A Biography*, p. 73.

20. *Ibid.*, p. 290.

21. Manuel, *A Portrait of Isaac Newton*, p. 96.

22. I. Bernard Cohen, *Franklin and Newton* (Philadelphia: 1956), p. 52.

23. More, *Isaac Newton, A Biography*, p. 81.

24. Florian Cajori, "Isaac Newton's Experiments on Light," *School Science and Mathematics*, vol. XXVIII, no. 6 (June, 1928), pp. 621–622.

25. Charles Coulston Gillispie, *The Edge of Objectivity: An Essay in the History of Scientific Ideas* (Princeton: 1960), p. 128.

26. Ralph M. Blake, "Sir Isaac Newton's Theory of Scientific Method," *The Philosophical Review*, vol. XLII, no. 5 (September, 1933), p. 462.

27. Manuel, *The Religion of Sir Isaac Newton*, p. 22.

28. Westfall, *Force in Newton's Physics*, p. 328.

29. *Ibid.*, p. 329 .

30. More, *Isaac Newton, A Biography*, pp. 246–247.

31. Quoted in Koyré, *From the Closed World to the Infinite Universe*, p. 185.

32. More, *Isaac Newton, A Biography*, p. 321.

33. Sir Isaac Newton, *Mathematical Principles of Natural Philosophy*, trans. by Andrew Motte in *The Great Books of the Western World*, vol. 34 (Chicago: 1952), pp. 370–371.

34. Bertrand Russell, *The Autobiography of Bertrand Russell, 1872–1914*, vol. I (London: 1967), p. 153.

35. Manuel, *A Portait of Isaac Newton*, p. 220.
36. More, *Isaac Newton, A Biography*, p. 382.
37. *Ibid.*, p. 385.

Chapter 9

1. Arthur Koestler, *The Act of Creation* (New York: 1967), p. 227.
2. Kuhn, *The Structure of Scientific Revolutions*, p. viii.
3. *Ibid.*, p. 103.
4. *Ibid.*, pp. 10–11.
5. *Ibid.*, p. 11.
6. *Ibid.*, p. 13.
7. *Ibid.*, p. 24.
8. *Ibid.*, p. 38.
9. *Ibid.*, p. 36.
10. *Ibid.*, p. 24.
11. *Ibid.*, p. 117.
12. *Ibid.*, p. 5.
13. *Ibid.*, p. 65.
14. *Ibid.*, p. 91.
15. Dudley Shapere, "The Paradigm Concept," *Science*, vol. 172 (May 14 1971), p. 707.
16. Kuhn, *The Structure of Scientific Revolutions*, p. 49.
17. *Ibid.*, p. 144.
18. Holton, *The Thematic Origins of Scientific Thought: Kepler to Einstein*, pp. 354–355.
19. Lewis S. Feuer, *Einstein and the Generations of Science*, p. 293.
20. Bronowski, *The Ascent of Man*, p. 247. For a somewhat different version of the question see Banesh Hoffman, *Albert Einstein: Creator and Rebel* (New York: 1972), p. 28.
21. Kuhn, *The Structure of Scientific Revolutions*, p. 10.
22. *Ibid.*, p. 11.
23. *Ibid.*, p. 137.
24. *Ibid.*, p. 138.
25. Sidney Pollard, *The Idea of Progress* (London: 1971), pp. 20–21.
26. Kuhn, *The Structure of Scientific Revolutions*, p. 164.
27. Gillispie, *The Edge of Objectivity: An Essay in the History of Scientific Ideas*, p. 10.
28. Kuhn, *The Structure of Scientific Revolutions*, pp. 84–85.
29. Kuhn, *The Copernican Revolution*, p. 265.

Selected Bibliography

Both the time period and subject matter covered in this study are so broad that they have produced a vast literature, only a very small part of which can be cited here. The following bibliography is highly selective and therefore excludes countless important studies. However it should serve as an adequate introduction for the reader new to the field.

ANDRADE, E. N. DA C. *Sir Isaac Newton: His Life and Work* (New York, 1954).

ARMITAGE, ANGUS. *Copernicus* (New York, 1957).

———, *Sun, Stand Thou Still* (New York, 1947).

BAUMGARDT, CAROLA. *Johannes Kepler: Life and Times* (New York, 1951).

BEER, A. AND STRAND, K. A., eds. *Copernicus* (Oxford & New York, 1975).

BOAS, MARIE. *The Scientific Renaissance, 1450–1630* (New York, 1962).

BREWSTER, DAVID. *The Life of Sir Isaac Newton* (London, 1831).

BRODERICK, JAMES. *Galileo: The Man, His Work, His Misfortunes* (New York, 1964).

BRONOWSKI, JACOB. *The Ascent of Man* (Boston, 1973).

BURTT, E. A. *The Metaphysical Foundations of Modern Physical Science*, rev ed. (New York, 1952).

BUTTERFIELD, HERBERT. *The Origins of Modern Science, 1300–1800*, rev. ed. (New York, 1957).

CASPAR, MAX. *Kepler,* trans. and ed. by C. Doris Hellman (London & New York, 1959).

COHEN, I. B. *Franklin and Newton* (Philadelphia, 1956).

———, *Introduction to Newton's Principia* (Cambridge, Massachusetts, 1971).

———, *The Birth of a New Physics* (Garden City, 1960).

COPERNICUS, NICOLAS. *On the Revolutions of the Heavenly Spheres,* trans. by Charles Glenn Wallis (Chicago, 1952).

———, *Three Copernican Treatises,* trans. with an Introduction by Edward Rosen, 3rd ed. (New York, 1971).

DIJKSTERHUIS, E. J. *The Mechanization of the World Picture,* trans. by C. Dijkshoorn (Oxford, 1961).

DOBBS, BETTY JO TEETER. *The Foundations of Newton's Alchemy* (Cambridge, England, 1975).

DRAKE, STILLMAN. *Discoveries and Opinions of Galileo* (New York, 1957).

DREYER, J. L. E. *A History of Astronomy From Thales to Kepler*, rev. ed. (1953).

———, *Tycho Brahe, A Picture of Scientific Life and Work in the Sixteenth Century*, 2nd ed. (New York, 1963).

FARRINGTON, BENJAMIN. *Greek Science* (London, 1953).

FERMI, LAURA AND BERNARDINI, GILBERTO. *Galileo and the Scientific Revolution* (New York, 1961).

FEUER, LEWIS S. *Einstein and the Generations of Science* (New York, 1974).

GADE, JOHN ALLYNE. *The Life and Times of Tycho Brahe* (New York, 1947).

GALILEI, GALILEO. *Dialogue Concerning the Two Chief World Systems—Ptolemaic and Copernican*, trans. by Stillman Drake, 2nd ed. (Berkeley and Los Angeles, 1967).

GILLISPIE, CHARLES COULSTON. *The Edge of Objectivity* (Princeton, 1960).

GINGERICH, OWEN, ed. *The Nature of Scientific Discovery* (Washington, D.C., 1975).

GOLINO, CARLO L., ed. *Galileo Reappraised* (Berkeley & Los Angeles, 1966).

GUTHRIE, W. K. C., ed. *Aristotle on the Heavens*, 4th ed. (Cambridge, Massachusetts, 1960).

HALL, A. R. *The Scientific Revolution, 1500–1800*, 2nd ed. (Boston, 1962).

HEATH, THOMAS. *Aristarchus of Samos: The Ancient Copernicus* (Oxford, 1959).

HOLTON, GERALD. *Thematic Origins of Scientific Thought: Kepler to Einstein* (Cambridge, Massachusetts, 1973).

KEPLER, JOHANNES. *Harmonice Mundi*, trans. by Charles Glenn Wallis (Chicago, 1952).

KOESTLER, ARTHUR. *The Sleepwalkers* (New York, 1959).

KOYRÉ, ALEXANDRE. *From the Closed World to the Infinite Universe* (Baltimore, 1957).

———, *Newtonian Studies* (Cambridge, Massachusetts, 1968).

———, *Metaphysics and Measurement: Essays in the Scientific Revolution* (Cambridge, Massachusetts, 1968).

KUHN, T. S. *The Copernican Revolution* (Cambridge, Massachusetts, 1957).

———, *The Structure of Scientific Revolutions*, 2nd ed. (Chicago, 1970).

LANGFORD, JEROME J. *Galileo, Science and the Church*, rev. ed. (Ann Arbor, 1971).

MANUEL, FRANK E. *A Portrait of Isaac Newton* (Cambridge, Massachusetts, 1968).

———, *The Religion of Isaac Newton* (Oxford, 1974).

McINTOSH, CHRISTOPHER. *The Astrologers and Their Creed* (New York, 1969).

McMULLIN, ERNAN, ed. *Galileo: Man of Science* (New York, 1957).

MORE, LOUIS T. *Isaac Newton: A Biography* (New York, 1934).

MUMFORD, LEWIS. *The Myth of the Machine: Technics and Human Development*, 2 vols. (New York, 1967).

NEWTON, ISAAC. *Mathematical Principles of Natural Philosophy*, trans. by Andrew Motte (Chicago, 1952).

PALTER, ROBERT, ed. *The Annus Mirabilis of Sir Isaac Newton: 1666–1966* (Cambridge, 1970).

PANNEKOEK, ANTONIE. *A History of Astronomy* (New York, 1961).

PANOFSKY, EDWIN. *Galileo as a Critic of the Arts* (The Hague, 1954).

POLLARD, SYDNEY. *The Idea of Progress* (London, 1971).

PTOLEMY, CLAUDIUS. *The Almagest*, trans. by R. Catesby Taliafero (Chicago, 1952).

ROSEN, EDWARD, ed. *Three Copernican Treatises*, 3rd ed. (New York, 1971).

RUDNICKI, JOZEF. *Nicholas Copernicus*, trans. by B. W. A. Massey (London, 1943).

DE SANTILLANA, GIORGIO. *The Crime of Galileo* (Chicago, 1955).

TOULMIN, STEPHEN AND GOODFIELD, JANE. *The Fabric of the Heavens* (New York, 1961).

WESTFALL, RICHARD S. *Force in Newton's Physics* (London & New York, 1971).

———, *Science and Religion in Seventeenth-Century England* (New Haven & London, 1958).

WHITEHEAD, ALFRED NORTH. *Science and the Modern World* (New York, 1925).

WILLEY, BASIL. *The Seventeenth Century Background*, 3rd ed. (New York, 1953).

WITTFOGEL, KARL A. *Oriental Despotism, A Comparative Study* (New Haven, 1957).

Index